Lasers in Applied and Fundamental Research

Lasers in Applied and Fundamental Research

compiled and introduced by

Stig Stenholm

Research Institute for Theoretical Physics
University of Helsinki

Adam Hilger Ltd, Bristol and Boston

© Adam Hilger Ltd 1985

All rights reserved. No part of this publication may be reproduced, stored in a retrieval system or transmitted in any form or by any means, electronic, mechanical, photocopying, recording or otherwise, without the prior permission of the publisher.

British Library Cataloguing in Publication Data

Lasers in applied and fundamental research.
  1. Lasers
  I. Stenholm, Stig
  621.36'6   TA1675

ISBN 0-85274-808-6

Consultant Editor: **Dr E R Pike** RSRE, Malvern

QC
689
.L38
1985

Published by Adam Hilger Ltd
Techno House, Redcliffe Way, Bristol BS1 6NX, England
PO Box 230, Accord, MA 02018, USA

Printed in Great Britain by J W Arrowsmith Ltd, Bristol

Contents

Foreword ix

Present Addresses of Contributors xiii

Introduction 1
S Stenholm

Interaction of Laser Radiation with Free Atoms 61
S Feneuille

1. Introduction	63
1.1 Structure of the review	64
2. Theoretical models for resonant phenomena	64
2.1 Parameters characterizing an optical resonance	65
2.2 Interaction of a two-level atom with a classical field	66
2.3 Interaction of a two-level atom with a quantum field	68
2.4 The Bloch–Siegert shift	71
2.5 Spontaneous emission and other relaxation processes	72
2.6 Steady-state and transient solutions of the Bloch equations including decay constants	74
2.7 Rate equations	76
2.8 Real systems	77
2.9 Concluding remarks	80
3. Response of an atomic system to resonant laser excitation	80
3.1 Atomic beams transversely illuminated	80
3.2 Vapours	88
3.3 Transient effects	93
4. Propagation effects	95
4.1 Incoherent saturation effects on the propagation of pulses	95
4.2 The quantum-mechanical 'area theorem'	96
4.3 Self-induced transparency	97
5. Multiphoton processes	101
5.1 Doppler-free multiphoton excitation	101
5.2 Resonant multiphoton ionization	103
6. Concluding remarks	105
Acknowledgments	105
References	105
Addendum	109

Optical Bistability and Related Devices   111
E Abraham and S D Smith

1. Introduction   113
    1.1 Aim of this review   113
    1.2 Historical remarks and general discussion   113
    1.3 Organisation of the review   115
2. Basic theory   115
3. Theory of optical bistability for atomic systems   119
    3.1 Maxwell–Bloch equations   120
    3.2 Steady state   123
    3.3 The mean field approximation   125
    3.4 Time-dependent phenomena   129
    3.5 Phase transition analogy   134
    3.6 Quantum-mechanical approach   137
4. Theory of alternative approaches to bistable systems   144
5. Non-linear refraction of semiconductors   146
    5.1 Non-linearity in InSb   146
    5.2 Non-linearity in GaAs   153
6. Experiments on intrinsic optically bistable systems   154
    6.1 Free atoms and molecules   155
    6.2 Kerr media   159
    6.3 Solids: semiconductors and ruby   160
    6.4 Non-linear interface and other geometries   164
7. Experiments on hybrid devices   166
8. Switching times and energies   173
9. Summary and conclusions   176
    Additional reading   176
    References   178
    Addendum   182

Non-classical Effects in the Statistical Properties of Light   185
R Loudon

1. Introduction   187
2. Classical degree of second-order coherence   188
3. Quantum degree of second-order coherence   193
4. Non-classical single-beam statistics   198
    4.1 Radiation by a single driven atom: special case   198
    4.2 Radiation by a single driven atom: general case   201
    4.3 Radiation by a distribution of atoms   204
    4.4 Non-linear optical processes   208
5. Non-classical double-beam coincidence statistics   211
    5.1 Two-photon cascade emission   211
    5.2 Non-linear optical processes   215
6. Conclusions   216
    Acknowledgments   217

Appendix. Solutions of the optical Bloch equations	217
References	220

Bell's Theorem: Experimental Tests and Implications 223
J F Clauser and A Shimony

1. Introduction	225
2. The Einstein–Podolsky–Rosen argument	227
3. Bell's theorem	228
3.1 Deterministic local hidden-variables theories and Bell (1965)	229
3.2 Foreword to the non-idealised case	231
3.3 Generalisation of the locality concept	233
3.4 Bell's 1971 proof	234
3.5 The proof by Clauser and Horne	236
3.6 Symmetry considerations	238
3.7 The proof by Wigner, Belinfante and Holt	239
3.8 Stapp's proof	240
3.9 Other versions of Bell's theorem	242
4. Considerations regarding a general experimental test	242
4.1 Requirements for a general experimental test	243
4.2 Three important experimental cases	244
5. Cascade-photon experiments	245
5.1 Predictions by local realistic theories	246
5.2 Quantum-mechanical predictions for a $J=0 \rightarrow 1 \rightarrow 0$ two-photon correlation	248
5.3 Description of experiments	250
5.4 Are the auxiliary assumptions for cascade-photon experiments necessary and reasonable?	254
6. Positronium annihilation and proton–proton scattering experiments	256
6.1 Historical background	256
6.2 The experiment by Kasday, Ullman and Wu	257
6.3 The experiments by Faraci *et al*, Wilson *et al* and Bruno *et al*	259
6.4 Proton–proton scattering experiment by Lamehi-Rachti and Mittig	259
7. Evaluation of the experimental results and prospects for future experiments	260
7.1 Two problems	260
7.2 Experiments without auxiliary assumptions about detector efficiencies	261
7.3 Preventing communication between the analysers	262
7.4 Conclusion	263
Appendix 1. Criticism of EPR argument by Bohr, Furry and Schrödinger	263
Appendix 2. Hidden-variables theories	264
Acknowledgments	268
References	268
Addendum	270

Foreword

Lasers have become central tools of physics investigations. Their use extends from purely technical applications to investigations into the basic principles of quantum mechanics. To collect a series of review articles exposing the range of applications of laser physics can hence be expected to be of interest to a wide range of readers.

In this book we have collected a series of four papers, which have been published in 'Reports on Progress in Physics'. They are:

1. S Feneuille: Interaction of laser radiation with free atoms

Rep.Prog.Phys. 40(1977) 1257-1304

2. E Abraham and S D Smith: Optical bistability and related devices

Rep.Prog.Phys. 45(1982) 815-885

3. R Loudon: Non-classical effects in the statistical properties of light

Rep.Prog.Phys. 43(1980) 913-949

4. J F Clauser and A Shimony: Bell's theorem: experimental tests and implications

Rep.Prog.Phys. 41(1978) 1881-1927

Of these paper #1 gives a general view of the use of lasers in spectroscopy. The basic theoretical tools are summarized in a manner detailed enough to serve as a pedagogical introduction. The complications appearing in real atomic systems are surveyed, and some fundamental experiments are discussed.

Paper #2 reviews the phenomenon of optical bistability, which has received a lot of interest recently. It has potential applications to optical computations, and it provided one of the roads into the very topical field of non-linear dynamics and chaotic behaviour of deterministic systems. The theory is based on that developed in paper #1, but usually realisations employ semiconductor materials.

In most problems concerned with the interaction between matter and laser light it suffices to use a classical description of the laser radiation, a semiclassical theory. In paper #3 the need for a quantum theory of light is discussed. The basic differences between the coherence properties of classical and quantum radiation are presented, and some experiments are discussed where the two types can be discerned.

Paper #4 gives a discussion of how lasers may be used to test the basic formulation of quantum mechanics itself. In spite of its tremendous success to describe experiments, many people feel that quantum mechanics is fundamentally an unsatisfactory theory. John Bell managed to derive certain relations which highlight the point where it appears offensive to our intuition. These relations are susceptible to experimental tests. Paper #4 presents the theoretical background and the experimental situation at the time of its writing. The work carried out later has only confirmed the picture obtained at that time; quantum mechanics predicts the correct experimental results and, when alternative theories differ, they are found to be wrong.

The aim of the present collection is two-fold. Firstly, it can act as a survey of the field of laser applications for physicists that work in other fields. Secondly its articles are detailed enough to present an introduction to the theory and observations of laser physics. Readers new to the field can penetrate into the details of the work and use the references to pursue interesting topics further.

The Introduction is written with both types of reader in mind. The basic theoretical questions are dealt with in the form of simple examples. These calculations are carried out in some detail and provide model examples for the more complicated cases in the reprinted papers. The discussions should enable the reader to consult the papers in any order desired, even if the subject matter appears to form a natural sequence as presented.

The Introduction also summarises the development in the field after the papers have been written. Newer reviews are referred to, and connections to other questions of topical interest are presented. In some cases there has been essential experimental progress, and then some key references are given. The authors have also been given the opportunity to add comments and references to their articles.

There is a problem of notation in this collection. The various articles use different symbols for the same quantities. I have chosen to handle this situation so that I use mainly the notation of paper #1 in my Introduction, and I provide a comparison of notations in table I.

In many applications the laser is but a glorious light source with a narrow spectral output and a high brightness. One interesting feature of laser spectroscopy is, however, that it has created many new methods and techniques for investigating the microscopic properties of atomic matter.

Many of these have developed from methods that were originally introduced in magnetic resonance spectroscopy and microwave spectroscopy. There are, however, some features of the interaction between matter and light that are new, and these lead to interesting phenomena. They are:

1. The wavelength is short. Any macroscopic sample extends over many waves. The atomic motion can take a particle through several wavelengths during the interaction period, and hence an appreciable Doppler shift appears.

2. The spontaneous emission at allowed transitions is fast because of the high frequency, and it contributes its own features to the spectra.

3. It is possible to detect a single quantum of the radiation using photon multipliers because of the high energy of the quanta. Then the information carried in the statistical properties of the photons is also accessible in contrast to the case for lower frequencies.

In this reprint collection we shall see examples of each of these features.

Table 1

Introduction: Paper	# 1	# 2	# 3
Lower-level frequency	ω_g	0 [2]	0 [2]
Upper-level frequency	ω_e	ω_0	ω_0
Laser frequency	ω	ω	ω
Detuning Δ	$\omega-(\omega_e-\omega_g)$	$\omega-\omega_0$	$\omega-\omega_0$
Dipole matrix element μ	$\langle e|\vec{\varepsilon}\cdot\vec{D}|g\rangle$ [1]	μ	$\hbar g\sqrt{2\varepsilon_0 V/\hbar\omega}$ [3]
Rabi frequency at resonance	$2K$	—	Ω
Rabi frequency with detuning	Ω	—	—
Decay rates	γ, Γ	γ, Γ	γ
Saturation parameter $\mu E/\hbar\gamma$	χ	—	—
Intensity parameter I	χ^2	I	—

1) $\vec{\varepsilon}$ = polarisation of radiation; $\vec{D}$ = dipole operator.

2) The lower level is given energy zero; the upper level energy $\omega_0 = (E_2-E_1)/\hbar$

3) V = the quantisation volume of the cavity.

Present Addresses of Contributors

Dr E Abraham
Physics Department
Heriot-Watt University
Riccarton
Edinburgh   EH14 4AS
UK

Dr J F Clauser
Lawrence Livermore National Laboratory
PO Box 808
Livermore
California 94550
USA

Dr S Feneuille
Lafarge Coppee
28 rue Emile Menier
BP 40
F-75782 Paris Cedex
FRANCE

Professor R Loudon
Physics Department
University of Essex
Wivenhoe Park
Colchester   CO4 3SQ
Essex
UK

Professor A Shimony
Physics Department
Boston University
Boston
Massachusetts 02215
USA

Professor S D Smith FRS
Physics Department
Heriot-Watt University
Riccarton
Edinburgh   EH14 4AS
UK

Professor S T Stenholm
University of Helsinki
Research Institute for Theoretical Physics
Siltavuorenpenger 20C
SF 00170 Helsinki 17
FINLAND

Introduction

S Stenholm

Research Institute for Theoretical Physics, University of Helsinki, Finland

1. COMMENTS ON PAPER #1

1.1 A two-level system in a strong field

In paper #1 the use of laser light for spectroscopic investigations is discussed, and hence we start by considering the utilisation of the strong field intensity. The basic ideas are presented in its section 2. Here, we present some aspects of this in a slightly simplified form to lead the reader into the subject. The archetype of a quantum atom is the two-level system of figure 1. The lower level $|g\rangle$ does not necessarily have to be the ground state. All other levels can be omitted from the discussion because, for atomic transitions, the energy differences are mostly such that laser light interacts resonantly with only one pair of levels at a time. In reality these would, of course, have fine structures, which we neglect here.

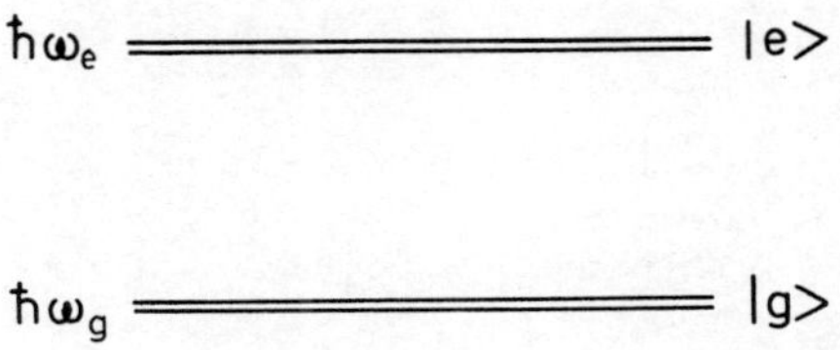

$$\hbar\omega_e = \!\!=\!\!=\!\!=\!\!=\!\!= |e\rangle$$

$$\hbar\omega_g = \!\!=\!\!=\!\!=\!\!=\!\!= |g\rangle$$

Fig.1. The simplest theoretical model in spectroscopy consists of two levels only. The lower one $|g\rangle$ has got the energy $\hbar\omega_g$ and the excited one $|e\rangle$ the energy $\hbar\omega_e$. The lower state may be the ground state, but it can be any state selected by a laser field nearly resonant with the frequency difference $\omega_e - \omega_g$.

If the laser light has frequency ω we introduce the quantum state in the form

$$|\psi\rangle = \alpha \exp(-i\omega_g t)|g\rangle + \beta \exp[-i(\omega+\omega_g t)]|e\rangle. \tag{1}$$

The coupling between the levels is assumed to be given by the dipole operator μ, and the interaction Hamiltonian is

$$H = -\mu E \begin{bmatrix} 0 & 1 \\ 1 & 0 \end{bmatrix} \cos \omega t. \tag{2}$$

Schrödinger's equation gives for the probability amplitudes of equation

(1) the time evolution

$$i\dot{\alpha} = -\frac{\mu E}{\hbar} \cos \omega t \, e^{-i\omega t} \beta \qquad (3a)$$

$$i\dot{\beta} + (\omega+\omega_g)\beta = \omega_e \beta - \frac{\mu E}{\hbar} \cos \omega t \, e^{i\omega t} \alpha. \qquad (3b)$$

In these we neglect rapidly oscillating components approximating

$$e^{\pm i\omega t} \cos \omega t \sim 1/2 \qquad (4)$$

which by analogy with magnetic resonance theory is called the 'rotating wave approximation'. The equations to be solved are then

$$i\dot{\alpha} = -K\beta \qquad (5a)$$

$$i\dot{\beta} = -\Delta\beta - K\alpha \qquad (5b)$$

where we have set

$$\Delta = \omega - \omega_e + \omega_g \qquad (6)$$

$$K = \frac{\mu E}{2\hbar}. \qquad (7)$$

The eigenvalues of this set are

$$\lambda_\pm = i\left[\Delta \pm \sqrt{K^2 + \Delta^2/4}\right] = i\left[\Delta \pm \Omega/2\right] \qquad (8)$$

and with the initial conditions

$$\alpha(0) = 1$$
$$\beta(0) = 0 \qquad (9)$$

we obtain the solution

$$\alpha(t) = e^{i\Delta t/2} \frac{2K^2}{\Omega} \left[\frac{e^{i\Omega t/2}}{\Omega + \Delta} - \frac{e^{-i\Omega t/2}}{\Omega - \Delta}\right] \qquad (10a)$$

$$\beta(t) = i \, e^{i\Delta t/2} \frac{2K}{\Omega} \sin(\Omega t/2). \qquad (10b)$$

This is a simplified form of the solution in equation (2.8) of paper #1.

The probability for occupation of the upper level becomes

$$|\beta(t)|^2 = \frac{4K^2}{\Omega^2} \sin^2(\Omega t/2)$$

$$= \frac{2K^2}{\Omega^2}\left[1 - \cos \Omega t\right] \tag{11}$$

and we see that it pulsates regularly between the upper and the lower level with the frequency

$$\Omega = (\Delta^2 + 4K^2)^{1/2} \tag{12}$$

which is called the 'Rabi flipping frequency' after its original discoverer Rabi(1937).

In many spectroscopic applications it is impossible to control the interaction time t with the accuracy required to apply equation (11). The measurement takes place in a steady state situation, but the individual atoms begin their interactions and end them at random times. It leads us to consider an ensemble of atoms introduced into the interaction region and removed after a random period. This corresponds to incoherent pumping of the atoms into the interacting level pair and the random termination of interaction through radiative decay to unobserved levels or quenching collisions. In these cases the expected lifetime of an interacting atom is described by the exponental distribution

$$W(t) = \gamma e^{-\gamma t} \tag{13}$$

and the observed signal becomes an average over an ensemble with this distribution.

Calculating the observed average from equation (11) we find that the upper state population becomes

$$P_e = \int \gamma e^{-\gamma t} |\beta(t)|^2 \, dt$$

$$= \frac{2K^2}{\Delta^2 + \gamma^2 + 4K^2} . \tag{14}$$

As a function of the detuning we thus obtain a Lorentzian line shape, but with the line width

$$\Gamma = \gamma\left(1 + \frac{4K^2}{\gamma^2}\right)^{\frac{1}{2}} \qquad (15)$$

where the flipping frequency 2K contributes to the broadening; this is called 'power broadening'. It can be interpreted through the uncertainty principle by saying that due to the flipping (11) there exists an uncertainty τ in the time spent at each level and hence the energy acquires the uncertainty or width

$$\hbar\Gamma \sim \Delta E \sim \frac{\hbar}{\tau} = \mu E. \qquad (16)$$

When the intensity becomes very large, the Rabi frequency goes to infinity, and the probability of finding the atom at the upper level becomes 1/2, which derives from the fact that the atom flips so often that it is found with equal probability at each level.

The field acting on the two-level system induces a dipole moment through the matrix element of μ. Using the state (1) we can calculate the expectation value of this in the form

$$P = N\langle\psi|\mu|\psi\rangle = N\mu[\alpha^*, \beta^* e^{i\omega t}]\begin{bmatrix} 0 & 1 \\ 1 & 0 \end{bmatrix}\begin{bmatrix} \alpha \\ \beta e^{-i\omega t} \end{bmatrix} \qquad (17)$$

$$= P^+ e^{-i\omega t} + P^- e^{i\omega t}$$

where the density of active atoms is denoted by N. The component P^+ follows from equation (17) in the form

$$P^+ = N\mu\,\alpha^*\beta$$

$$= \frac{2N\mu K^3}{\Omega^2}\left[\frac{e^{-i\Omega t/2}}{\Omega + \Delta} + \frac{e^{i\Omega t/2}}{\Omega - \Delta}\right]\sin\left(\frac{\Omega t}{2}\right) \qquad (18)$$

which displays the detailed time dependence. If we again apply the argument from above, we must average this result with the distribution function equation (13) and obtain

$$P^+ = N\mu \int_0^\infty \gamma e^{-\gamma t} \alpha^*(t)\beta(t)dt$$

$$= \frac{N\mu K}{\Delta^2 + \gamma^2 + 4K^2} (\Delta + i\gamma). \tag{19}$$

In section 3 we shall find that the imaginary and real parts of this serve to determine the gain and dispersion of the medium. Here we notice that the main dependence on the detuning is again the power-broadened Lorentzian just as in the expression for the upper state occupation equation (14).

The power broadened line width can be written as

$$\Gamma = \gamma(1+I)^{1/2} \tag{20}$$

where the dimensionless saturation parameter I is

$$I = \chi^2 = \left[\frac{\mu E}{\hbar\gamma}\right]^2. \tag{21}$$

The physical significance of this is that it gives the ratio of the resonant Rabi flipping rate $\mu E/\hbar$, i.e. the rate at which the probability is transferred coherently from one level to the other, to the incoherent decay rate γ. When this ratio is larger than unity the coherent flipping becomes dominant. It is hence an appropriate measure of the degree of saturation.

1.2. The Bloch equations

In quantum electronic calculations it is often an advantage to use the density matrix

$$\rho_{ij} = \overline{c_i c_j^*} \tag{22}$$

where the bar denotes an ensemble average. For this it is simple to introduce phenomenological rate processes like pumping and decay. A detailed discussion of the use of the density matrix in laser spectroscopy is found in Stenholm (1984). Here we list only a few aspects.

The relaxation terms can be written as

$$\frac{d}{dt} \rho_{ee} = -\gamma_{ee} \rho_{ee} + \ldots$$

$$\frac{d}{dt} \rho_{gg} = -\gamma_{gg} \rho_{gg} + \ldots \qquad (23)$$

$$\frac{d}{dt} \rho_{eg} = -\gamma_{eg} \rho_{eg} + \ldots$$

where the decay rates satisfy

$$\gamma_{eg} \geq \frac{1}{2}(\gamma_{ee} + \gamma_{gg}). \qquad (24)$$

The off-diagonal elements may disappear faster than the average rate of disappearance of the diagonal elements, because they are affected by all processes that tend to mix up the phases. Especially atomic collisions have this influence and hence the inequality (24) is often said to derive from these. However, other phase-interrupting processes exist, for instance laser field fluctuations.

When the two levels are coupled by a dipole a spontaneous decay appears between them. This cannot be described at the level of Schrödinger's equation but can be written into the equation of motion for the density matrix. For the derivation we refer to Cohen-Tannoudji (1977) or Stenholm (1984). We have

$$\frac{d}{dt} \rho_{ee} = -\Gamma \rho_{ee} + \ldots$$

$$\frac{d}{dt} \rho_{gg} = \Gamma \rho_{ee} + \ldots$$

$$\frac{d}{dt} \rho_{eg} = -\tfrac{1}{2}\Gamma \rho_{eg} + \ldots \qquad (25)$$

and the decay rate Γ is given by

$$\Gamma = \frac{1}{4\pi\varepsilon_0} \frac{4\mu^2 \omega_{eg}^3}{3\hbar c^3}. \qquad (26)$$

For the diagonal elements the effect is as expected, but for the off-diagonal element half the rate only is seen. This is against our naive expectations because usually the off-diagonal element decays faster than the populations on the diagonal. From equation (17) we can see that the off-diagonal element gives the expectation value of the dipole moment and hence it is often called the induced dipole.

In many applications the real variables

$$u = \rho_{eg} + \rho_{ge}$$
$$v = i(\rho_{eg} - \rho_{ge}) \qquad (27)$$
$$w = \rho_{ee} - \rho_{gg}$$

are used. They form a three component vector which obeys the equation of motion

$$\frac{d}{dt}(u\hat{e}_1 + v\hat{e}_2 + w\hat{e}_3) = (2K\hat{e}_1 - \Delta\hat{e}_3) \times (u\hat{e}_1 + v\hat{e}_2 + w\hat{e}_3)$$

$$- \frac{u\hat{e}_1 + v\hat{e}_2}{T_2} - \frac{(w - w^0)\hat{e}_3}{T_1} \qquad (28)$$

where $\hat{e}_i$ are orthogonal unit vectors. This is called the optical Bloch equation and its decay parameters T_1 and T_2 are called the longitudinal and transverse decay times, respectively. Giving an intuitive picture of the behaviour of the two-level system, closely related to the magnetic resonance case, the Bloch vector is widely used in quantum electronics (see e.g. Allen and Eberly 1975).

1.3. Spectroscopic aspects

Using the dressed atom picture paper #1 (sections 2 and 3) shows that the transition in a two-level system is split into a doublet. This can be seen when the level pair is coupled to a third level as, for example, shown in figure 2. Then a doublet spectrum is seen. The splitting is proportional to the Rabi flipping rate (2). This can be understood by considering a dipole oscillating at the imposed optical frequency ω but precessing at the rate Ω. This puts sidebands on the central frequency separated just by Ω. This is called the Autler-Townes or AC Stark effect, and it is discussed in some detail in paper #1. It is a manifestation of the same physics which is seen as a power broadening in equation (15). Another case where we can see the effect is in the spectrum of the spontaneous decay to the lower level, but this will be considered separately in section 1.4.

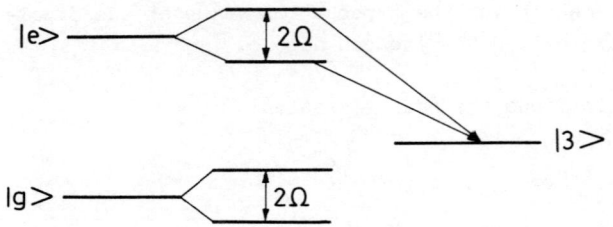

Fig.2 In a strong field each level of a two-level system is split up into a doublet. This splitting can be observed directly if either level (here the upper one) is coupled to a third level. Then a doublet is observed.

In order to be able to investigate an isolated atom we must use either an atomic beam or a gas-filled cell. In the latter case the motion of the atoms causes their resonance frequencies to be Doppler shifted to

$$\omega'_{eg} = \omega_{eg} \frac{1 + v/c}{\sqrt{1-(v/c)^2}} \cong \omega_{eg}(1 + \frac{v}{c} + \ldots)$$

$$\cong \omega_{eg} + qv \qquad (29)$$

because the atom traverses several wavelengths during its interaction with the field. Here q is the wave vector of the light, ω/c. The resonance denominator in equation (14) becomes shifted, and the populations on the two levels are given by

$$P_e = \frac{2K^2}{(\Delta+qv)^2 + \gamma^2 + 4K^2} \qquad (30a)$$

$$P_g = 1 - P_e . \qquad (30b)$$

When this occurs at the top of a Maxwellian velocity distribution the population distribution takes the form shown in figure 3. Each atomic velocity group has got its own resonance frequency, and the field can interact with a group of a width equal to the line width of the two-level transition. This type of broadening is called 'inhomogeneous'.

The holes 'burned' in the velocity distribution are analogous to holes in the inhomogeneous lines in solids. In the optical context they are

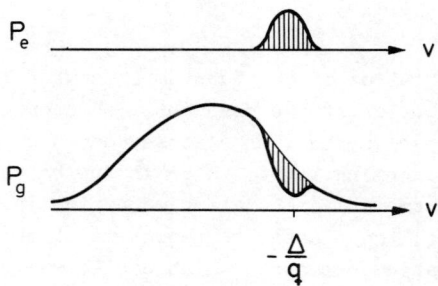

Fig. 3. The population probability in our atomic gas is found entirely in the lower level with a Maxwellian velocity distribution. When a laser field with detuning Δ couples the two levels, those atoms with the right velocity to Doppler compensate the detuning are transferred to the upper level over a velocity range determined by the width of the interacting resonance.

called Bennett holes (Bennett 1962). For a laser based on a standing wave in a cavity it means that the two counter-propagating waves can interact with two velocity groups which compensate the detuning

$$v = \pm \frac{|\omega - \omega_{eg}|}{q} . \tag{31}$$

These contribute to the gain. When the laser cavity becomes tuned to the centre of the atomic line, both holes overlap and only one group remains to give gain. Hence in a gas laser the output shows a dip at the centre of the tuning curve. Its width is the line width of the individual atoms, the homogeneous line width.

The dip at the centre of the tuning curve was found theoretically by Lamb (1963), and is hence called the Lamb dip. He understood it to relate to Bennett's hole and together they managed to report its experimental verification (McFarlane et al 1963). The details of its theory were reported in Lamb (1964).

The Lamb dip gives a picture of the line shape of the individual atom, albeit a distorted one. The large power and the relatively high pressure in the laser cavity distorts the line shape. Later the same type of experiments have been conducted outside the laser cavity; this is the work on saturation spectroscopy, which is discussed in paper #1.

1.4. Resonance fluorescence

An interesting manifestation of the quantum theory of light is presented by the spectral redistribution of the resonantly scattered light from a strongly driven two-level system. An elastic component appears at exactly the frequency of the incoming light, which is due to the radiation from the precessing dipole. If the only relaxation mechanism is spontaneous decay between the two levels, the elastic component is of zero width. Of more interest is the inelastic component, which can be shown to dominate for large intensities. The shape of the spectrum is discussed in section 3.1 of paper #1.

The spectrum of the scattered light can be obtained from the Fourier transform of the dipole correlation function

$$C(t) = \langle \mu^+(t) \mu^-(0) \rangle \qquad (32)$$

where the components of the dipole are obtained from the Pauli matrices $\hat{\sigma}_i$ through

$$\bar{\mu} = \mu_1 \hat{\sigma}_1 + \mu_2 \hat{\sigma}_2 + \mu_3 \hat{\sigma}_3 \qquad (33a)$$

$$\mu^\pm = \mu_1 \pm i\mu_2 . \qquad (33b)$$

The simplest way to obtain the spectrum is to evoke the 'quantum regression theorem' of Lax (1968). This states that the correlation function $\langle \mu(t)\mu(0) \rangle$ shows the same time evolution as the expectation value $\langle \mu(t) \rangle$ with the appropriate initial condition.

As the dipole moment components (33) are the same as the components of the Bloch vector (28) we find that we have to solve the equations

$$\frac{d\mu_1}{dt} = -\frac{1}{2} \Gamma \mu_1 + \Delta \mu_2 \qquad (34a)$$

$$\frac{d\mu_2}{dt} = -\frac{1}{2} \Gamma \mu_2 - \Delta\mu_1 - 2K\mu_3 \qquad (34b)$$

$$\frac{d\mu_3}{dt} = -\Gamma \mu_3 + 2K\mu_2 \qquad (34c)$$

as follows from equation (28). For simplicity we choose the resonant case $\Delta=0$. As the initial condition we take

$$\mu^+(0) = \mu_1(0) + i\mu_2(0) = 1 \qquad (35a)$$

which gives

$$\mu_1(0) = \frac{1}{2} \qquad \mu_2(0) = -\frac{i}{2}. \tag{35b}$$

The eigenvalues λ of the time evolution operator are determined by

$$\begin{vmatrix} \lambda + \frac{1}{2}\Gamma & 0 & 0 \\ 0 & \lambda + \frac{1}{2}\Gamma & 2K \\ 0 & -2K & \lambda+\Gamma \end{vmatrix} = (\lambda + \frac{1}{2}\Gamma)\left[\lambda^2 + \frac{3\Gamma}{2}\lambda + \frac{1}{2}\Gamma^2 + 4K^2\right] = 0$$

(36)

and if we assume $K \gg \Gamma$ we find

$$\lambda_1 = -\frac{1}{2}\Gamma \tag{37a}$$

$$\lambda_{2,3} = -\frac{3\Gamma}{4} \pm 2iK. \tag{37b}$$

The solution satisfying the initial conditions is thus

$$\mu_1(t) = \frac{1}{2} e^{-\Gamma t/2}$$

$$\mu_2(t) = -\frac{i}{2} e^{-3\Gamma t/4} \cos 2Kt \tag{38}$$

$$\langle \mu^+(t) \rangle = \frac{1}{2}\left[e^{-\Gamma t/2} + e^{-3\Gamma t/4} \cos 2Kt\right].$$

Calculating the Fourier transform and taking the real part we find the spectrum

$$\Phi(\nu) = \frac{1}{2}\left[\frac{\Gamma/2}{\nu^2 + \Gamma^2/4} + \frac{1}{2}\frac{3\Gamma/4}{(\nu-2K)^2 + (3\Gamma/4)^2} + \frac{1}{2}\frac{3\Gamma/4}{(\nu+2K)^2 + (3\Gamma/4)^2}\right].$$

(39)

This is valid in the limit of a large field intensity when the elastic peak is already negligible.

The surprising feature of equation (39) is the line width of the sidebands, $3\Gamma/4$, which is not obtainable from simple perturbation theory. As its consequence the peak value of the sideband is only 1/3 of the central peak. This has been calculated first by Burshtein (1966), and Mollow (1969); see also Cohen-Tannoudji (1977) and Stenholm (1984). The experimental verification is discussed in paper #1.

1.5. Transient phenomena

There exists a wide range of time dependent phenomena in quantum electronics. They are used to obtain a multitude of data for both atomic systems and solids. The brief description in paper #1 gives only part of the picture. Many more applications can be found in the book by Allen and Eberly (1975).

If the interaction period is so short that no relaxation processes have time to take place as in equation (5), we can omit the relaxation terms and at exact resonance we find the solution

$$|\beta(t)|^2 = \sin^2(Kt). \tag{40}$$

This shows that when we have

$$Kt = \pi/2 \tag{41}$$

the population is found entirely in the upper level. This is called a $\pi/2$ pulse. For any integer n the condition

$$Kt = \pi n \tag{42}$$

guarantees that the population is returned to the initial lower level. Thus the pulse has not affected the energy of the atomic system, its own energy must be unaffected and no absorption takes place. This is the simplest case of self-induced transparency.

The pulse shape is, however, affected by the presence of the medium, and a consideration of the propagation effects gives equation (4.5) of paper #1

$$\frac{d\theta}{dz} = -\frac{1}{2}\alpha \sin \theta \tag{43}$$

for the integrated pulse area

$$\theta(t,z) = 2 \int_{-\infty}^{t} K(t',z)dt' . \tag{44}$$

By separating the variables we can solve the equation to obtain the pulse shape

$$\tan[\tfrac{1}{2}\theta(z)] = \tan[\tfrac{1}{2}\theta_0] \, e^{-\alpha z/2} \tag{45}$$

which shows that from the value θ_0 at $z = 0$ we obtain the value $2\pi n$ at plus infinity. This is one example of the general theory discussed in section 4.3 of paper #1.

The theory of optical pulse propagation is just a special case of the general theory of non-linear propagation in a medium. This is connected with the concept of soliton pulses and their properties, which has attracted much interest recently. Several reviews are available: Whitham (1974), Scott et al (1973) and Bishop and Schneider (1978).

The range of transient effects in optics is larger, there are important aspects connected with optical nutations, free induction decay and photon echoes. These are not discussed in paper #1 and the reader is referred to the text book by Allen and Eberly (1975) for details.

2. COMMENTS ON PAPER #2

2.1 The response of matter

For quantum electronics the non-linearities of the material response have always been an essential point. As is often found to happen in non-linear systems the steady state conditions prove to be non-unique for some parameter ranges. The bistabilities found in optical systems suggest applications to computing machines. Similarly the bistabilities of electronic devices form the basis for memory circuits in ordinary electronics.

There are several different types of bistabilities in quantum electronics, and in order to understand their differences we shall look at them in their most simple setting. They are discussed in greater detail in paper #2.

We use the results of section 1.1 for a two-level system, which, from equation (19), gives the polarisation in the form

$$P = N\langle\mu\rangle = P^+ e^{-i\omega t} + P^- e^{i\omega t}$$

$$= \frac{N\mu K}{(\Delta - i\gamma)} \frac{e^{-i\omega t}}{(1 + IL(\Delta))} + c.c. \quad (46)$$

where the detuning of the laser from resonance is Δ, half of the Rabi flipping frequency is K, the dimensionless saturation parameter is

$$I = \left[\frac{2K}{\gamma}\right]^2 \quad (47)$$

and the Lorentzian is given by

$$L(\Delta) = \frac{\gamma^2}{\Delta^2 + \gamma^2}. \quad (48)$$

The real part of the polarisation gives the dispersive properties of the medium

$$\text{Re } P^+ = \frac{N\mu \, K\Delta}{\Delta^2 + \gamma^2(1+I)} \quad (49)$$

and shows the familiar dispersion shape but with a power-broadened Lorentzian denominator. The imaginary part

$$\text{Im } P^+ = \frac{N\mu \, K\gamma}{\Delta^2 + \gamma^2(1+I)} \tag{50}$$

gives the loss or gain introduced by the matter. In our case we assume the absorptive case to begin with, later we shall also discuss the case with an initially inverted population, the laser amplifier.

2.2. The absorptive bistability

Let us make a simple model of the absorptive bistability based on the results we have obtained so far. The medium is supposed to be situated in a Fabry-Perot cavity as shown in figure 4. Through the mirror we inject a

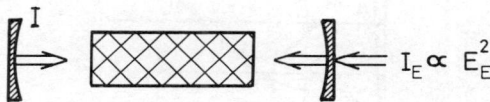

Fig.4 A non-linear medium is situated between the mirrors of a Fabry-Perot resonator. It is excited by an externally imposed laser field of amplitude E_E, which penetrates through the mirror and drives the dipoles inside the cavity with a power proportional to $T \, E_E^2 = T \, I_E$, where T is the mirror transmission coefficient.

signal from an external laser, which is seen from the inside of the laser as $\sqrt{T} \, E_E$, where T is the transmission coefficient of the mirror. If we write an equation for the time derivative of the field inside the laser, this is a term causing the field to grow; the damping due to the medium, equation (50), and the cavity decay time τ provide the losses. We multiply the equation by $\mu/2\hbar$ to write it in terms of the Rabi parameter K and find the equation

$$\frac{dK}{dt} = \sqrt{T} \, K_E - \frac{K}{\tau} - \frac{N\mu\gamma}{\Delta^2 + \gamma^2 + \gamma^2 I} \, K \; . \tag{51}$$

Introducing the scaled variables

$$x = \sqrt{I} \qquad y = \frac{2\sqrt{T}\tau}{\gamma} K_E \qquad t' = t/\tau \qquad C = \frac{\tau N\mu}{2\gamma} \tag{52}$$

we find, at resonance, the equation

$$\frac{dx}{dt'} = y - x - \frac{2Cx}{1+x^2} \qquad (53)$$

with the steady state given by

$$y = f(x) = x\left(1 + \frac{2C}{1 + x^2}\right). \qquad (54)$$

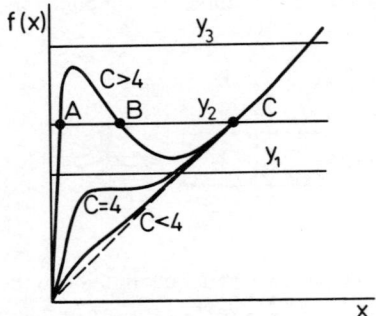

Fig.5 The function $f(x)$ defined in equation (54) plotted for three values of the parameter C. For $C > 4$ the function has got a maximum whereas for $C < 4$ it rises monotonically.

The function $f(x)$ is plotted in figure 5 and we can see that for $C>4$ it shows a non-monotonic behaviour. The solution to equation (54) is obtained by drawing a 'load line' horizontally corresponding to the injected signal y. For y_1 there is only one solution at a very small value of x, for y_3 again only a large value of x is possible. For the intermediate case y_2 however, there are three different intersections. The central one will be seen to be unstable and hence only two values for x are possible; the system is bistable. In state A the field inside the Fabry-Perot is small, the losses in the absorber are large and prevent the intensity from growing. If, however, the other operating point C can be reached, the absorber is saturated and the decreased absorption allows the larger intensity to prevail. This is the basis for the bistability.

The stability analysis can be based on a linearisation of the equation (53), as is done in paper #2, section 3.3, but the main physics can be seen directly. If the intensity grows at the operating point A the function f(x) becomes larger than the injected signal y, and consequently the losses become larger than the inflow of energy; the amplitude tends to decrease. Similarly a decrease leads to a compensating growth. The operating point A is found to be stable. Also at the point C the operation is stable. The same discussion applied to the point B proves it to be unstable; any deviation takes the operation towards the stable points A or C.

These calculations are based on the assumption that the field can be described by only one value. There are no propagation effects or spatial variations. In spectroscopy this has often been called the 'thin medium approximation', and in the present context it has been introduced as the 'mean field theory'; see the work by Bonifacio and Lugiato (1978).

Many calculations have been performed to consider the detailed spatial and temporal behaviour of the fields inside the bistable cavity. The details of these calculations and the relevant references are found in section 3 of paper #2.

The absorptive bistability was first suggested by Lee et al (1968), who considered a system with the absorber inside a laser cavity; the suggestion to use an absorbing Fabry-Perot was given by Szöke et al (1969). Using the absorber inside a laser cavity does not change the argument given above, it only makes the theory more complicated because one needs to establish the operating point selfconsistently. The experimental investigations of the laser with an intracavity absorber did indeed show bistable behaviour, Lisitsin and Chebotaev (1968). These are the first observed optical bistabilities, and still the main ones based on saturable absorption. Most of the recent experimental work has been based on the non-linear dispersion of the medium to be discussed in the next section.

2.3. Dispersive bistability

When laser light enters a Fabry-Perot cavity, as shown in figure 4, its frequency is determined at its distant source relating it to the wavelength by

$$\omega = cq = c\frac{2\pi}{\lambda}. \tag{55}$$

When the cavity is filled with a medium of refractive index n the velocity of light becomes c/n, and in order to keep the frequency, equation (55), constant the wavelength reduces to λ/n. This changes the wave-vector of light q and makes the round-trip phase in the cavity depend on the refractive index n by

$$\Phi = 2Lq = \frac{4\pi}{\lambda} Ln. \tag{56}$$

From the theory for Fabry-Perot resonators, Yariv (1976), the ratio between the intensity I inside the cavity, and the externally impinging light intensity I_E is

$$\frac{I}{I_E} = \frac{\Lambda}{1 + F \sin^2(\Phi/2)} \tag{57}$$

where the finesse F of the cavity is given in terms of the energy transmission coefficient T of the mirrors as

$$F = \frac{4(1-T)}{T^2}. \tag{58}$$

The coefficient Λ depends only on the transmission coefficient T.

Now, if the refractive index n in the cavity depends on intensity according to

$$n(I) = n_0 + n_2 I \tag{59}$$

we find that equation (57) becomes a non-linear equation for determining the steady state intensity I inside the cavity, once the external light intensity I_E is given. The situation is shown in figure 6, and we can see that for the parameters shown in case 2, bistable operation is possible. The argument here follows closely the one used for the absorptive case in the preceding section.

The dispersive bistability was first observed in a gas-filled cell by Gibbs et al (1976), but later work has mainly concentrated on semiconductor materials. These offer a more promising component for optical computers and integrated optics, see Abraham et al (1983), and provide large non-linear coefficients. The mechanisms for these are briefly discussed in paper #2 and a more detailed description of the theory of optical processes in semiconductors can be found in Haug and Schmitt-Rink (1984).

Some work has also been carried out on so called 'hybrid systems' where part of the feedback providing the means of achieving selfconsistency is purely electronic (see e.g. Derstine at al 1982 and 1983). Unless this is a

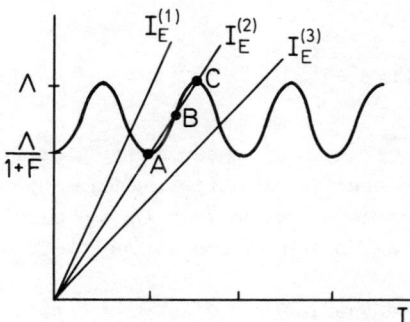

Fig.6 This diagram solves the nonlinear equation (57) for the dispersive bistability as the one in Fig. 5 solves the case of absorptive bistability. The transmission curve of the Fabry-Perot varies between Λ and $\Lambda/(1+F)$. The solution is obtained by the intersection of this curve with a linearly rising one with slope I_E. For $I^{(1)}$ and $I^{(3)}$ only one intersection appears, but for $I^{(2)}$ three intersections are seen. Of these the points A and C are stable and point B is unstable. The discussion of this case is similar to the one concerning figure 5.

digitally realised component it does not destroy the system's character of being a non-linear physical device. As a consequence, the border between a system and its analogue simulation becomes amusingly blurred. The various ways of achieving bistable operation are briefly discussed in paper #2. More recent results can be found from the Rochester conference (Bowden et al 1984).

The non-linear part of the refractive index in equation (59) can be related to the non-linear properties of the physical system used in section 1.1 as an example. We take the real part of the polarisation equations (49) and expand in the light intensity parameter I to obtain the non-linear susceptibility $\chi(I)$ as follows

$$\text{Re } P^+ = \frac{N\mu K\Delta}{\Delta^2+\gamma^2+\gamma^2 I}$$

$$= \frac{N\mu K\Delta}{\Delta^2+\gamma^2} \left(1 - \frac{\gamma^2 I}{\Delta^2+\gamma^2} + \ldots \right) \qquad (60)$$

$$= \text{Re } (\chi^{(1)}+\chi^{(3)}I) \, E$$

where we can write

$$\text{Re } \chi^{(3)} = -\frac{N\mu^2\gamma^2\Delta}{2\hbar(\Delta^2+\gamma^2)^2}. \tag{61}$$

This is a simple case of the non-linear susceptibilities introduced by Blombergen (1965) into non-linear optics. When several fields are present the susceptibility becomes a tensor, but in our simple case it can be expressed as a series expansion in the intensity I.

According to electrodynamics the electric displacement vector D is written (in SI units)

$$D = \varepsilon_0 E + \text{Re}P = \varepsilon_0 \left[1 + \frac{\text{Re } \chi^{(1)} + \text{Re } \chi^{(3)} I}{\varepsilon_0} \right] E = \varepsilon_0 \varepsilon_r E \tag{62}$$

which defines the relative dielectric constant

$$\varepsilon_r = 1 + \frac{\text{Re } \chi^{(1)}}{\varepsilon_0} + \frac{\text{Re } \chi^{(3)}}{\varepsilon_0} I. \tag{63}$$

Because the magnetic properties of our medium do not differ appreciably from those of the vacuum, the velocity of light in the medium is

$$\frac{c}{n} = \frac{1}{\sqrt{\varepsilon\mu}} = \frac{c}{\sqrt{\varepsilon_r}} \tag{64}$$

which gives the relation

$$(n_0 + n_2 I)^2 \cong n_0^2 + 2n_0 n_2 I + \ldots = 1 + \frac{\text{Re } \chi^{(1)}}{\varepsilon_0} + \frac{\text{Re } \chi^{(3)}}{\varepsilon_0} I. \tag{65}$$

The part of the refractive index which does not depend on intensity, n_0, derives from $\chi^{(1)}$, and consequently the non-linear coefficient n_2 can be expressed through the non-linear susceptibility

$$n_2 = \frac{\text{Re } \chi^{(3)}}{2n_0 \varepsilon_0}. \tag{66}$$

The relation (5.6) of paper #2 differs because of the choice of units.

2.4. Time delay instabilities

We consider a non-linear optical device which transforms its input signal V_{in} into its output signal V_{out} according to

$$V_{out} = f(V_{in}). \tag{67}$$

In bistable systems, the feedback is often achieved by turning the light beam back onto itself through the use of mirrors or light guides. Such a system is shown in figure 7. If the delay time through the loop can be neglected this makes V_{out} and V_{in} equal, but in other cases we must write

$$V_{in}(t + \tau) = V_{out}(t) = f[V_{in}(t)]. \tag{68}$$

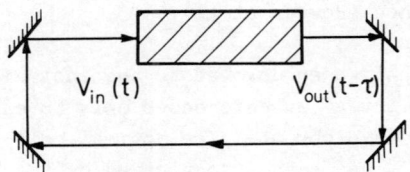

Fig. 7. A non-linear optical device transform its input signal V_{in} into its output signal V_{out}. Through an optical feedback loop the output is taken to the input of the device but with a delay time τ. Then we must set $V_{in}(t) = V_{out}(t-\tau)$ as seen in equation (68).

This provides a delay time of length τ in the loop, and from control theory it is known that this may introduce instabilities. Many an engineering creation has lost its stability thanks to a sluggish feedback mechanism.

If we look only at the input signal, we can obtain it at the successive times $t + n\tau$ by iterating the non-linear transformation

$$V(n+1) = f[V(n)]. \tag{69}$$

This can show unstable behaviour, which was first pointed out by Ikeda (1979). Many optically bistable systems do, in fact, show oscillational instabilities and a pulsating or ringing behaviour, as discussed in paper #2. The optical signal V is complex, it has got two components, but the more general equation of type (69) has come to play a very important role in the recent theory of chaotic behaviour in deterministic systems.

It was a major revolution in the thinking of physicists, when it was realised that purely deterministic systems could show a practically unpredictable behaviour; for a review and discussion see e.g. Helleman (1980). Sampling the motion according to some well defined method leads to a set of transformations similar to equation (69), which in this context is called a Poincaré map. The variable V may have any number of components but the ordinary case is a point in the two-dimensional plane.

Feigenbaum (1979, 1980) showed that even simple one-dimensional mappings have very interesting properties, characterised by a universality which is similar to that found in the theory of critical points of phase transitions. The onset of chaotic motion follows after a series of thresholds for subharmonic generation at half the frequency, the period doubling. This is one way to reach chaos in a non-linear system. Many of the optical devices used for bistability also display chaotic outputs, and the various routes to chaos have been investigated. A useful reference is the Rochester conference (Bowden et al 1984).

Much work has recently been devoted to the instabilities in optical systems. Here I wish to list a few references only to enable the reader to trace the more recent developments. The optical self-pulsing, discussed by Bonifacio et al (1979), has been investigated by the hybrid devices (Derstine et al 1982, 1983). Related work has been published by Hopf et al (1982), Ikeda et al (1982), Mandel and Kapral (1983), and Gao et al (1983a, 1983b).

Relations to low frequency noise have been presented by Arecchi et al (1982), Geisel and Nierwetberg (1982), Grossman and Fujisaka (1982), and Procaccia and Schuster (1983).

Casperson (1978) suggested that an operating laser system may turn unstable. This has been investigated experimentally by Weiss and King (1982), Gioggia and Abraham (1983, 1984), and Halas et al (1983). Recent theoretical investigations include Graham and Cho (1983) and Lugiato et al (1983). For a general overview of bistability and chaos in various systems the reader should consult Abraham et al (1984).

The stability properties of equations of type (69) can be seen in the one-dimensional case as discussed by Kadanoff (1983), Feigenbaum (1980), Eckmann (1981) and Ott (1981). In figure 8 we show the function f mapping $V(n)$ into $V(n+1)$. The next iteration can be obtained by going to the dotted diagonal line and taking this as the new value for $V(n)$. In this way we generate a zig-zag line which represents the consecutive results of the iteration. In figure 8(a) the motion approaches the point V^* where it becomes stable, we have a fixed point of the iteration

$$V^* = f(V^*). \tag{70}$$

In figure 8(b) the iteration becomes unstable and leaves the neighbourhood of the starting point; what happens later depends on the details of the problem. We can approach another fixed point, find simple oscillatory behaviour or obtain an unpredictable motion, which is termed chaotic. For details we refer to the references.

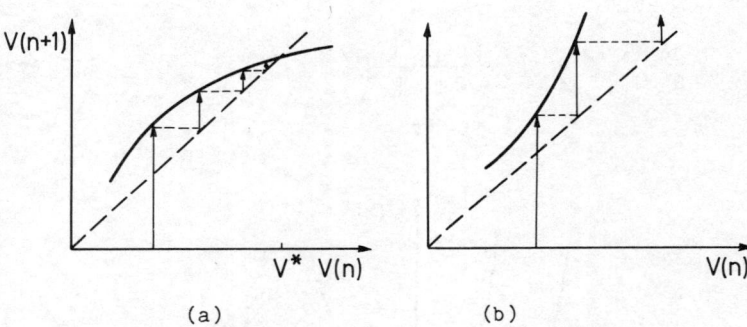

(a)   (b)

Fig.8 The graphic method to follow subsequent iterations of the equation (69) is illustrated in these figures. The case (a) shows the approach to a fixed point V^*, whereas in the case (b) the iteration rapidly takes us out from the region shown.

2.5. Connection with phase transitions

The steady state condition (54) for the non-linear medium can be written as a condition for the extremum of a potential

$$0 = y - x(1 + \frac{2C}{1+x^2}) = -\frac{\partial U}{\partial x} \tag{71}$$

with

$$U(x) = -yx + \frac{1}{2} x^2 + C \log (1+x^2)$$

$$\cong -yx + \frac{1}{2} (1+2C)x^2 - \frac{1}{2} C x^4 + \ldots$$

$$= -yx + \frac{1}{2} A x^2 + \frac{1}{4} B x^4 + \ldots, \tag{72}$$

From the series expansion we can see that this shows a certain similarity with the free energy of the Landau theory for a second order phase transition (Lifshitz and Pitaevskii 1980). Here the field x assumes the role of the order parameter. In a laser system the medium is inverted, the coefficient C is negative and we have amplification and gain. In this case the coefficient A can change sign and the free energy behaves as shown in figure 9 for the case when the injected signal y is zero. When A<0 the equilibrium value is given by

$$x = \left[\frac{-A}{B}\right]^{1/2} = \left[\frac{2|C|-1}{2|C|}\right]^{1/2} \qquad (73)$$

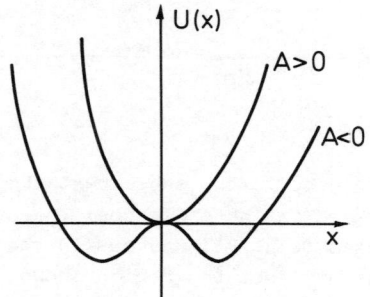

Fig. 9. The potential U(x) obtained from the absorptive bistability condition. It is similar to the free-energy curve in the Landau theory of phase transitions. For positive A the minimum energy occurs at x=0, but when A changes sign the order parameter x tends to localise itself near one of the new minima.

which gives the Landau order parameter x as a function of the critical parameter A. The analogy is discussed in some detail in section 3.5 of paper #2. The exponent in equation (73) is 1/2 and is called the critical exponent. For A < 0 two different minima occur, and the system will have to choose one of them; the symmetry becomes broken.

The injected signal y corresponds to an external field and for the case A > 0 it only tilts the free energy to one side; the minimum becomes displaced as shown in figure 10. For small values of y the displacement is small, and we can neglect the non-linear term to find

$$x \approx \frac{y}{A} = \frac{y}{1-2|C|} \qquad (74)$$

which shows that the linear response of the system (its susceptibility) diverges at the critical point $A = 1-2|C| = 0$. For the case $A < 0$, we can linearise the free energy around its minimum and calculate the response in

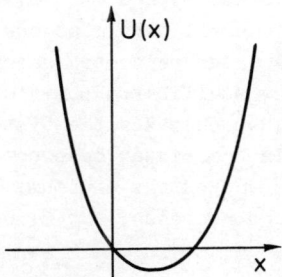

Fig. 10. When $A > 0$ the presence of an externally injected signal ($y \neq 0$) tilts the potential and displaces the minimum to a small but nonzero value of the order parameter x.

a similar way. We find that the symmetry is broken by the presence of the external field and the response becomes singular also when approaching the critical point from this side.

Taking the analogy with thermophysics seriously we must assume that the equilibrium distribution function is given by a Boltzmann factor of the form

$$P \sim \exp[-\frac{U(x)}{kT}] . \qquad (75)$$

For $A > 0$ and $y = 0$ this can be written

$$P \sim \exp(-\frac{Ax^2}{2kT}) \qquad (76)$$

and it gives the mean square fluctuations of the order parameter the value

$$\langle x^2 \rangle = \frac{\int P(x)x^2 dx}{\int P(x) dx} = \frac{kT}{A} . \qquad (77)$$

In our case the parameter x corresponds to the field amplitude in the laser cavity, and its square is proportional to the thermal energy in the cavity mode. From statistical mechanics we know that this is of the form

$$E_{cav} = \hbar\omega \langle n \rangle = \frac{\hbar\omega}{e^{\hbar\omega/kT} - 1} \approx kT \qquad (78)$$

where we have used the high temperature expansion. As we see the result is in agreement with equation (77).

When the critical point is approached, $A \to 0$, the fluctuation energy in equation (77) diverges, the thermal perturbations on the system cause ever increasing deviations in the order parameter which finally, at the critical point, goes over into its new equilibrium position. The value $A = 0$ gives the potential a flat bottom, which makes the system very sensitive to perturbations, as was seen in the linear response equation (74). By linearising the fluctuations around the minimum, one can carry out a similar calculation in the ordered case, $A < 0$, and find the analogous results.

It is unclear exactly when the analogy between a phase transition and a laser was first discovered. One of the early papers by Haken (1964) already displays the similarity clearly but without mentioning it explicitly. In 1966 the Brandeis summer school (Cretien et al 1968) was devoted to phase transitions but it also contained lectures on laser theory. The relationship was discussed explicitly by DeGiorgio and Scully (1970) and Graham and Haken (1970). These papers discussed the ordinary laser and a second order phase transition, but soon Kazantsev and Surdotovich (1970) saw the relationship between a laser with a saturable absorber and a first order phase transition. This was also discussed further by Scott et al (1975) and Salomaa and Stenholm (1977).

Suggestive as the analogy may be, it should not be pushed too far. It rests mainly on the effective field descripton of a phase transition, or its equivalent form of a Landau theory. It is now known that there are important shortcomings in this approach when it is applied to a real thermodynamic transition. The critical fluctuations are not described properly and the critical exponents come out incorrectly. There has been a large number of investigations devoted to the theory of critical phenomena, and the results can be found e.g. in Ma (1976).

3. COMMENTS ON PAPER #3.

3.1. Quantisation of fields

There actually was a quantum theory of radiation before there was one for matter. The treatment of the black-body radiation by Planck and Einstein was originally concerned solely with the propertes of the fields. Only the work by Bohr on the hydrogen atom, followed by de Broglie's concept of matter waves, included matter in the quantisation. For the early history the reader is referred to the reprint volume by ter Haar (1967).

Later developments of the theory have shown that some of the originally compelling reasons for field quantisation have become only suggestive, see e.g. Lamb and Scully (1969), Schrödinger (1927) and Boyer (1980). There are, however, no reasons to dismiss the quantum theory of radiation even though most electrodynamic phenomena can be adequately treated by a classicaltheory of light. It is the topic of paper #3 to stress the quantum features of our understanding of radiation phenomena.

The single most insistent argument for field quantisation is the emission of atomic excitation energy into an empty mode of the vacuum of the radiation field, that is spontaneous emission. If a dipole moment exists, it will radiate energy out into space; any radio transmitter does that. This fact prevented the stability of the classical planetary model of atoms and provided the reason for Bohr's original theory of the atom. In quantum theory an excited state even without a dipole moment drops to the lower level at a rate given by equation (26).

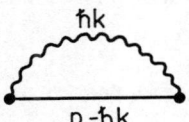

Fig. 11. An excited atom propagates with the momentum p. It emits a virtual photon of momentum $\hbar k$ and re-absorbs it later. Its propagator will acquire a self-energy, which to lowest order is given by this diagram.

Let us briefly review the result of quantum eletrodynamics. We consider an excited atom propagating in its upper level. It will decay spontaneously to the lower level and the atomic propagator will have a contribution to the self-energy Σ given in figure 11. The perturbation theory expression for this diagram is

$$\Sigma = \int d^3q \, \frac{|\gamma(q)|^2}{\omega_{eg} - cq + i\varepsilon} \, . \tag{79}$$

The imaginary part of this is given by

$$\text{Im } \Sigma = -\pi \int d^3q \, |\gamma(q)|^2 \delta(\omega_{eg} - cq) = -\frac{1}{2}\Gamma \tag{80}$$

and this is the decay rate of the quantum mechanical probability amplitude of the excited state. Hence Γ is the decay rate of the upper level and an evaluation of the expression (80) gives the result (26); see e.g. Stenholm (1984).

The real part of equation (79) gives an energy shift, which is found to be divergent when the integration is extended to large photon momenta q:

$$\text{Re } \Sigma = \int d^3q \, \frac{|\gamma(q)|^2}{\omega_{eg} - cq}$$

$$\sim \int d^3q \, \frac{|\gamma(q)|^2}{q} \sim \int dq \, . \tag{81}$$

This was the problem that Bethe (1947) solved when he suggested the renormalisation scheme, which has developed into a calculational machinery capable of making sense of divergent expressions in quantum field theory. For a simple discussion see Feynman (1962) or for a modern one see Itzykson and Zuber (1980). From this perturbation theory one derives an expression for the level shift (81), the Lamb shift, which agrees with experiments to an astonishing degree.

The point here is to stress the fact that both the presence of an energy shift and the spontaneous decay derive ultimately from the atom's ability to emit its excitation energy into empty space. This feature can only be discussed satisfactorily from the point of view of a quantised field theory. In this the field cannot be exactly zero, its degrees of freedom must have zero point fluctuations which can induce transitions even if the field is in its ground state. The considerations of vacuum fluctuations do indeed give a qualitatively correct picture of the quantum effects, as shown by Welton (1948). This point of view has, however, recently been criticised by Ginzburg (1983). A modern discussion of the description of quantum fluctuations is given by Cohen-Tannoudji (1982).

Introduction

The quantum theory of light, see e.g. Loudon (1983), writes the electromagnetic field variables in terms of a set of classical field modes $U_i(\bar{r})$ in the form

$$E(\bar{r},t) = \sum_i \left[a_i^\dagger(t) U_i(\bar{r}) + h.c. \right] \quad (82)$$

where the operators $a_i^\dagger$ act on the states of the field, and are said to create a photon into the mode $U_i(\bar{r})$. If there are n photons in this mode we write the state as $|n\rangle$ giving the relations

$$a_i^\dagger |n\rangle = \sqrt{n+1} \; |n+1\rangle$$

$$a_i |n+1\rangle = \sqrt{n+1} \; |n\rangle \quad (83)$$

$$a_j a_i^\dagger - a_i^\dagger a_j = \delta_{ij}$$

which characterise boson operators. The normalised eigenstates of the number operator $a^\dagger a$ are found to be

$$|n\rangle = \frac{1}{\sqrt{n!}} (a_i^\dagger)^n |0\rangle \quad (84)$$

and the Hamiltonian is written

$$H = \sum_i \hbar\omega_i (a_i^\dagger a_i + \frac{1}{2}). \quad (85)$$

The whole formalism of quantising the fields contains some subtlety but is essentially based on the reduction of Maxwell's equations to a set of independent harmonic oscillators, which then are quantised in the ordinary fashion. The algebraic relations in equations (83)-(85) are just the usual ones for the occupation number representation of the harmonic oscillator. The theoreticians' habit of referring to photons as particles need not be taken too seriously. The quantum number may just be taken to label the excitation state of a quantised field mode. This point of view is advocated by van Vleck and Huber (1977), and Scully and Sargent (1977) discuss the necessity and the misuse of the concept of a photon in quantum electronics.

The theory of measurements is usually complicated in quantum mechanics, but for photon detectors it can be formulated in a surprisingly simple way, see Glauber (1965). The detector is an absorber which is situated far away from the region of the physical action. Here only the field variables are accessible and we can describe the process as a simple photon destruction. As an example let us consider the detection of a photon in mode 1 at time t_1 and another in mode 2 at time t_2. The matrix element from an initial state to a final one is for this process

$$V_{fi} = \langle f | a_2(t_2) a_1(t_1) | i \rangle . \qquad (86)$$

The quantum mechanical rate of absorption, the counting rate, is proportional to the square of this. The final state cf the field is, however, not recorded and there appears an average over the initial states with the weight P_i. As a result the observation gives a result proportional to the quantity

$$\begin{aligned}
W &= \sum_{if} P_i \; | \langle f | a_2 a_1 | i \rangle |^2 \\
&= \sum_{if} P_i \; \langle i | a_1{}^\dagger a_2{}^\dagger | f \rangle \langle f | a_2 a_1 | i \rangle \\
&= \sum_i P_i \langle i | a_1{}^\dagger a_2{}^\dagger a_2 a_1 | i \rangle \\
&= \langle a_1{}^\dagger(t_1) a_2{}^\dagger(t_2) a_2(t_2) a_1(t_1) \rangle .
\end{aligned} \qquad (87)$$

This result was derived by Glauber (1965) and can easily be generalised to any number of photons. The important point is that the operators are normally ordered, i.e. the creation operators $a^\dagger$ all stand to the left of the annihilation operators a. This implies that the second absorption process has to take place in a situation where the first photon has already been absorbed. The quantum nature of the states then takes care of the combinatorial factors.

For a general field we separate the absorbing and emitting parts by setting

$$E(\bar{r},t) = \sum_q [C_q \, a^\dagger{}_q \, U_q(\bar{r}) + \text{h.c.}]$$

$$= E^{(-)}(\bar{r},t) + E^{(+)}(\bar{r},t) . \qquad (88)$$

Generalising the result (87) we obtain the probability for finding a photon at point $\bar{r}$ at time 0 from

$$\begin{aligned}
W &= \langle E^{(-)}(\bar{r}) E^{(+)}(\bar{r}) \rangle \\
&= \sum_{qq'} U_q(\bar{r}) U_{q'}{}^*(\bar{r}) \langle a_q^\dagger{}_{q'} a_q \rangle .
\end{aligned} \qquad (89)$$

If the field is expressed in terms of the eigenfunctions of the energy (85) we find

$$W = \sum_q |U_q(\bar{r})|^2 \, n_q . \qquad (90)$$

This result shows the nature of the quantum theory of light. The factors n_q give the probability of absorbing photons from the field, but the spatial probability is determined by the classical mode functions $U(\bar{r})$, and hence all results of classical diffraction theory can be found in the quantum theory as well. Such quantum calculations were carried out in the paper by Fermi (1932), just to show the agreement with the previously known phenomena.

Dirac (1927) introduced the annihilation operator through the relation

$$a_q = \theta_q \, n_q^{1/2} . \qquad (91)$$

From the properties of the operators a and $a^\dagger$ we derive the relations

$$\theta_q \, \theta^\dagger_q = 1$$

$$\theta_q^\dagger \, \theta_q = 1 - |0\rangle\langle 0| \qquad (92)$$

which shows that the operator θ_q is unitary except when acting on the vacuum $|0\rangle$. The commutation relation is given by

$$\theta_q \, n_q - n_q \, \theta_q = \theta_q . \qquad (93)$$

If we neglect the fact that θ_q is not exactly unitary, we can introduce the phase operator ϕ_q through the relation

$$\theta_q = \exp(-i\phi_q) \qquad (94)$$

which defines ϕ_q for states having a neglible admixture of the vacuum. If the phase is small the relation (93) is equivalent to

$$[n_q, \phi_q] = 1 \qquad (95)$$

which shows that the phase is the conjugate variable to the occupation number. This gives the uncertainty relation

$$\Delta n_q \, \Delta \phi_q \geq \frac{1}{2} \qquad (96)$$

which implies complementarity between knowing the phase of the field and its photon number. This discussion is not exact but forms the basis for more detailed discussions; see Carruthers and Nieto (1968) and Fain (1967).

For bosonic variables there are no restrictions on the occupation numbers, and hence its uncertainty can be made arbitrarily large. Consequently the phase can be well defined and the classical limit of a boson system can be a classical field. A similar discussion for fermions gives that the uncertainty of the occupation number must not exceed one and hence the phase must always remain uncertain. The classical limit of a fermion field is a particle description and not a field; see the discussion by Peierls (1979).

3.2. Classical states of the field

In the states $|n\rangle$ the phase is unknown and there exists no field variable. This is seen for a single mode field (82)

$$E = \epsilon (a U + a^\dagger U^*) \tag{97}$$

which has the expectation value

$$\langle n|E|n\rangle = 0. \tag{98}$$

To describe the behaviour of nearly classical fields Glauber (1963) introduced the states

$$|z\rangle = \exp[-\frac{1}{2}|z|^2] e^{za^\dagger} |0\rangle$$

$$= \exp[-\frac{1}{2}|z|^2] \sum_{n=0}^{\infty} \frac{z^n}{\sqrt{n!}} |n\rangle \tag{99}$$

which are called coherent states. They are eigenstates of the annihilation operator

$$a|z\rangle = z|z\rangle \tag{100}$$

and the probability of finding n photons in it is a Poissonian distribution

$$P_n = \frac{|z|^{2n}}{n!} e^{-|z|^2} \tag{101}$$

with the average photon number

$$\langle z|n|z\rangle = \sum_{n=0}^{\infty} n\, P_n = |z|^2 \qquad (102)$$

and the Poissonian dispersion

$$\langle \Delta n^2 \rangle = \langle n^2 \rangle - \langle n \rangle^2 = \langle n \rangle . \qquad (103)$$

Glauber wrote the density matrix of the field system in the form

$$\rho = \int P(z)\, |z\rangle\, dz\, \langle z| \qquad (104)$$

where $P(z)$ is a quasi-probability distribution. This is not always positive, and hence it cannot be interpreted as the weight function for the state $|z\rangle$. This state occurs with the probability

$$\langle z|\rho|z\rangle = \int e^{-|z-\xi|^2} P(\xi)\, d\xi \qquad (105)$$

which is always positive. The gaussian weight in equation (105) smears the distribution P over a unit circle in the complex z-plane, because in the state $|z\rangle$ the phase related to the operator a is well defined. We cannot know the photon number arbitrarily precisely, and hence the structure of P within each circle of radius unity can signify no classically observable property.

To see the appearance of the classical result we evaluate the action of the creation and annihilation operators on the Glauber P function. By direct calculation we find

$$a\,|z\rangle\langle z| = z|z\rangle\langle z|$$

$$a^\dagger\,|z\rangle\langle z| = \left(\frac{\partial}{\partial z} + z^*\right)|z\rangle\langle z| \qquad (106)$$

which implies for the density matrix

$$a\rho = \int z P(z)|z\rangle\, dz\, \langle z|$$

$$a^\dagger \rho = \int \left(z^* - \frac{\partial}{\partial z}\right) P(z)|z\rangle\, dz\, \langle z| \qquad (107)$$

where a partial integration has been carried out.

The state of the field can be regarded as classical when the average photon number n is much larger than one, and the Glauber function P does not vary much over the unit circle. The last requirement eliminates the possibility of large non-classical effects. These conditions require the function P to satisfy

$$\left|\frac{\partial P}{\partial z}\right| \cdot 1 \ll |P| \ll |zP| \tag{108}$$

over most of the area where it is non-zero. The relations can be written

$$\frac{1}{|zP|} \cdot \left|\frac{\partial P}{\partial z}\right| = \frac{\partial \log P}{\sqrt{n}\, \partial\sqrt{n}} \sim \frac{\partial \log P}{\partial n} \ll 1. \tag{109}$$

For a Poisson distribution of the type (101) we apply Stirling's approximation and find

$$\log P = n \log \langle n \rangle - n(\log n - 1). \tag{110}$$

The condition for the validity of the classical description becomes

$$\frac{\partial \log P}{\partial n} = \log \frac{\langle n \rangle}{n} \approx \log \frac{\langle n \rangle}{\langle n \rangle - \sqrt{\langle n \rangle}} \sim \frac{1}{\sqrt{\langle n \rangle}} \tag{111}$$

where the width of the distribution is taken as a typical value for n. These considerations are found in Kazantsev and Surdotovich (1969). They imply that when (109) and (110) are satisfied the derivative term in equation (107) can be omitted and for these states the creation and annihilation operators can be replaced by c numbers; the theory becomes indistinguishable from the classical one. The field (97) is then given by

$$E = \varepsilon (z U + z^* U^*). \tag{112}$$

These considerations are easily extended to several modes of the radiation field. The relation of quantum oscillators to their classical description is discussed also by Mollow (1967).

3.3. The correlation functions

The main point of paper #3 is that there are non-classical effects to be found in the correlation functions of electromagnetic fields. Such functions have been widely used to discuss the coherence properties of light, see Mandel and Wolf (1965). Of special interest is the two-photon correlation function, which according to the result (87) must be written as the normally ordered expression

$$g^{(2)}(t) = \frac{\langle a^\dagger(0) a^\dagger(t) a(t) a(0)\rangle}{|\langle a^\dagger a\rangle|^2} . \qquad (113)$$

The corresponding classical coherence function is discussed in section 2 of paper #3.

Using the commutation properties for the boson operators we find the initial value

$$g^{(2)}(0) = \frac{\langle a^\dagger [a^\dagger, a] a\rangle + \langle (a^\dagger a)^2\rangle}{|\langle a^\dagger a\rangle|^2} \qquad (114)$$

$$= 1 + \frac{Q}{\langle n\rangle}$$

where the deviation from Poisson statistics is given by the Mandel parameter (Mandel 1979)

$$Q = \frac{\langle n^2\rangle - \langle n\rangle^2 - \langle n\rangle}{\langle n\rangle} . \qquad (115)$$

This is zero for a Poisson distribution and hence $g^{(2)}(0) = 1$. For a broader distribution, a super-Poissonian one, its value is larger. Using the Glauber P representation (104) of the density matrix we find

$$g^{(2)}(0) = \frac{1}{\langle |z|^2\rangle^2} \int P(z) |z|^4 \, dz$$

$$= 1 + \int P(z) \frac{(|z|^2 - \langle |z|^2\rangle)^2}{\langle |z|^2\rangle^2} . \qquad (116)$$

In the classical case, the P function is always positive and we see that the statistical distribution must be super-Poissonian

$$g^{(2)}_{cl}(0) > 1 . \qquad (117)$$

The photons tend to appear in bunches. For a quantum field, on the other hand, the distribution function $P(z)$ can be non-positive and we may have

$Q < 0$

$$g^{(2)}(0) < 1. \tag{118}$$

This shows that the sub-Poissonian statistical distribution is a genuine quantum effect, which cannot be found in a classical theory. A review of photon antibunching is given by Paul (1982).

The standard example is the photon field emitted in resonance fluorescence, as discussed in section 1.4. To relate the present approach to our earlier treatment we must express the observable (113) in terms of the atomic dipole operator (33). From electrodynamics we find the relation between the field at a distant observation point r from a radiating dipole μ

$$E(r) = \frac{1}{4\pi\varepsilon_0 c^2 r} \mu(\tau) \tag{119}$$

where the retarded time τ is given by

$$\tau = t - r/c. \tag{120}$$

Because the relationship between the dipole and the field is linear, the same relationship can be taken over directly in the quantum theory.

Generalising the result (113) and using the relation (119) we find

$$g^{(2)}(t) = \frac{\langle E^{(-)}(0)E^{(-)}(t)E^{(+)}(t)E^{(+)}(0)\rangle}{\langle E^{(-)}E^{(+)}\rangle^2}$$

$$\sim \langle \mu^-(0)\,\mu^-(t)\,\mu^+(t)\,\mu^+(0)\rangle. \tag{121}$$

The correlation function in equation (121) can now be evaluated by the quantum regression theorem as in section 1.4. We use the expansion (33a) of the dipole. The time dependence of the correlation function is the same as that of the operator

$$\mu^-(t)\mu^+(t) \sim \sigma^-(t)\sigma^+(t)$$

$$= \frac{1}{2}[1 + \sigma_3(t)] \tag{122}$$

with the appropriate initial condition.

In the case of resonance fluorescence the spontaneous emission of a photon must be followed by the restriction of the atomic state to the lower level. Solving the equations (34) for the Bloch vector with the initial condition

$$\langle \mu_3(0) \rangle = -1 \tag{123}$$

and using the approximation (37) valid for large field intensities, we find

$$\langle \mu_3(t) \rangle = -e^{-3\gamma t/2} \cos(2\,Kt) \tag{124}$$

from which we obtain

$$g^{(2)} \sim \left[1 - e^{-3t\gamma/2} \cos(2\,Kt)\right]. \tag{125}$$

At the initial time t=0, this function gives zero, which is an extreme case of violation of the classical behaviour (117). The result (125) is given in equation (4.37) of paper #3 as an example. Its experimental verification is also discussed further in the paper.

The physical reason for the behaviour of the correlation function is easy to see. Once a photon is emitted, the atom is found in the lower state and it takes the field some time to re-excite the atom to the upper level, from which the next photon can start. On average this is of the order of the Rabi period $(2K)^{-1}$ as can be seen from equation (125). Hence the outgoing, spontaneously emitted photons have a tendency to appear regularly, which gives them a smaller variance than a perfectly random, Poissonian distribution. They are antibunched.

We should, however, be wary of interpreting the antibunching as too closely related to the state of the radiation field. From the result (121) we can see that the property is solely due to the characteristics of the radiating atomic system and not particularly characteristic of the field. This only acts as the carrier which transfers the quantum properties of the radiating dipole to the distant detector. If we could observe the dipole directly we would see the same behaviour. In paper #3 there are other, more complicated examples of non-classical correlation functions. It seems to me that the same argument applies to all of these, the quantum behaviour takes place locally in the radiating system, and these properties are passively carried by the field to the observer. In this sense the radiation field can be said to display non-classical correlations, but the examples appear not to give directly an illustration of the violation of the inequality (117) due to the properties of the P function. The antibunching in two-photon absorption seems to be more promising, see section 5.2 of paper #3. The paper also discusses some of the details connected with experiments on non-classical aspects of light.

3.4. Squeezed states

One very recent development in the quantum theory of light fields is not covered in paper #3. This concerns a set of states which differ from classical ones even in the limit of a large amplitude. They are called squeezed states; they were introduced by Stoler (1970, 1971) and Yuen (1976). We cannot here completely review all recent publications, but a brief presentation of the basic idea and some representative references will be given.

To understand the idea we must remember that the basic quantum mechanical restriction

$$\Delta p \, \Delta x \geq \tfrac{1}{2}\hbar \qquad (126)$$

does not prevent us from making the one uncertainty as small as we please, if only the quantum fluctuations are taken up by the other uncertainty. The most classical situation is one where the uncertainty is distributed equally between the two, both position and momentum are reasonably well defined, and a classical path does exist approximately. For the equality sign to hold, we must use the well-known minimum uncertainty wave packets.

For the coherent states (99) we have an uncertainty in the photon number of roughly the order $\sqrt{\langle n \rangle}$, but the phase is well defined. If we divide up the operator a in its Hermitian components by setting

$$a = X_1 + iX_2, \qquad (127)$$

we find that in the complex z-plane the expectation values of X_1 and X_2 can be used to label the axes. In figure 12 we show the circle of highest probability for a coherent state.

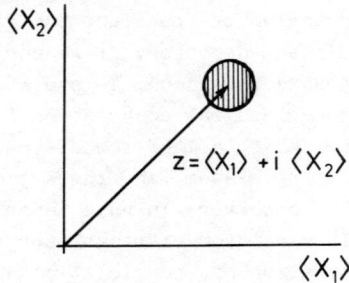

Fig. 12. The coherent state distributes its uncertainties equally between the two components X_1 and X_2.

If we use the commutation rules from (83) we find

Introduction

$$[X_1, X_2] = \frac{i}{2} \qquad (128)$$

which implies the uncertainty product

$$\Delta X_1 \, \Delta X_2 \geq \frac{1}{4}. \qquad (129)$$

From figure 12 we can see that, in a coherent state, the uncertainty is symmetrically divided between X_1 and X_2. This equal distribution of the quantum noise is the basis for estimates of quantum limits on the performance of optical devices.

If the uncertainty could be unevenly distributed between the components X_1 and X_2 one of them could be read with a greater accuracy than the quantum limit, at the expense of the accuracy in the other one. Such a state is shown in figure 13 centred at the origin. Here

$$\Delta X_1 \ll \Delta X_2. \qquad (130)$$

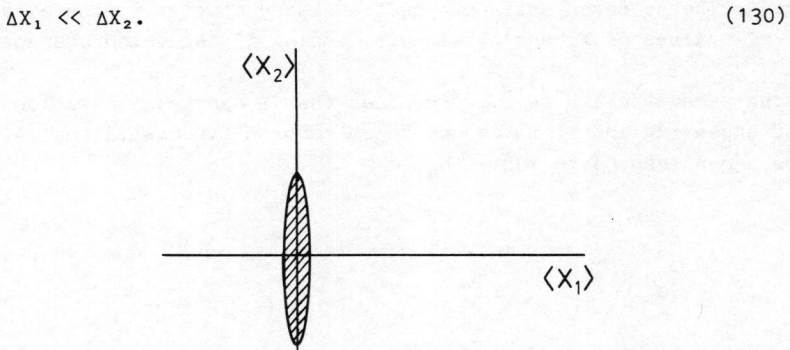

Fig.13. We can decrease the uncertainty in the one direction (X_1) at the expense of that in the other one (X_2); this is a squeezed state.

Such a state can be generated by the squeezing operator

$$S(\xi) = \exp\left[\tfrac{1}{2}\xi^* a^2 - \tfrac{1}{2}\xi a^{\dagger 2}\right]. \qquad (131)$$

The coherent state can be generated from the vacuum state by the displacement operator

$$D(z) = e^{-\frac{1}{2}|z|^2} e^{za^\dagger} e^{-z^* a}$$
$$= \exp(za^\dagger - z^* a) \qquad (132)$$

as we can see from equation (99). This displacement operator can also be used to take the squeezed state generated at the origin by (131) to any place z in the complex plane. We thus define the squeezed states

$$|z,\xi\rangle = D(z)S(\xi)|0\rangle \qquad (133)$$

and their probability distribution is shown in figure 14. If we write the

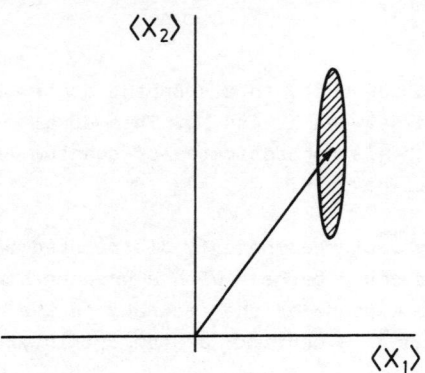

Fig.14. The squeezed state can be translated to give finite expectation values of X_1 and X_2 without losing its squeezed character.

squeezing parameter $\xi = re^{i\theta}$, we can see that θ denotes a rotation of the axis of squeezing and it is chosen to be zero in figures 13 and 14. It can then be shown (see Caves 1981) that

$$\Delta X_1 = \tfrac{1}{2}e^{-r}$$
$$\Delta X_2 = \tfrac{1}{2}e^{r}. \qquad (134)$$

The squeezing is hence indicated by the parameter r; its value r=0 denotes the limiting case of equal distribution of uncertainty.

The squeezing is a quantum characteristic of the states, which lacks a classical interpretation, but it retains its meaning when the state is displaced arbitrarily far into the complex plane by z. This shows that we can have arbitrarily large photon numbers in the field for any given value of r; in the state (133) the average photon number is

$$\langle n \rangle = z^*z + \sinh^2 r. \qquad (135)$$

Consequently the squeezed states are of great interest from a fundamental point of view. Their existence is a rigorous implication of the formalism, but it remains to be seen if the actual electromagnetic field can be prepared in such states.

The squeezed states are also of great practical interest for applications. They have been suggested for optical information transmission (Yuan and Shapiro 1978, 1979, 1980 and Shapiro et al 1979) with minimum noise. For the detection of the feeble signals caused by gravitational waves circumvention of the conventional quantum noise limitations may be essential (Hollenhorst 1979, Bragensky et al 1980 and Coves et al 1980).

There have been many suggestions for the production of squeezed states. The simplest is the parametric process where one photon is split into two according to figure 15 (the process inverse to second harmonic generation). If the incoming field is strong enough to be describable

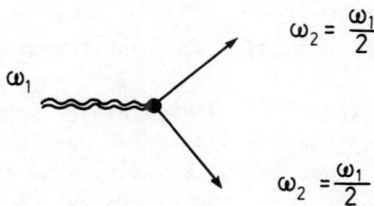

Fig. 15. In this parametric process a photon of frequency ω_1 is split in to two new photons of equal frequency.

Classically the photon generation can be represented by the Hamiltonian

$$H = \hbar[Va^2 + V^*a^{\dagger 2}] .\qquad (136)$$

At exact resonance this is the correct operator in the interaction picture. It gives rise to the time evolution operator

$$U(t) = \exp[i(Va^2+V^*a^{\dagger 2})t] \qquad (137)$$

which according to equation (131) produces a squeezed state out of the vacuum with the squeezing parameter

$$r = 2|V|t .\qquad (138)$$

This straightforward method does, however, not seem to offer a realistic possibility. For a review of various suggestions the concise and clear article by Walls (1983) should be consulted. Later discussions have been provided by Loudon (1984a, 1984b).

The experimental verification of the presense of a squeezed state is not obvious. In homodyn detection the squeezing manifests itself in the photon counting statistics (Mandel 1982 and Loudon 1984b). The effect of photon statistics on interferometric experiments is discussed by Heidmann et al (1985). The non-classical nature of the squeezed states does not necessarily manifest itself in photon antibunching as discussed in section 3.3. When the light intensity is large ($|z|^2 \gg \sinh^2 r$ in equation (135)), the correlation function $g^{(2)}(0)$ in equation (114) becomes

$$g^{(2)}(0) = 1 + \frac{2}{|z|^2}(e^{-2r} - 1) \qquad (139)$$

which is larger or smaller than unity depending on the sign of r (see Walls 1983). In the limit of small intensity, $|z| \ll 1$, the result is

$$g^{(2)}(0) = 1 + (1 + \operatorname{ctgh}^2 r) > 1 \qquad (140)$$

and the 'squeezed vacuum' can never show antibunching.

In conclusion I note that the squeezed states denote an interesting formal property of the vacuum oscillators we use to describe the quantised electromagnetic fields. Whether it is possible to prepare such states and utilise them in optical detection and communication schemes remains a question to be settled in the future. I want to thank Professor Loudon for discussions and suggestions concerning my treatment of paper #3.

4. COMMENTS ON PAPER #4

4.1. The nature of quantum mechanics

It was recognised early that quantum mechanics implied a total revision of our manner of describing nature. It is not a theory of physical reality as much as a theory of how we obtain knowledge about phenomena in nature. Paraphrasing Bohr we can say that the question is not so much how the physical objects are, but how we can obtain unambiguously communicable information about them.

Quantum mechanics arose in an atmosphere of positivistic philosophy, and many hailed the development as a natural consequence of the new theory of knowledge. Others claimed that the supremacy of mind over matter had finally been vindicated; no purely materialistic theory could henceforth be taken seriously. Sixty years of philosophical evolution and experimental progress have made cautious scientists much less categorical in claiming philosophical implications of our contemporary theoretical understanding.

It has never been clear what kind of reality we should attribute to the quantum objects between the acts of observations. Only recorded measurements are unambiguously objective, and according to Bohr the rest is a purely formal apparatus useful for predictions only. If we try to attach to an object all the results of possible measurements, we clearly create a paradoxical situation.

The most straightforward consequence of the quantum mechanical formalism is that it does not allow a local description of events. The main lesson of wave mechanics is that interference appears between spatially separated parts of the experimental setup. This predicts a definite quantum behaviour, which shows that it is more than merely smeared classical mechanics. The non-local character is most trivially seen in the standard scattering experiment of figure 16. The flux of incoming particles is re-directed in space in the form of outgoing spherical waves. If we look at only one particle the wave determines its detection probability in space. An array of detectors D_i are used to verify this, and once the particle observation has been secured in one of them, D_1 say, the detection probability drops to zero for the others. This is a non-local effect, and it is accepted by every one understanding the basis of quantum theory.

The counter-intuitive character of the quantum mechanical non-locality is not always obvious. The point was emphasised by Einstein et al (1935) in what has been called the EPR paradox. After the initial period of discussions, the rapid and successful deployment of quantum theory to new phenomena relegated the problem to the physicists' coffee table discus-

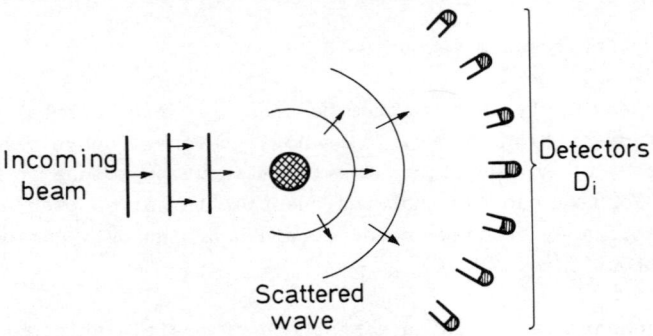

Fig. 16. The scattering of an incoming collimated particle beam by a local target. The outgoing wave reaches a battery of detectors, each one receiving a finite probability to record the particles; when one detector records the particle the probability drops to zero for an observation at all other detectors.

sions. Only when Bell (1964, 1966) showed that the counter-intuitive content of quantum mechanics could be expressed in the form of exact mathematical relationships did the interest flare up again. Its predictions could be subjected to experimental tests, and it turned out that certain very natural assumptions are not in agreement with experiments.

Here we do not intend to enter the discussion of the interpretation of quantum theory or its history. The interested reader is referred to the detailed exposition by Jammer (1974). Paper #4 discusses the possibilities to test the theory from a fundamental point of view, and in this section we shall briefly present the conceptual frame of the approach. Some key references are reprinted in Wheeler and Zurek (1983).

4.2. Einstein, Podolsky and Rosen

The original discussion by Einstein et al (1935) was concerned with the spatial properties of a pair of particles. A conceptually simpler version was introduced by Bohm (1951). He considered the decaying system at O in figure 17, which emits two spin 1/2 particles A_1 and A_2 in opposite directions. If the total spin is zero (S state) the wavefunction is

$$\psi = \frac{1}{\sqrt{2}} (\alpha_1 \beta_2 - \beta_1 \alpha_2) \qquad (141)$$

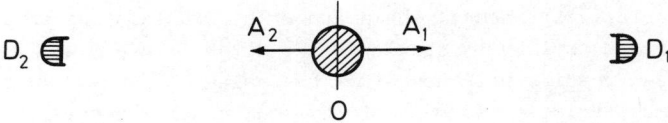

Fig. 17. A system in a spin singlet state decays at the origin O into particles A_1 and A_2, and their spinstates are eventually recorded in the detectors D_1 and D_2.

where the two-level states α and β are defined by

$$\alpha = \begin{bmatrix} 1 \\ 0 \end{bmatrix} \quad \beta = \begin{bmatrix} 0 \\ 1 \end{bmatrix}. \tag{142}$$

The subscripts refer to particles 1 and 2 respectively.

According to quantum mechanics, if the detector D_1 finds that particle A_1 is in the spin state α then the detector D_2 must see particle A_2 in the state β and vice versa. Thus the result obtained at one observation point restricts the values possible at another one however large their separation. It should be stressed, however, that only a correlation appears and there is no question of transmitting any influence across the distance between the observation points. No information is carried across space.

There exists a similar non-local correlation in the classical case. If, by accident, you pack only one shoe, and when arriving in Paris you find it to be the left one, you know immediately that the shoe remaining at home is the right one. The quantum mechanical correlation is, however, more complex. According to the theory, an S state is rotationally invariant and the reasoning is independent of the direction chosen for the spin measurement. We can arrange the detectors in figure 17 to switch between measuring different components of the spin at random, see d'Espagnat (1979). If the two experimental records are compared, it is found that when, by accident, the two detectors have measured the same spin direction, there appears the correlation described above, but when they have been measuring in orthogonal directions no special connection can be found between the results. This is in accordance with quantum theory but not with any classical theory known today. This correlation between the outcomes of multiple-choice experiments is a manifestation of the non-local correlations in quantum mechanics.

Recently the instantaneous character of the appearance of the correlations has received much attention. This may be of interest from the point of view of relativity theory. The argument by Einstein was a different one. He characterised reality by stating: 'If, without in any way disturbing a system, we can predict with certainty the value of a physical quantity, then there must exist an element of physical reality corresponding to this physical quantity.' In the spin-correlation experiment, if we know the value of the x component at D_1 we know the corresponding value at D_2 without taking any action there; the same holds for the components y and z. Thus all these components must have physical reality according to Einstein. Because quantum mechanics does not allow this, the theory must be an incomplete description of physical reality. This is Einstein's reality, which has been discussed extensively later; see section 2 of paper #4.

The paradoxical formulation of the problem can be partly removed by using the density matrix description by Cantrell and Scully (1978). The reduced density matrix describing particle A_1 in the state (141) becomes

$$\rho_2 = Tr_1 \ |\psi\rangle\langle\psi|$$

$$= \frac{1}{2} \begin{bmatrix} 1 & 0 \\ 0 & 1 \end{bmatrix}. \quad (143)$$

Looking at the spin in a direction differing from the one chosen in writing equation (141), we must perform a rotation, i.e. a unitary transformation, on the states. It is easily seen that the density matrix (143) is invariant with respect to any unitary transformation, and hence the statistical properties of particle A_2 are independent of direction. When A_1 has been recorded the situation changes, of course. We must then project (143) on the observed direction and, if we find + 1/2 at D_1, we must have the reduced density matrix

$$\rho_2 = \begin{bmatrix} 0 & 0 \\ 0 & 1 \end{bmatrix} \quad (144)$$

which is no longer invariant under rotation of the axes. The direction along which the correlation appears becomes fixed by the first recorded observation.

4.3 The Bell inequalities

During the fifties there was a revival of attempts to derive quantum theory from an underlying theory without some of the less satisfactory features of quantum mechanics; these were called 'hidden variable theories' (Belinfante 1973). John Bell got involved in this work, and in 1964 he wrote a paper where he stated: 'It is urged that in further examination of this problem an interesting axiom would be that mutually distant systems are independent of each other.' Owing to accidental reasons, this paper was published only in 1966 (Bell 1966), so Bell had time to publish his own investigation (Bell 1964) into this problem, where he established the first of a series of inequalities now associated with his name.

The Bell inequalities are usually assumed to derive from the assertions:

1. The objects of nature are real and have their properties irrespective of our observational acts. The outcomes of all possible measurements are uniquely determined by these properties.

2. The properties of an object are determined by its local conditions and not by distant objects or events.

To see the type of argument that is used, we derive one of the simplest inequalities and show it to be in conflict with quantum theory, see Wigner (1970). We take the case of two particles with spin 1/2 travelling away from a region where they were prepared in an S state like (141), see figure 17. At the point D_1 we measure a spin-direction a and denote the event of obtaining +1/2 by 'a' and the event of obtaining -1/2 by '$\bar{a}$'. The probabilities for these two events must satisfy

$$P(a) + P(\bar{a}) = 1. \tag{145}$$

Because we have assumed that the two spins always give opposite results we do not need to display the spin value at the point D_2 explicitly, because it is fixed by that at D_1. The realism assumed guarantees that these values have an existence independent of any observations.

We now introduce another direction b of the spin. The assumptions of realism and locality suggest that we may introduce a joint probability function P(a,b) for the case when a is obtained if measuring in this direction, but b if measuring in direction b. By '$\bar{b}$' we again denote the value -1/2 in the direction b. These satisfy

$$P(a,b) + P(a,\bar{b}) = P(a). \tag{146}$$

In P(a,$\bar{b}$) the bar signifies that we have +1/2 in the direction a at D_1 and +1/2 in the direction b at D_2 (because we have -1/2 in direction b at D_1).

To conclude the argument we choose a third, independent direction c and the joint probabilities P(a,b,c). Remember that specifying the spin in any of the directions a,b or c we establish the value of the spin in just the opposite direction at the other detector. We have the relation

$$P(a,b,c) + P(a,b,\bar{c}) = P(a,b). \tag{147}$$

Because probabilities are positive we can write

$$P(a,\bar{b},c) + P(\bar{a},b,\bar{c}) \geq 0$$

$$P(a,\bar{b},c) + P(a,\bar{b},\bar{c}) + P(\bar{a},b,\bar{c}) + P(a,b,\bar{c})$$

$$\geq P(a,\bar{b},\bar{c}) + P(a,b,\bar{c})$$

$$P(a,\bar{b}) + P(b,\bar{c}) \geq P(a,\bar{c}). \tag{148}$$

This is a Bell inequality establishing a relation between three, by assumption independent, spin measurements.

To test the result obtained we need to calculate the quantum predictions for the probabilities P. We choose to do this for three polarisation states of two photons described by the S state (141). Then no longitudinal component of the spin is present. A detector rotated by an angle ϕ tests for the presence of the state

$$|\phi\rangle = \cos(\frac{1}{2}\phi)|\alpha\rangle + \sin(\frac{1}{2}\phi)|\beta\rangle \tag{149}$$

which can be verified by calculating the expectation value of the z

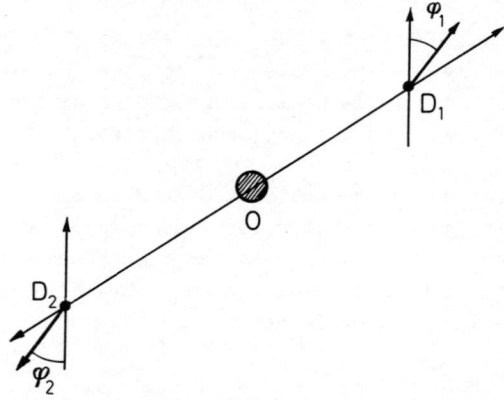

Fig. 18. The angles ϕ_1 and ϕ_2 used to define the projectors used to describe the quantum mechanical situation discussed in the text.

component of the spin

$$\langle\phi|\sigma_z|\phi\rangle = \cos^2(\tfrac{1}{2}\phi) - \sin^2(\tfrac{1}{2}\phi) = \cos\phi. \tag{150}$$

A filter turned by the angle ϕ_1, see figure 18, projects the incoming states on the state $|\phi_1\rangle$ and this projector is

$$Q_1 = \begin{bmatrix} \cos(\tfrac{1}{2}\phi_1) \\ \sin(\tfrac{1}{2}\phi_1) \end{bmatrix} [\cos(\tfrac{1}{2}\phi_1), \sin(\tfrac{1}{2}\phi_1)]$$

$$= \begin{bmatrix} \cos^2(\tfrac{1}{2}\phi_1) & \cos(\tfrac{1}{2}\phi_1)\sin(\tfrac{1}{2}\phi_1) \\ \cos(\tfrac{1}{2}\phi_1)\sin(\tfrac{1}{2}\phi_1) & \sin^2(\tfrac{1}{2}\phi_1) \end{bmatrix}. \tag{151}$$

The detector at D_2 must be denoted by the angle $\pi+\phi_2$ because the direction a at D_1 implies the direction $\bar{a}$ at D_2 and so on. The projector for states $|\phi_2\rangle$ at D_2 is thus given by

$$Q_2 = \begin{bmatrix} \cos^2(\tfrac{1}{2}\phi_2) & \cos(\tfrac{1}{2}\phi_2)\sin(\tfrac{1}{2}\phi_2) \\ \cos(\tfrac{1}{2}\phi_2)\sin(\tfrac{1}{2}\phi_2) & \sin^2(\tfrac{1}{2}\phi_2) \end{bmatrix}. \tag{152}$$

If we select the angle ϕ_1 at detector D_1 and the angle ϕ_2 at D_2 we test for the state

$$|\psi_{12}\rangle = Q_2 Q_1 |\psi\rangle. \tag{153}$$

The probability to observe a positive value in both these directions is given by the norm of the state (153) and we find

$$P_{12} = \langle\psi|Q_1 Q_2 Q_2 Q_1|\psi\rangle = \langle\psi|Q_2 Q_1|\psi\rangle. \tag{154}$$

Evaluating this result for the state (141) we obtain

$$P_{12} = \frac{1}{2} \left[(\alpha, Q_1 \alpha)(\beta, Q_2 \beta) + (\beta, Q_1 \beta)(\alpha, Q_2 \alpha) \right.$$

$$\left. - (\alpha, Q_1 \beta)(\beta, Q_2 \alpha) - (\beta, Q_1 \alpha)(\alpha, Q_2 \beta) \right] . \tag{155}$$

The first two terms give the sum of the expectation values in the two states and the next two give the interference terms characteristic of quantum theory. They contain the essential non-local features of the theory. The calculation gives

$$P_{12} = \frac{1}{2} \left[\cos(\frac{1}{2}\phi_1) \sin(\frac{1}{2}\phi_2) - \cos(\frac{1}{2}\phi_2) \sin(\frac{1}{2}\phi_1) \right]^2$$

$$= \frac{1}{2} \sin^2(\frac{\phi_1 - \phi_2}{2}) . \tag{156}$$

If we now choose the three cases

$$P(a, \bar{b}) = \text{Prob} \{0° \text{ at } D_1 \text{ and } 45° \text{ at } D_2\} \tag{157a}$$

$$P(b, \bar{c}) = \text{Prob} \{45° \text{ at } D_1 \text{ and } 90° \text{ at } D_2\} \tag{157b}$$

$$P(a, \bar{c}) = \text{Prob} \{0° \text{ at } D_1 \text{ and } 90° \text{ at } D_2\} \tag{157c}$$

we find from (148) the result

$$\frac{1}{2} \sin^2(22.5°) + \frac{1}{2} \sin^2(22.5°) \geq \frac{1}{2} \sin^2(45°)$$

$$0.1464 \geq 0.25 \tag{158}$$

which is a contradiction. The above derivation of the inequality (148) hence contain some inadmissible assumption. In my opinion this is the locality postulate, and no far reaching conclusions concerning reality needed. Others have, however, reacted differently, see e.g. d'Espagnat (1979).

The argument above is perhaps too naive. Even in the classical theory we cannot assume three polarisation directions in a plane to posess independent probabilities. If we choose a spin 1/2 massive particle the three orthogonal directions are much better candidates, and even if the

the argument becomes more involved, disagreement with quantum theory is still found. In addition an experimental investigation of the present simple inequality is difficult to envisage. Paper #4 discusses in some detail other inequalities and the possibility of their experimental verifications.

The proof above was essentially given by Wigner (1970). It is not exactly clear how it utilises the assumption about locality, and it also rests, to some extent, on the rotational invariance of the spin states. Hence its conclusions are not as compelling as those of the proofs by Bell in 1971 and Clauser and Horne but these are discussed in some detail in paper #4. Professor Shimony has pointed out to me that these proofs work also against hidden variable theories with stochastic time evolution. Such theories are more general than the deterministic ones, and the experiments must be able to exclude those too. I thank Professor Shimony for clarification on this point, and for further developments I refer the reader to the remarks added by Professors Clauser and Shimony to paper #4.

4.4. The verdict by experiments

At the time of the writing of paper #4 only one experiment, that of Holt and Pipkin, was known to disagree with quantum mechanics and agree with Bell's inequalities. The reasons for this singular observation are still unclear, but it is assumed that it derives from an experimental error. This is supported by the fact that other similar experiments have turned out in agreement with quantum mechanics.

Recently Aspect et al (1981, 1982a, 1982b) have carried out careful experiments to check Bell's inequalities. Their results appear to satisfy most physicists that the predictions of quantum theory are valid and the type of reasoning lying behind Bell's inequalities leads to incorrect predictions. In Aspect's experiments the polarisation directions measured were changed in a random manner. Possible loopholes have to be of the exotic type discussed in Sec. 7.3 of paper #4. The distance between the detectors was extended to 13 m, which starts to have implications for the possible rate of propagation of information and the theory of relativity.

The main impact of the discussion around Bell's inequalities has been to focus the attention on those spots where the non-local character of quantum mechanics becomes acute. No future hidden variable theory can overcome the violation of the inequalities now well established both experimentally and theoretically. It seems to me that new theories must depend on non-local hidden variables, and then much of the original motivation is lost. Whether these theories have any implications for our concept of reality remains to be seen. As such the present results do not deny the objective existence of any of our familar physical objects but they are stranger than we can imagine indeed.

A simple discussion of the alternative interpretations of the present situation can be found in d'Espagnat (1979) and Mattuck (1982a, 1982b). For details the readers are referred to these.

REFERENCES

Braham N B, Gollup J P and Swinney H L 1984 Physica **11D** 252-264

Braham E, Seaton C T and Smith S D 1983 Scientific American **248** 3-71

Allen L and Eberly J H 1975 Optical Resonances and Two-Level Atoms (New York: Wiley)

Arecchi T T, Meucci R, Puccioni G and Tredicce J 1982 Phys.Rev.Lett. **9** 1217-1220

Aspect A, Dalibard J and Roger G 1982a, Phys.Rev.Lett. **49** 1804-1807

Aspect A, Grangier P and Roger G 1981 Phys.Rev.Lett. **47** 460-463

Aspect A, Grangier P and Roger G 1982b Phys.Rev.Lett. **49** 91-94

Belinfante F J 1973 A Survey of Hidden-Variable Theories (Oxford: Pergamon)

Bell J S 1964 Physics **1** 195-200

Bell J S 1966 Rev.Mod.Phys. **38** 447-452

Bennett W R Jr 1962 Phys.Rev. **126** 580-593

Bethe H A 1947 Phys.Rev. **72** 339-341

Bishop A R and Schneider T (eds) 1978 Solitons and Condensed Matter Physics (Heidelberg: Springer)

Blombergen N 1965 Nonlinear Optics (New York: Benjamin)

Bohm D 1951 Quantum Theory (London: Prentice-Hall)

Bonifacio R, Gronchi M and Lugiato L A 1979 Opt. Commun. **30** 129-133

Bonifacio R and Lugiato L A 1978, Phys.Rev. **A18** 1129-1144

Bowden C M, Gibbs H M and McCaLL S L (eds) 1984 Optical Bistability 2 (New York: Plenum)

Boyer T H 1980 in Foundations of Radiation Theory and Quantum Electrodynamics ed A O Barut 49-63 (New York: Plenum)

Braginsky V B, Vorontsov Yu I and Thorne K S 1980 Science **209** 547-557

Burshtein A I 1966 Sov.Phys.-JETP **22** 939-947

Cantrell C D and Scully M O 1978 Phys.Rep. **C43** 499-508

Carruthers P and Nieto M M 1968 Rev.Mod.Phys. **40** 411-440

Casperson L W 1978 IEEE J.Quant.Electron. QE-14 756-761

Caves C M 1981 Phys.Rev. D23 1693-1708

Caves C M, Thorne K S, Drever R W P, Sandberg V D and Zimmerman M 1980 Rev.Mod.Phys 52 341-392

Cohen-Tannoudji C 1977 in Frontiers in Laser Spectroscopy, LesHouches Summer School 1975 eds R Balian, S Haroche and S Liberman 5-104 (Amsterdam: North-Holland)

Cohen-Tannoudji C 1982 Lectures at LesHouches 1982

Cretien M, Gross E P and Deser S 1968 (eds) Statistical Physics, Phase Transitions and Superfluidity (New York: Gordon & Breach)

DeGeorgio V and Scully M O 1970 Phys.Rev. A2 1170-1177

Derstine M W, Gibbs H M, Hopf F A and Kaplan D L 1982 Phys.Rev. A26 3720-3722

Derstine M W, Gibbs H M, Hopf F A and Kaplan D L 1983 Phys.Rev. A27 3200-3208

Dirac P A M 1927 Proc.R.Soc. A114 243-265

Eckman J-P 1981 Rev.Mod.Phys. 53 643-654

Einstein A, Podolsky B and Rosen N 1935 Phys.Rev. 47 777-780

d'Espagnat B 1979 Sci.Am. 241 258-281

Fain V M 1967 Sov.Phys.-JETP 25 1027-1030

Feigenbaum M I 1979 J.Stat.Phys. 21 669-706

Feigenbaum M I 1980 Los Alamos Science 1 4-27

Fermi E 1932 Rev.Mod.Phys 4 87-132

Feynman R P 1962 Quantum Electrodynamics (New York: Benjamin)

Gao J Y, Yuan J M and Narducci L M 1983a Opt.Commun. 44 201-206

Gao J Y, Narducci L M, Schulman L S, Squicciarini M and Yuan J M 1983b Phys.Rev. A28 2910-2914

Geisel T and Nierwetberg J 1982 Phys.Rev.Lett. 48 7-10

Gibbs H M, McCall S L and Venkatesan T N 1976 Phys.Rev.Lett. 36 1135-1148

Ginzburg V L 1983 Sov.Phys. Usp. 26 713-719

Boggia R S and Abraham N B 1983 Opt.Comm. **47** 278-282

Boggia R S and Abraham N B 1984 Phys.Rev. **A29** 1304-1309

Glauber R J 1963 Phys.Rev. **131** 2766-2788

Glauber R J 1965 in Quantum Optics and Electronics, LesHouches 1964 eds C DeWitt, A Blandin and C Cohen-Tannoudji 63-185 (New York: Gordon & Breach)

Graham R and Cho Y 1983 Opt.Commun. **47** 52-56

Graham R and Haken H 1970 Z. Physik **237** 31-46

Grossman S and Fujisaka H 1982 Phys.Rev. **A26** 1779-1782

ter Haar D 1967 The Old Quantum Theory (Oxford: Pergamon)

Haken H 1964 Z.Physik. **181** 96-124

Halos N J, Liu S-N and Abraham N B 1983 Phys.Rev. **A28** 2915-2920

Haug H and Schmitt-Rink S 1984 Prog. Quantum Electron. **9**, 3-100

Heidmann A, Reynand S and Cohen-Tannoudji C 1984 Opt.Commun. **52**, 235-240

Helleman R H G 1980 in Fundamental Problems in Statistical Mechanics vol 5, ed E G D Cohen 169-233 (Amsterdam: North-Holland)

Hollenhorst H N 1979 Phys.Rev. **D19** 1669-1679

Hopf F A, Kaplan D L, Gibbs H M and Schoemaker R L 1982 Phys.Rev. **A25** 2172-2182

Ikeda K 1979 Opt.Commun. **30** 257-261

Ikeda K, Kondo K and Akimoto O 1982 Phys.Rev.Lett. **49** 1467-1470

Jammer M 1974 The Philosphy of Quantum Mechanics (New York: Wiley)

Kadanoff L P 1983 Phys. Today **36** 46-53

Kazantsev A P and Surdotovich G I 1969 Sov.Phys.JETP **29** 1075-1083

Kazantsev A P and Surdotovich G I 1970 Sov.Phys.JETP **31** 133-137

Lamb W E Jr 1963 in Quantum Electronics and Coherent Light ed P A Miles 78-110 (New York: Academic)

Lamb W E Jr 1964 Phys.Rev. **A134** 1429-1450

Lamb W E Jr and Scully M O 1969 in Polarization, Matter and Radiation, Jubilee volume in honour of A Kastler, 363-369, (Paris: Presse Universitaires de France)

Lifshitz E M and Pitaevski L P 1980 Statistical Physics, 3rd Edition, Part 1 (Oxford: Pergamon)

Lax M 1968 Phys.Rev. 172 350-361

Lee P H, Schoefer P B and Barker W P 1968 Appl.Phys.Lett. 13 373-375

Lisitsyn V N and Chebotaev V P 1968 JETP Letters 7 1-3

Loudon R 1983 The Quantum Theory of Light (Oxford: Oxford university Press)

Loudon R 1984a Opt.Commun. 49 24-28

Loudon R 1984b Opt.Commun. 49 67-70

Lugiato L A, Narducci L M, Bandy D K and Pennise C A 1983 Opt.Commun. 46 64-68

Ma S-K 1976 Modern Theory of Critical Phenomena (Reading: Benjamin)

Mandel L 1979 Opt.Lett. 4 205-207

Mandel L 1982 Phys.Rev.Lett. 49 136-138

Mandel P and Kapral R 1983 Opt.Commun. 47 151-156

Mandel L and Wolf E 1965 Rev.Mod.Phys. 37 231-287

Mattuck R D 1982a Eur.J.Phys. 3 107-112

Mattuck R D 1982b Eur.J.Phys. 3 113-118

McFarlane R A, Bennett W R and Lamb W E Jr 1963 Appl.Phys.Lett. 2 189-190

Mollow B R 1967 Phys.Rev. 162 1256-1273

Mollow B R 1969 Phys.Rev. 188 1969-1975

Ott E 1981 Rev.Mod.Phys. 53 655-671

Paul H 1982 Rev.Mod.Phys. 54 1061-1102

Peierls R 1979 Surprises in Theoretical Physics (Princeton: Princeton University Press)

Procaccia I and Schuster H 1983 Phys.Rev. A28 1210-1212

Rabi I I 1937 Phys.Rev. 51 652-654

Salomaa R and Stenholm S 1977 Appl.Phys. 14 355-360

Schrödinger E 1927 Annalen der Physik 82 257-265

Scott A C, Chiu F Y F and Mclaughlin D W 1973 Proc.I.E.E.E. 61 1443-1481

Scott J F, Sargent M III and Cantrell C D 1975 Opt.Commun. 15 18-10

Scully M O and Sargent M III 1972 Phys.Today 25 38-47

Shapiro J H, Yuen H P and Machado Mata J A 1979 IEEE Trans.Inform.Theory IT-25 179-192

Stenholm S 1984 Foundations of Laser Spectroscopy (New York: Wiley)

Stoler D 1970 Phys.Rev. D1 3217-3219

Stoler D 1971 Phys.Rev. D4 1925-1926

Szöke A, Danew V, Goldhar J and Kurnit N A 1969 Appl.Phys.Lett. 15 376-379

van Vleck J H and Huber D L 1977 Rev.Mod.Phys. 49 939-959

Walls D F 1983 Nature 306 141-146

Weiss C O and King H 1982 Opt.Commun. 44 59-61

Welton T A 1948 Phys.Rev. 74 1157-1167

Wheeler J A and Zurek W H 1983 Quantum Theory and Measurement (Princeton: Princeton University Press)

Whitham G B 1974 Linear and Nonlinear Waves (New York: Wiley)

Wigner E P 1970 Am.J.Phys. 38 1005-1009

Yariv A 1976 Introduction to Optical Electronics (New York: Holt, Rinehard and Winston)

Yuen H P 1976 Phys.Rev. A13 2226-2243

Yuen H P and Shapiro J H 1978 IEEE Trans.Inform.Theory IT-24 657-672

Yuen H P and Shapiro J H 1979 Opt.Lett. 4 334-336

Yuen H P and Shapiro J H 1980 IEEE Trans.Inform.Theory IT-26 78-92

Interaction of laser radiation with free atoms

SERGE FENEUILLE

Laboratoire Aimé Cotton, Université Paris-Sud, CNRS II, Bâtiment 505,
91405 Orsay, France

Abstract

The object of this review is to describe and to interpret the basic phenomena recently observed in the interaction between free atoms and laser radiation. Some of these have been known for a long time in magnetic resonance but, in the optical range, the situation is strongly modified because of spontaneous emission and the Doppler effect. Therefore, particular attention is paid to resonance fluorescence and Doppler-free spectroscopy, but without dwelling on applications. Transient effects are also discussed, but we ignore all the phenomena which characterize, not specifically the interaction between atoms and fields, but the free evolution of atoms after coherent excitation. On the other hand, since propagation effects lead in the visible range to completely new aspects of the role of saturation in the response of atomic systems to resonant monochromatic excitation, pulse propagation experiments are described in detail. Finally, multiphoton processes are considered, with special emphasis on Doppler-free multiphoton excitation and resonant multiphoton ionization.

1. Introduction

Atomic and molecular physics are certainly the oldest subfields of quantum physics and everybody knows the major role that they played in the 1920s when the first principles of quantum mechanics were elaborated. After this 'golden age', many physicists considered research in these subfields essentially complete and in fact, for several decades, activity in atomic and molecular physics decreased steadily in comparison with the large effort directed towards nuclear and high-energy physics. The advent of lasers, and more precisely of tunable lasers, changed this situation and for five years now, atomic and molecular physics, gradually intermingling with laser physics, has become an area where more and more activity is contributing substantially to the understanding of many phenomena.

The main advances resulting from the use of lasers fall essentially into three categories. The first, related to the monochromaticity of laser radiation, consists of a considerable increase in resolution in atomic and molecular spectroscopy, thanks in particular to Doppler-free techniques. This not only leads to much better precision of spectroscopic data but also offers the possibility of studying non-radiative processes (collisions and energy transfers, for example) through their influence on spectral line profiles. The second advance comes from the selective character of laser excitation. The most important technical application might turn out to be laser isotope separation, but more basically the essential interest of such a selective character is the possibility of studying independently the various (radiative and non-radiative) relaxation processes of an atom or a molecule in a well-defined state. The origin of the third advance is the high spectral density of laser excitation. This allows one either to work with very small numbers of atoms (which is essential, for example, in the study of spectroscopic properties of isotopes far from the valley of stability) or to excite atoms through transitions with extremely small transition probabilities. This latter possibility plays a crucial role in optical experiments testing weak interactions through parity violation in atoms. Furthermore, atomic Rydberg states can now be significantly populated even for very high values of principal quantum number. A completely new area of atomic physics is now appearing through the study of these highly excited bound states.

However, all these new trends in atomic and molecular physics do not constitute the main subject of this review. In fact, it must not be supposed that the advantages of using lasers in atomic and molecular physics reduce to a quantitative extension of conventional spectroscopic techniques without introducing new phenomena or new concepts. On the contrary, the interaction between intense monochromatic light and absorbing media presents many new aspects which do not appear in broadband and low-power excitation. The corresponding phenomena do not depend on the details of atomic and molecular structures. Therefore, the electron–nuclear systems which are the most adapted for their study are the simplest, that is to say free atoms. This explains the title of this review but, of course, we shall not refrain from reporting on experiments on free molecules when they illustrate a significant point of the interaction between light and matter.

The interaction between atoms and laser radiation can be investigated essentially in two different ways. The first consists of looking for the response of the atomic system during the interaction; the second is based on the study of the free evolution of the atoms after the interaction is finished. Here, we shall consider only the first possibility, that is to say, we shall ignore all the phenomena which time-resolved spectroscopy is based on. Moreover, we shall describe only the effects which can be interpreted in terms of the interaction between light and a single atom, even when they are observed on a collection of atoms. In other words, collective phenomena will not be considered.

In spite of these limitations, this review covers a very broad area of research and thus cannot be exhaustive. Indeed, we have chosen to describe the most basic phenomena and to report only a few experiments which are particularly demonstrative. As a consequence, the list of references quoted is far from being complete and the choice between such and such theoretical approach or between such and such experiment is somewhat arbitrary. The reader will be able to find exhaustive lists of references in the numerous more specialized reviews written recently on this subject: for the most part, these are quoted here.

1.1. Structure of the review

Most of the phenomena appearing in the resonant interaction between atoms and monochromatic light can be understood in the two-level approximation. Section 2 is devoted to the description of simple theoretical models based on this approximation. Classical and quantum descriptions of the electromagnetic field are successively introduced and lead respectively to the Bloch equations (initially developed for magnetic resonance theory) and to the dressed-atom model. Particular attention is paid to spontaneous emission which plays a crucial role in the optical range. Validity conditions of the two-level atom and of the monochromatic approximation in real systems are also discussed.

In §3, we characterize the response of an atomic system to resonant laser excitation in selected experimental situations using either atomic beams transversely illuminated or vapours. In the first case power-dependent effects such as the dynamical Stark effect are especially considered, while for vapours we lay the stress on Doppler-free techniques based on hole-burning or saturation phenomena.

While most of the basic phenomena introduced in §§2 and 3 were known in magnetic resonance a long time before the advent of lasers, propagation effects, which are described in §4, are specific to the optical range. We shall pay particular attention to self-induced transparency which certainly is the most fascinating conceptual contribution to quantum optics during the last few years.

Section 5 is devoted to a brief discussion of multiphoton processes with special emphasis on Doppler-free multiphoton excitation and resonant multiphoton ionization.

Concluding remarks are finally given to try to sketch out the main perspectives in the field.

2. Theoretical models for resonant phenomena

The theoretical description of the various effects appearing in the interaction between a collection of atoms and light comes up against many difficulties in the general case. However, most of the basic phenomena can be understood by using exactly soluble models which represent good approximations for particular experimental conditions. These models separate into two classes according to the representation chosen for the driving electromagnetic field which, of course, can be described either as a classical field or a quantum field. In fact, up to now, no difference in observable quantities has appeared between these two treatments, which are thus totally equivalent. From a practical point of view, however, they are rather different and in the description of a particular experiment the choice between the two possible descriptions of the field is not really unimportant. So, in many resonance experiments, the quantum approach, which leads to a time-independent interaction, provides simple physical explanations of the phenomena more easily, while propagation effects, for example, can be studied by using a classical description of the field only. In the following the two representations will be considered in turn and their respective advantages will be discussed for each particular problem.

Furthermore, collective phenomena, such as super-radiance, appear under very special circumstances and will be ignored here. Then all the effects observed from a collection of atoms can be interpreted from the results obtained for a single atom. Lastly, since this review is mainly concerned with resonant phenomena appearing when 'a quasi-monochromatic field almost coincides in frequency with a particular Bohr frequency of the considered atom, the two-level atom approximation will be used throughout this section. The validity conditions of these approximations are discussed in §2.1.

2.1. Parameters characterizing an optical resonance

The situation that we discuss here is schematized in figure 1. The two-level atom is characterized by a ground state $|g\rangle$ and an excited state $|e\rangle$, the energies of these states being respectively $\hbar\omega_g$ and $\hbar\omega_e$. The power spectrum of the light is assumed to be a peak function centred on ω with a width equal to $\delta\omega$. The detuning Δ is defined by:

$$\Delta_0 = \omega - (\omega_e - \omega_g).$$

The strength of the interaction between the atom and the field is usually measured by the so-called Rabi frequency at exact resonance ($2K_0$) which is proportional to

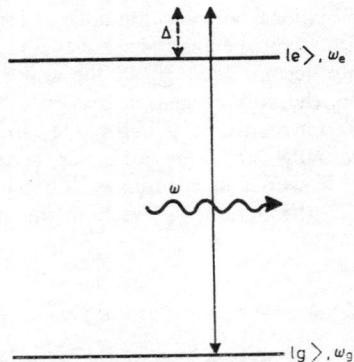

Figure 1. The two-level atom and monochromatic approximations.

the square root of the energy density flux (itself proportional to the amplitude of the electric field) and to the matrix element of the atomic electric dipole between the states $|g\rangle$ and $|e\rangle$. To characterize fully the resonance concerning a single atom, two other types of parameters must be introduced: first, various relaxation constants denoted by Γ related to atomic dipole and population decay times; and secondly, the inverse of the mean interaction time, τ, between the atom and the field. For continuous excitation, τ is the time for the atom to cross the laser beam; for pulsed excitation, τ is the pulse duration if this quantity is smaller than the previous time. For a fully coherent wave packet, τ^{-1} can be included, by Fourier transform, in $\delta\omega$.

All the previous parameters are identical for all the atoms and therefore they are called homogeneous quantities. For a collection of atoms, however, other parameters due to inhomogeneous effects play a role in resonant processes. For a vapour in a cell, the only inhomogeneous effect is the Doppler effect which introduces a velocity-dependent detuning:

$$\Delta_v = \Delta_0 - (\boldsymbol{k}\cdot\boldsymbol{v})$$

where $\boldsymbol{k}$ is the light wavevector and $\boldsymbol{v}$ is the velocity of the atom. The only inhomogeneous parameter is thus the Doppler width, Δ_D, of the transition.

The parameter playing the most important role is certainly the spectral width of the incident light. In fact, two extreme cases are to be considered. In classical resonance experiments using spectral lamps for example, $\delta\omega$ is large with respect to all the other parameters, and under these conditions the resonance characteristics can be deduced from the well-known Einstein rate equations. On the other hand, if one utilizes CW single-mode lasers, $\delta\omega$ is quite small and the monochromatic approximation is a very good one. We shall use it throughout the rest of this section. For coherent pulsed excitation, this approximation will still be used, pulses being described in the time domain.

The situation corresponding to K_0 being very small with respect to all the Γ is called the weak-field limit or the linear regime since the interaction between the atom and the field can be treated by using standard perturbation theory to first order. The opposite situation ($K_0 \gg$ all the Γ) is called the strong-field limit and in this case relaxation constants can often be ignored in predicting the essential features of the phenomena. However, most of the experimental situations correspond to intermediate cases.

Lastly, two extreme situations can be contemplated for the inhomogeneous parameter. In experiments with atomic beams transversely illuminated, no inhomogeneous effect has to be taken into account. Thus, all the atoms can be assumed to be at rest. Besides, in this case, the only relaxation process is spontaneous emission that introduces a single relaxation constant γ. However, in optical experiments on a vapour in a cell, Δ_D is usually very large with respect to all the other parameters and in most cases can be considered as infinite. The specific role of the Doppler effect will be studied in §3: throughout this section, the atom will be assumed to be at rest.

2.2. Interaction of a two-level atom with a classical field

In this subsection, the field emitted by a single-mode laser is written in a classical way:

Interaction of laser radiation with free atoms 67

$$E(t) = 2^{-1/2} \mathscr{E}(t)(\exp\{i[\omega t + \phi(t)]\} + \exp\{-i[\omega t + \phi(t)]\})\boldsymbol{\epsilon} \tag{2.1}$$

$\boldsymbol{\epsilon}$ is the polarization unit vector of the field; $\mathscr{E}(t)$ and $\phi(t)$ are slowly varying functions of time; $\dot{\mathscr{E}}(t), \dot{\phi}(t) \ll \omega$ for any time, and $\langle \dot{\phi}(t) \rangle$, the mean value in time of $\dot{\phi}(t)$, is assumed to be zero. With the two-level atom approximation, the atomic Hamiltonian reduces to

$$H_A = \hbar\{\omega_g |g\rangle\langle g| + \omega_e |e\rangle\langle e|\} \tag{2.2}$$

and, since we are concerned with electric dipole transitions, the coupling between the atom and the field is taken to be:

$$H_{AF} = \hbar \mathscr{K}(t)|e\rangle\langle g| + \hbar \mathscr{K}^*(t)|g\rangle\langle e| \tag{2.3}$$

where

$$\mathscr{K}(t) = \hbar^{-1}\langle e|E(t)\cdot D|g\rangle = K(t)(\exp\{i[\omega t + \phi(t)]\} + \exp\{-i[\omega t + \phi(t)]\}) \tag{2.4}$$

$(K(t) = 2^{-1/2}\mathscr{E}(t)\langle e|\boldsymbol{\epsilon}\cdot D|g\rangle$ can be assumed to be real without loss of generality), D being the electric dipole operator. Then, if the state of the system $|\Psi\rangle$ is written:

$$|\Psi(t)\rangle = \alpha(t)\exp\{\tfrac{1}{2}i[\omega t + \phi(t)]\}|g\rangle + \beta(t)\exp\{-\tfrac{1}{2}i[\omega t + \phi(t)]\}|e\rangle \tag{2.5}$$

the Schrödinger equation becomes:

$$i\dot{\alpha}(t) - \tfrac{1}{2}\alpha(t)[\omega + \dot{\phi}(t)] = \omega_g \alpha(t) + K(t)\beta(t)(1 + \exp\{-2i[\omega t + \phi(t)]\}) \tag{2.6(a)}$$

$$i\dot{\beta}(t) + \tfrac{1}{2}\beta(t)[\omega + \dot{\phi}(t)] = \omega_e \beta(t) + K(t)\alpha(t)(1 + \exp\{2i[\omega t + \phi(t)]\}). \tag{2.6(b)}$$

The usual way to simplify these equations is to neglect the rapidly varying terms at frequencies $\pm 2\omega$ which correspond to non-resonant processes. These terms are automatically suppressed if H_{AF} is replaced by:

$$\tilde{H}_{AF} = K(t)(\exp\{-i[\omega t + \phi(t)]\}|e\rangle\langle g| + \exp\{+i[\omega t + \phi(t)]\}|g\rangle\langle e|) \tag{2.7}$$

and, as we shall see later, this approximation corresponds to the well-known rotating-wave approximation in magnetic resonance theory (Bloch 1946).

Now, for continuous excitation ($K(t) = K_0$, $\phi(t) = \phi_0$), equations (2.6) are exactly soluble and by assuming that, for $t = t_0$, the atom is in the ground state, one gets:

$$\alpha(t) = \exp\left[-\tfrac{1}{2}i(\omega_g + \omega_e)(t - t_0)\right]\left(\frac{2K_0}{\Omega - \Delta}\exp\left[-i\tfrac{1}{2}\Omega(t - t_0)\right]\right.$$

$$\left. + \frac{2K_0}{\Omega + \Delta}\exp\left[i\tfrac{1}{2}\Omega(t - t_0)\right]\right)\frac{K_0}{\Omega} \tag{2.8(a)}$$

$$\beta(t) = \exp\left[-\tfrac{1}{2}i(\omega_g + \omega_e)(t - t_0)\right]\{\exp\left[-i\tfrac{1}{2}\Omega(t - t_0)\right] - \exp\left[i\tfrac{1}{2}\Omega(t - t_0)\right]\}\frac{K_0}{\Omega} \tag{2.8(b)}$$

where $\Omega(\Delta) = (4K_0^2 + \Delta^2)^{1/2}$ is the Rabi frequency. This leads to the famous Rabi solution (Rabi 1937) for the population of the excited level:

$$n_e = \langle g|\Psi(t)\rangle\langle\Psi(t)|g\rangle = \frac{4K_0^2}{\Omega^2}\sin^2 \tfrac{1}{2}\Omega(t - t_0). \tag{2.9}$$

Now, for a collection of atoms, if there exists any process such that t_0 takes all the values within a domain which is very broad with respect to Ω^{-1}, then n must be averaged over time and reduces to

$$\bar{n}_e = \frac{1}{2}\frac{4K_0^2}{4K_0^2+\Delta^2} = \frac{1}{2}\left(\frac{\Omega(0)}{\Omega(\Delta)}\right)^2. \qquad (2.10)$$

At exact resonance $\Delta = 0$, $\bar{n} = \frac{1}{2}$; in other words, the field equalizes the respective populations of the ground and the resonant excited states.

Instead of using the preceding representation, we can, of course, use a density matrix formalism. At this stage, the two languages are strictly equivalent. If the density matrix is defined between atomic steady states, the equations of motion are

$$\dot{\rho}_{gg} = -\dot{\rho}_{ee} = \mathrm{i}[K(t)\rho_{ge} - K^*(t)\rho_{eg}] \qquad (2.11(a))$$

$$\dot{\rho}_{eg} = \dot{\rho}_{ge}^* = -\mathrm{i}\rho_{eg}(\omega_e - \omega_g) - \mathrm{i}K(t)(\rho_{gg}-\rho_{ee}). \qquad (2.11(b))$$

After making the resonant approximation ($H_{AF} \to \tilde{H}_{AF}$), these equations can be written in a rather compact form:

$$\dot{w} = 2K(t)v \qquad (2.12(a))$$

$$\dot{v} = -(\Delta+\dot{\phi})u - 2K(t)w \qquad (2.12(b))$$

$$\dot{u} = (\Delta+\dot{\phi})v \qquad (2.12(c))$$

where the population inversion, w, is equal to $(\rho_{ee}-\rho_{gg})$ and v and u are defined by

$$u - \mathrm{i}v = 2\rho_{eg}\exp\{-\mathrm{i}[\omega t + \phi(t)]\}.$$

Formally, equations (2.12) are identical to the Bloch equations developed initially for magnetic resonance (Bloch 1946). This is not surprising since obviously a spin $\frac{1}{2}$ in a static magnetic field is a particular two-level system. More precisely, one can identify w, v and u respectively to $2\langle S_z \rangle$, $2\langle S_y \rangle$ and $2\langle S_x \rangle$, and with this picture the vector model can be used to predict most of the phenomena (see, for example, Allen and Eberly 1975, Cohen-Tannoudji 1975a). In particular, it appears that the resonant approximation defined previously is equivalent to decomposing the linear oscillating field into two components rotating in opposite directions around the static magnetic field and keeping only the component rotating as a right-handed screw as t increases (see figure 2). The validity conditions of this 'rotating-wave' approximation will be discussed later.

2.3. Interaction of a two-level atom with a quantum field

Instead of describing monochromatic light by a classical field, one can utilize a quantum description of the field (see, for example, Heitler 1954). This approach is really only convenient for continuous (or square pulsed) excitation but, in this case, it is very attractive since the energy of the total system (atom plus field) is no longer time-dependent. In the single-mode case, the field Hamiltonian reduces to $H_F = \hbar\omega a^+ a$ where a^+ and a are respectively creation and annihilation operators for a photon with frequency ω, polarization ϵ and wavevector $\mathbf{k}$. The state of the free field $|f\rangle$ can be written in a general way:

$$|f\rangle = \sum_n \langle n|f\rangle|n\rangle = \sum_n \rho_n|n\rangle \qquad (2.13)$$

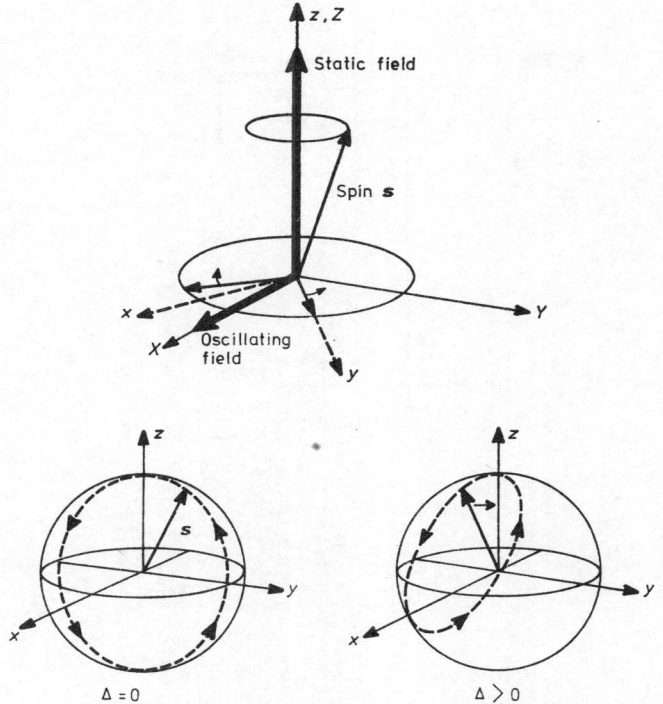

Figure 2. The rotating-wave approximation in the Bloch vector model.

where $|n\rangle$ is an eigenstate of H_F defined by:

$$H_F|n\rangle = n\hbar\omega|n\rangle. \qquad (2.14)$$

The Hamiltonian characterizing the interaction between the atom and the field now becomes:

$$H_{AF} = \hbar\kappa[a^+ + a](|e\rangle\langle g| + |g\rangle\langle e|) \qquad (2.15)$$

where κ is a coupling parameter proportional to $\langle e|\boldsymbol{\epsilon}.\boldsymbol{D}|g\rangle$.

If we ignore this interaction, the eigenstates of the system are of the type $|g, n\rangle$ or $|e, m\rangle$, the corresponding energies being respectively $\hbar(\omega_g + n\omega)$ and $\hbar(\omega_e + m\omega)$. If Δ is much smaller than ω, then the spectrum of $H_A + H_{AF}$ is a sequence of nearly degenerate doublets $\{|g, n\rangle, |e, n-1\rangle\}$ separated in energy by $\hbar\omega$ (see figure 3). Now, if $n\kappa$ is much smaller than $\omega_e - \omega_g = \omega$, near a resonance ($\Delta \ll \omega$), first-order perturbation theory can be used to take H_{AF} into account. Under this approximation, if we denote by $|n_\pm\rangle$ the eigenstates of the total Hamiltonian and by $\hbar\lambda_\pm$ the corresponding energies, one gets in a straightforward way:

$$\lambda_\pm = \tfrac{1}{2}[\omega_g + \omega_e + (2n-1)\omega] \pm \tfrac{1}{2}\Omega_n \qquad (2.16(a))$$

$$|n_\pm\rangle = \left(\frac{\Omega_n \pm \Delta}{2\Omega_n}\right)^{1/2}|g, n\rangle \pm \left(\frac{\Omega_n \mp \Delta}{2\Omega_n}\right)^{1/2}|e, n-1\rangle \qquad (2.16(b))$$

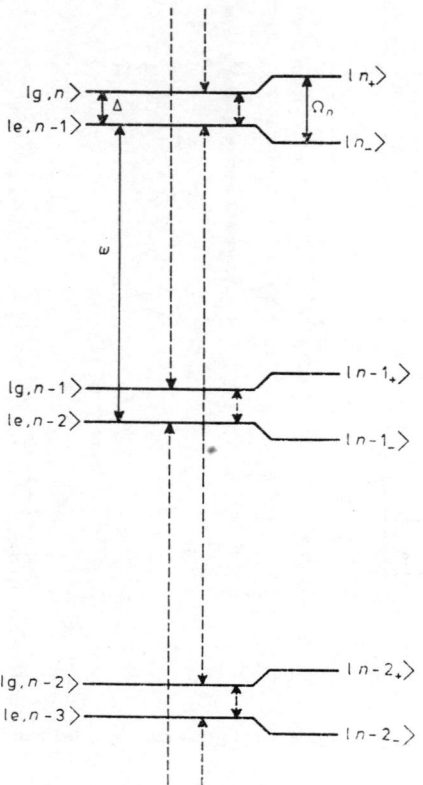

Figure 3. The dressed-atom energy spectrum.

Ω_n being defined by $\Omega_n = 4n\kappa^2 + \Delta^2$. Furthermore, if one assumes for $t < t_0$ that there is no interaction between the atom and the field and that the atom is in its ground state, the state of the system for $t > t_0$ is given by:

$$|\Psi\rangle = \sum_n \rho_n \left[\left(\frac{\Omega_n + \Delta}{2\Omega_n}\right)^{1/2} \exp\left[-i\lambda_+(t-t_0)\right] |n_+\rangle \right.$$
$$\left. + \left(\frac{\Omega_n - \Delta}{2\Omega_n}\right)^{1/2} \exp\left[-i\lambda_-(t-t_0)\right] |n_-\rangle \right]. \quad (2.17)$$

Therefore, the atomic population in the excited state is

$$n_e = \sum_n \rho_n^* \rho_n \frac{4n\kappa^2}{\Omega_n^2} \sin^2 \tfrac{1}{2} \Omega_n(t-t_0). \quad (2.18)$$

At first sight, this result is rather different from the Rabi solution given in equation (2.9). However, the laser field is known to be a quasi-classical one, and therefore the normalized distribution ρ_n of the number n of photons has a dispersion Δn around the mean value $\bar{n}$ which is large in absolute value but small in relative value ($\Delta n \simeq (\bar{n})^{1/2} \ll \bar{n}$) (Glauber 1963). This means that one can take $\rho_n = \delta(\bar{n}, n)$ without introducing a significant error. It is clear that, under this condition, equation (2.18) reduces to equation (2.9) if we identify $\bar{n}\kappa^2$ to K_0^2.

So one gets exactly the same result by using either a quantum or a classical approach to describe the driving laser field, but the corresponding physical pictures are quite different. In the classical description, the Hamiltonian is time-dependent and no energy level can be actually defined. In other words, some notions such as the concept of light shifts of energy levels (see, for example, Cohen-Tannoudji 1968, Series 1970, Stenholm 1973a) have, strictly speaking, no meaning for resonant phenomena since, in this case, perturbation theory cannot be used. On the contrary, in the quantum approach, the Hamiltonian is not time-dependent and the system has well-defined energy levels and stationary states, but one must always remember that this system is no longer an isolated atom but an atom dressed by photons according to the terminology introduced by Cohen-Tannoudji and Haroche (1969).

2.4. The Bloch–Siegert shift

It is well known in magnetic resonance theory that the rotating-wave approximation is valid only if the intensity of the oscillating field is sufficiently small. Otherwise, the effect of the counter-rotating field has to be taken into account and this was first investigated by Bloch and Siegert (1940) who utilized a classical approach to the field. However, the phenomenon can be described more easily with the dressed-atom picture (Walls 1972, Armstrong and Feneuille 1973). In fact, the rotating-wave approximation corresponds exactly to the use of perturbation theory to first order to take into account, in the quantum scheme, the interaction between the atom and the field, or in other words, to the replacement of H_{AF} by

$$\tilde{H}_{AF} = \hbar\kappa(a^+|g\rangle\langle e| + a|e\rangle\langle g|). \quad (2.19)$$

The contribution of non-resonant terms can be considered as resulting from higher orders of perturbation theory. Now, the matrix elements of H_{AF} can be written in the general way:

$$\langle e, m|H_{AF}|g, n\rangle = \hbar\kappa[(n)^{1/2}\,\delta(m, n-1) + (n+1)^{1/2}\,\delta(m, n+1)]. \quad (2.20)$$

This means that the states $|n_\pm\rangle$ are connected through H_{AF} to $|(n\pm 2)_\pm\rangle$ (see figure 3). Therefore, to second order, the correction to the energies of the system is given by:

$$\lambda_\pm^2 = \pm \frac{\bar{n}\kappa^2\Delta}{2\omega\Omega} = \pm \frac{K_0^2\Delta}{2\omega\Omega} \quad (2.21)$$

which leads to the following result for the population of the excited level:

$$n_e = \frac{4K_0^2}{\Omega^2}\left(1 + \frac{\Delta}{2\omega}\frac{4K_0^2}{\Omega^2}\right)\sin^2 \tfrac{1}{2}\Omega\left(1 - \frac{\Delta}{4\omega}\frac{4K_0^2}{\Omega^2}\right)t. \quad (2.22)$$

Thus the resonance is no longer obtained for $\Delta = 0$ but for $\Delta = K^2/\omega$. This is nothing else but the Bloch–Siegert shift. During the last few years, new interest has been paid to this shift and more complete expressions have been given by various authors (Shirley 1965, Cohen-Tannoudji et al 1973, Hannaford et al 1973, Stenholm 1973b). Now the question arises of the importance of this shift in the optical range. For resonance transitions of alkali atoms, for example, K_0 is of order of 10 MHz for an energy density flux equal to 1 W cm^{-2}, and therefore Δ is less than 0.2 Hz, a quantity certainly too small to have been observed. However, in some experiments such as, for example, resonant multiphoton ionization studies, very high energy density fluxes of the order of one GW cm^{-2} are commonly used, and under these conditions, the Bloch–Siegert shift may not be negligible.

2.5. Spontaneous emission and other relaxation processes

A complete theory of spontaneous emission is outside the scope of this review, and for details the reader is referred to the lecture notes recently published by Cohen-Tannoudji (1975b). The aim of this subsection is only to describe heuristically the various ways to introduce spontaneous emission in the equations of motion which have been already given.

If we ignore neoclassical theories of spontaneous emission (see, for example, Jaynes 1973) that still give rise to much controversy, field quantization is the common starting point of the various presentations of spontaneous decay of excited atomic levels. However, in the final result, this quantization can be somehow hidden. This is the case, for example, in all the theories describing the damping phenomenon as resulting from the coupling of the atom with a bath approximated at any time by its initial vacuum state $|0\rangle_B$. This allows one to approximate for the system of atom and bath at time t by using the product

$$\rho(t) = \rho_a(t) |0\rangle_B {}_B\langle 0|$$

where $\rho_a(t)$ is a reduced density operator for the atom. With this assumption, one gets equations of motions involving $\rho_a(t)$ only. For a two-level atom, this leads to the well-known result (Argyres and Kelley 1964):

$$\dot\rho_a(t) = \gamma |g\rangle\langle g| \langle e| \rho_a(t) |e\rangle - (\tfrac{1}{2}\gamma + i\delta)|e\rangle\langle e| \rho_a(t) - (\tfrac{1}{2}\gamma - i\delta)\rho_a(t)|e\rangle\langle e| \quad (2.23)$$

where γ is the decay constant of the upper level and δ is a frequency shift which is nothing else but the Lamb shift and which can be included in the experimental frequency of the transition $\omega_e - \omega_g$. The generalization to a multilevel atom is straightforward.

So, equation (2.23) shows that introducing spontaneous emission in the interaction between a two-level and a classical driving field can be achieved by replacing equations (2.12) by:

$$\dot w = 2K(t)v - \gamma(w+1) \qquad (2.24(a))$$

$$\dot v = -(\Delta + \phi)u - 2K(t)w - \tfrac{1}{2}\gamma v \qquad (2.24(b))$$

$$\dot u = (\Delta + \phi)v - \tfrac{1}{2}\gamma u. \qquad (2.24(c))$$

In these equations, spontaneous emission and coupling with an external field appear to be completely independent phenomena. This is a consequence of the vacuum approximation for describing spontaneous emission. It has been recently noticed that this approximation could be invalid when the atom interacts with a strong driving field since, in this case, some modes of the electromagnetic field are far from being empty (C Cohen-Tannoudji 1977, private communication). Equations (2.24) are a particular case ($\Gamma_e = 2\Gamma_{eg} = \gamma$, $w_0 = -1$) of the familiar Bloch equations (Bloch 1946) including phenomenological decay constants and pumping terms:

Interaction of laser radiation with free atoms

$$\dot{w} = 2K(t)v - \Gamma_e(w - w_0) \qquad (2.25(a))$$

$$\dot{v} = -(\Delta + \dot{\phi})u - 2K(t)w - \Gamma_{eg}v \qquad (2.25(b))$$

$$\dot{u} = (\Delta + \dot{\phi})v - \Gamma_{eg}u. \qquad (2.25(c))$$

In fact, if we consider a collection of atoms in a cell, spontaneous emission is not the only damping process. In particular, some types of atomic collisions can affect the atomic dipole oscillations without disturbing its energy, and therefore there is no longer a definite relation between Γ_e and Γ_{eg}. In the very low pressure limit, however, the relation $\Gamma_e = 2\Gamma_{eg} = \gamma$ must be satisfied. On the contrary, in the strong collision limit, energy decay and dipole phase interruption are simultaneously induced by collisions, and thus Γ_e and Γ_{eg} become nearly equal. w_0 is the equilibrium value of w in the absence of the external field. For isolated atoms, w_0 is obviously equal to -1, but in a cell it can be larger than this limiting value if the atoms receive some incoherent energy by means of an electric discharge, for example. So equations (2.24) are essentially valid for atoms in an atomic beam, while equations (2.25) must be generally utilized for atoms in a cell. At this point, a remark must be made. The only decay constants that appear in the Bloch equations are homogeneous quantities and characterize the free evolution of a single atom. Now, for a macroscopic collection of atoms, macroscopic polarization damping depends not only on the homogeneous Γ but also on the Doppler width Δ_D, since the free evolution of the atomic dipoles depends on the detuning, that is to say, on their velocities. Since Δ_D is usually much larger than the Γ, the decay time of macroscopic polarization of a collection of atoms is of the order of Δ_D^{-1}, a much shorter time than $1/\Gamma_e$ or $1/\Gamma_{eg}$.

Now, instead of considering the damping resulting from spontaneous emission and the interaction with the driving field separately, one can treat them simultaneously but, of course, only if a quantum description of the field is chosen. In this case, it is natural to introduce a reduced density operator for the dressed atom, $\sigma(t)$ (Oliver et al 1971). If the other electromagnetic field modes are again treated as a bath coupled to the dressed atom, with the same approximations as previously described, $\sigma(t)$ satisfies the following equation of motion:

$$\dot{\sigma}(t) = -i\hbar^{-1}[H, \sigma] + \gamma(|g\rangle\langle e|\sigma|e\rangle\langle g| - \tfrac{1}{2}|e\rangle\langle e|\sigma - \tfrac{1}{2}\sigma|e\rangle\langle e|). \qquad (2.26)$$

On the basis of the unperturbed states $|e, m\rangle$ and $|g, n\rangle$ the matrix elements of $\sigma(t)$ defined by:

$$\langle e, m-1|\sigma|e, m-1+p\rangle = \sigma_{eem}{}^p \qquad (2.27(a))$$

$$\langle e, m-1|\sigma|g, m+p\rangle = \sigma_{egm}{}^p \qquad (2.27(b))$$

$$\langle g, m|\sigma|e, m-1+p\rangle = \sigma_{gem}{}^p \qquad (2.27(c))$$

$$\langle g, m|\sigma|g, m+p\rangle = \sigma_{ggm}{}^p \qquad (2.27(d))$$

obey the system of differential equations:

$$\dot{\sigma}_{eem}{}^p = (ip\omega - \gamma)\,\sigma_{eem}{}^p + i\kappa[(n+p)^{1/2}\,\sigma_{egm}{}^p - (n)^{1/2}\sigma_{gem}{}^p] \qquad (2.28(a))$$

$$\dot{\sigma}_{egm}{}^p = [i(p\omega + \Delta) - \tfrac{1}{2}\gamma]\,\sigma_{egm}{}^p + i\kappa[(n+p)^{1/2}\,\sigma_{een}{}^p - (n)^{1/2}\sigma_{ggn}{}^p] \qquad (2.28(b))$$

$$\dot{\sigma}_{gem}{}^p = [i(p\omega - \Delta) - \tfrac{1}{2}\gamma]\,\sigma_{ggm}{}^p + i\kappa[(n+p)^{1/2}\,\sigma_{ggm}{}^p - (n)^{1/2}\sigma_{eem}{}^p] \qquad (2.28(c))$$

$$\dot{\sigma}_{ggm}{}^p = ip\omega\sigma_{ggm}{}^p + \gamma\sigma_{eem+1}{}^p + i\kappa[(n+p)^{1/2}\,\sigma_{gem}{}^p - (n)^{1/2}\sigma_{egm}{}^p] \qquad (2.28(d))$$

where the rotating-wave approximation has been used. These equations separate according to a definite value of p but do not separate according to a definite value of m. This obviously comes from cascade effects. However, as already noticed, the normalized distribution $\rho_n(t=0)$ of the number of photons in the laser mode has a dispersion Δn around the mean value $\bar{n}$ which is large in absolute values ($\Delta n \gg 1$) and small in relative values ($\Delta n \simeq (\bar{n})^{1/2} \ll \bar{n}$). This allows us in equations (2.28) to replace $\sigma_{eem+1}{}^p$ by $\sigma_{eem}{}^p + (\partial/\partial n) \sigma_{eem}{}^p \simeq \sigma_{eem}{}^p$ and to write for any time t, $\sigma(n,t) = \rho_n \rho(t)$. With these approximations, for $p=0$, the matrix elements of $\rho(t)$ satisfy the following equations:

$$\dot{\rho}_{ee}{}^0 = -\gamma \rho_{ee}{}^0 + i\kappa(\bar{n})^{1/2}(\rho_{eg}{}^0 - \rho_{ge}{}^0) \qquad (2.29(a))$$

$$\dot{\rho}_{eg}{}^0 = (i\Delta - \tfrac{1}{2}\gamma)\,\rho_{eg}{}^0 + i\kappa(\bar{n})^{1/2}(\rho_{ee}{}^0 - \rho_{gg}{}^0) \qquad (2.29(b))$$

$$\dot{\rho}_{gg}{}^0 = \gamma \rho_{ee}{}^0 + i\kappa(\bar{n})^{1/2}(\rho_{ge}{}^0 - \rho_{eg}{}^0) \qquad (2.29(c))$$

which are fully equivalent to equations (2.24) if one makes the same identifications as in equations (2.12) and if one still identifies $\kappa(\bar{n})^{1/2}$ with K_0. Once again, one obtains exactly the same results in the classical and in the quantum approaches, but it must be noticed that this is true only because the term $(\partial/\partial n)\sigma_{eem}{}^p$ has been neglected. The role of this term is to produce a drift of ρ_n towards the lower values of n with a velocity of the order of γ. It can be ignored only if the interaction time is sufficiently small so that the corresponding decrease in the mean number of photons is much smaller than the width Δn of ρ_n. This condition, which is satisfied in most actual experiments, is equivalent to considering that the external field is not affected after interacting with the atom, an assumption which is always implied in the classical approach.

For the study of resonance fluorescence, the quantum approach to the driving field is more powerful, however, since one can obtain directly the evolution of the off-diagonal elements of $\sigma(t)$, corresponding to $p=1$ for example, and therefore one can derive in a simple way the evolution of the mean dipole moment. This has been recently discussed in detail by Cohen-Tannoudji and Reynaud (1977a), who have furthermore generalized this formalism to multilevel atoms.

2.6. Steady-state and transient solutions of the Bloch equations including decay constants

Solutions to equations (2.24) were given many years ago by Torrey (1949) for the particular case of continuous excitation: $K(t)=K_0$, $\phi=0$. The long-time steady-state solutions can be easily obtained by putting $\dot{w}=\dot{v}=\dot{u}=0$ in equations (2.25). The results are:

$$\bar{w}_\infty(\Delta) = w_0 \frac{1+(\Delta/\Gamma_{eg})^2}{1+(\Delta/\Gamma_{eg})^2 + (4K_0{}^2/\Gamma_e\Gamma_{eg})} \qquad (2.30(a))$$

$$\bar{v}_\infty(\Delta) = w_0 \frac{-2K_0/\Gamma_{eg}}{1+(\Delta/\Gamma_{eg})^2 + (4K_0{}^2/\Gamma_e\Gamma_{eg})} \qquad (2.30(b))$$

$$\bar{u}_\infty(\Delta) = -w_0 \frac{\Delta K_0/\Gamma_{eg}{}^2}{1+(\Delta/\Gamma_{eg})^2 + (4K_0{}^2/\Gamma_e\Gamma_{eg})} \qquad (2.30(c))$$

and, of course, they do not depend on initial conditions. For $\Delta=0$, $w_0=-1$ and $2\Gamma_{eg}=\Gamma_e$ (spontaneous emission limit), one gets:

$$\bar{w}(0) = -\frac{1}{1+2(2K_0/\Gamma_e)^2} = -\frac{1}{1+\chi} \tag{2.31}$$

where $\chi = 8^{1/2} K_0/\Gamma_e$ is the so-called saturation parameter. This parameter characterizes the importance of power effects on an optical resonance with monochromatic light. Since Γ_e is proportional to the square of the dipole matrix element, χ is proportional to $\Gamma_e^{-1/2}$. In other words, a transition is the more saturable the smaller the corresponding transition probability. More precisely, χ can be rewritten as

$$\chi = \left(\frac{3\lambda^3}{4\pi^2} \frac{I}{\hbar c \Gamma_e}\right)^{1/2}$$

where λ is the wavelength of the transition and I is the light power per unit area. In the visible range, Γ_e is of the order of 10^7 s^{-1} and therefore, for $\lambda = 500$ nm, one obtains a saturation parameter equal to one for I of the order of 0·1 W cm^{-2}. Now let us compare the foregoing result with the population inversion that would be obtained by using broad-band excitation:

$$w_{BB} = -\frac{1}{1+2B\rho(\nu)/A} \tag{2.32}$$

where A and B are the usual Einstein coefficients; A is equal to Γ_e, and we see that the two results become equivalent if we identify $4K_0^2$ and $AB\rho(\nu)$. Thus, for resonant monochromatic excitation, one can define an effective value of $\rho(\nu)$ which is, of course, proportional to the light energy density flux, but also to Γ_e^{-1}. In other words, everything happens as if the energy of the light was uniformly distributed inside the homogeneous width of the transition.

Now, let us go back to the transient solutions of equations (2.25). The general form of these solutions can be written:

$$w(t, \Delta) = w^0 \exp(-\lambda_0 t) + w^+ \exp(-\lambda_+ t) + w^- \exp(-\lambda_- t) + \bar{w} \tag{2.33(a)}$$

$$v(t, \Delta) = v^0 \exp(-\lambda_0 t) + v^+ \exp(-\lambda_+ t) + v^- \exp(-\lambda_- t) + \bar{v} \tag{2.33(b)}$$

$$u(t, \Delta) = u^0 \exp(-\lambda_0 t) + u^+ \exp(-\lambda_+ t) + u^- \exp(-\lambda_- t) + \bar{u} \tag{2.33(c)}$$

where λ_0, λ_+ and λ_- are the complex solutions of the cubic equation:

$$(-\lambda+\Gamma_{eg})[(\lambda-\Gamma_{eg})(\lambda-\Gamma_e)+4K_0^2] + \Delta^2(-\lambda+\Gamma_e) = 0. \tag{2.34}$$

This equation has simple solutions in three particular cases only: $\Delta = 0$ (exact resonance), $\Gamma_e = \Gamma_{eg}$ (strong collision limit) and $K_0 \gg \Gamma_{eg}$ (strong-field limit). The corresponding expressions of $w(t, \Delta)$, $v(t, \Delta)$ and $u(t, \Delta)$ can be found in the literature (see, for example, Allen and Eberly 1975) and we discuss here the strong-field limit only. If K_0 is much larger than Γ_{eg}, it is clear that $K_0 \gg \Gamma_e$ and $K_0 \gg \Gamma_{eg}' = \Gamma_e - \Gamma_{eg}$. Equation (2.34) can be rewritten:

$$(-\lambda+\Gamma_{eg})[\Delta^2 + 4K_0^2 + (\Gamma_{eg}-\lambda)^2 + \Gamma_{eg}'(-\lambda+\Gamma_{eg})] = -\Gamma_{eg}'\Delta^2. \tag{2.35}$$

Ignoring terms of the order of $(\Gamma_{eg}'/K_0)^2$, one gets:

$$\lambda_0 = \Gamma_{eg} + \Gamma_{eg}' \frac{\Delta^2}{\Delta^2 + 4K_0^2} \qquad (2.36(a))$$

$$\lambda_\pm = \Gamma_{eg} + \tfrac{1}{2}\Gamma_{eg}' \frac{4K_0^2}{\Delta^2 + 4K_0^2} \pm i(\Delta^2 + 4K_0^2)^{1/2} = \lambda^0 \pm i\Omega(\Delta) \qquad (2.36(b))$$

which leads to the following expression for the population inversion:

$$w(t, \Delta) = a \exp[-\lambda_0(t - t_0)]$$
$$+ \exp[-\lambda^0(t - t_0)][b \cos \Omega(\Delta)(t - t_0) + c \sin \Omega(\Delta)(t - t_0)] + \bar{w}(\Delta) \quad (2.37)$$

where a, b and c are time-independent coefficients depending on initial conditions. So, in the strong-field limit, $w(t, \Delta)$ (and, of course, $v(t, \Delta)$ and $u(t, \Delta)$) exhibit a damped oscillation at the Rabi frequency, which corresponds to spin nutation in the Bloch vector model. If the condition $K_0 \gg \Gamma_{eg}$ is no longer fulfilled, the nutation frequency remains the Rabi frequency in the strong collision limit only, and in the weak-field limit ($K_0 < \Gamma_{eg}$), the oscillation disappears completely.

These transient solutions provide the response of a two-level atom not only to continuous excitation (the interaction time being defined by the geometry of the experiment and most often being very large with respect to all the other parameters characterizing the resonance) but also to square-pulsed excitation. In this latter case, the interaction time is usually defined by the pulse duration and it can be much smaller than the other resonance parameters. Of course, pulsed lasers do not provide square pulses and the question arises of the role of the pulse shape. This can be studied by a numerical integration of equations (2.25) but, from the corresponding calculations, it is difficult to derive general conclusions.

2.7. Rate equations

The difficulties encountered in solving the Bloch equations in a general way mainly come from the interdependence of the populations and coherences. However, a few assumptions allow one to derive, from the Bloch equations, rate equations for the populations only (see, for example, Allen and Eberly 1975). In particular, if Γ_{eg} is very large, one can assume that u and v have already reached their steady-state values while w is still varying in time. Under this condition, equation (2.25(a)) reduces for $K(t) = K_0$, $\phi = 0$, to

$$\dot{w}(t, \Delta) = 2K_0 v_\infty(\Delta) - \Gamma_e(w - w_0)$$

$$= -2K_0 \frac{2K_0/\Gamma_{eg}}{1 + (\Delta/\Gamma_{eg})^2} w(t, \Delta) - \Gamma_e[w(t, \Delta) - w_0]$$

$$= -\Gamma_e\left[\left(\frac{4K_0^2}{\Gamma_e \Gamma_{eg}} \frac{1}{1 + (\Delta/\Gamma_{eg})^2} + 1\right) w(t, \Delta) - w_0\right] \qquad (2.38)$$

which can be solved explicitly for any time without any difficulty. Let us remark that equation (2.38) is very similar to the well-known Einstein equation:

$$\dot{w}(t, \Delta) = -2B\rho w - A(w - w_0) \qquad (2.39)$$

the only difference coming from the Lorentzian factor $(1+\Delta/\Gamma_{eg})^{-1}$ which distinguishes monochromatic from broad-band excitation.

Since the general rate equation (2.38) is much simpler than the optical Bloch equations, it is generally used as a starting point to investigate theoretically some complicated problems such as the theory of gas lasers, saturation spectroscopy or pulse propagation which arise in quantum optics. This is justified since in many cases it appears that the results obtained from a rate equation approach do not differ greatly from those of a more sophisticated treatment. In particular, it has been demonstrated by Lamb (1964) that this is true for a gas laser well above threshold. All the phenomena which cannot be predicted by rate equations are usually called 'coherence effects' since their origin clearly lies in the characteristic time evolution of atomic coherences.

2.8. Real systems

Throughout this section, the two-level atom and monochromatic radiation approximations have been used. Several questions now arise about the validity conditions of these approximations in real systems: is it actually possible to isolate two-level systems in atomic spectra? What is the influence of the other levels? What is the role of atomic degeneracy? How good is the monochromatic approximation for real lasers? Etc..... It is difficult to give general answers to these questions and each individual case must be separately discussed but we shall try nevertheless to provide some general indications and to illustrate them with simple examples.

2.8.1. Real atoms. After a look at the famous tables compiled by Moore (1949) giving all the known atomic energy levels in simple spectra, it appears that well-isolated two-level systems in atomic spectra are quite exceptional. *A priori*, atomic levels having a hyperfine structure do not represent a good case, especially in the strong-field limit, and if, further, we wish to avoid decay to other levels leading to optical pumping effects, only a few resonance transitions can be retained. Unfortunately, if one takes into account these two conditions most of the possible transitions lie in the ultraviolet range, a wavelength domain which is not yet well covered by conventional tunable single-mode lasers. For some resonance transitions however, such as the transition $6s^2\ {}^1S_0 \rightarrow 6s6p\ {}^1P_1$ of Ba I ($\lambda = 553\cdot 7$ nm), branching ratios characterizing the various decay channels of the upper level are such that for reasonable interaction times ($\tau < 100/\gamma$), optical pumping is negligible and the two-level approximation is valid. A similar situation appears in Yb I but, paradoxically, these transitions have been studied for spectroscopic purposes; only, that is to say, in the weak-field limit. In fact, most of the experiments on non-linear interactions between free atoms and laser radiation have been performed on resonance transitions of alkali atoms or on transitions between excited states of noble gases, two situations which *a priori* are not favourable for defining well-isolated two-level systems but which are well adapted to the use of dye lasers.

In alkali atoms, a particular choice of hyperfine transitions can eliminate optical pumping problems. This is the case for the resonance transitions

$$nS_{1/2}F = I + \tfrac{1}{2} \rightarrow nP_{3/2}F = I + 3/2,$$

but two difficulties remain:
  (i) The presence of other hyperfine sublevels which cannot be considered as

far from the sublevels under consideration since the distance between two hyperfine components is often only a few times larger than the natural linewidths (see figure 4).

(ii) Atomic degeneracy and the coexistence of several Rabi frequencies. The obvious way to solve these difficulties is to put the atoms in a strong magnetic field to separate the levels and to remove atomic degeneracy. This has been done in particular by Gibbs and Slusher (1970) on an atomic beam of Rb. The magnetic field was adjusted to bring the atomic transition into resonance with the fixed frequency of a Hg II pulsed laser. By using tunable lasers there is another, much simpler, way to isolate a two-non-degenerate-level system. The principle is to take advantage of optical pumping within magnetic sublevels. In particular, for the sodium resonance transition $3S_{1/2}F=2\rightarrow 3P_{3/2}F=3$, it is clear from figure 5 that, if σ^+ circularly polarized light is used, optical pumping concentrates the atoms in the ground magnetic sublevel $M_F=2$ and then the only possible transition is $3S_{1/2}F=2$, $M_F=2\rightarrow 3P_{3/2}F=3$, $M_F=3$. Originally suggested by Hartig et al

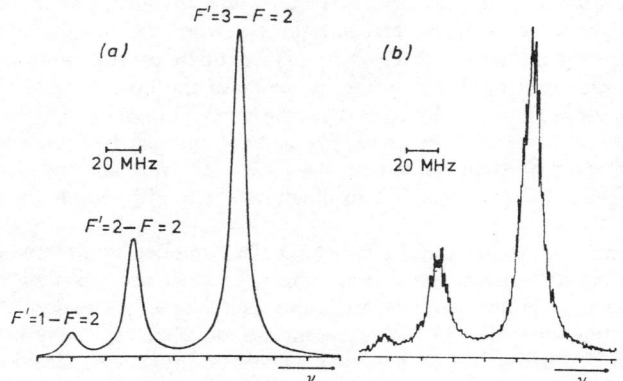

Figure 4. Part of the homogeneous spectrum of the D_2 line of sodium. (a) Calculated signal with natural linewidth; (b) recorded signal (from Lange et al 1973).

(1976), this technique has been actually utilized in resonance fluorescence experiments (see §3).

In experiments using noble-gas atomic beams, the only transitions which can be studied must start from metastable levels: $(np)^5 (n+1)s$ $^3P_0$, $^3P_2$. Here again, by an appropriate choice of the transition, optical pumping effects can be avoided. This is the case in neon, for example, for the transition $2p^53s$ $^3P_2\rightarrow 2p^53p$ $^3D_3$ ($\lambda=640$ nm) and here again, difficulties connected with atomic degeneracy can be suppressed by using circularly polarized light.

For atoms in a cell, optical pumping does not play any role, and in a discharge, any transition can be studied in principle. From a certain point of view, the problem is simpler and any transition without hyperfine structure can be described in the two-level approximation. However, atomic degeneracy effects are more difficult to suppress. This is nevertheless possible in some peculiar cases by a proper choice of polarizations and fortunately, in many cases, atomic degeneracy effects do not strongly perturb the phenomena under study.

So, though it seems *a priori* difficult to isolate well-defined two-level systems

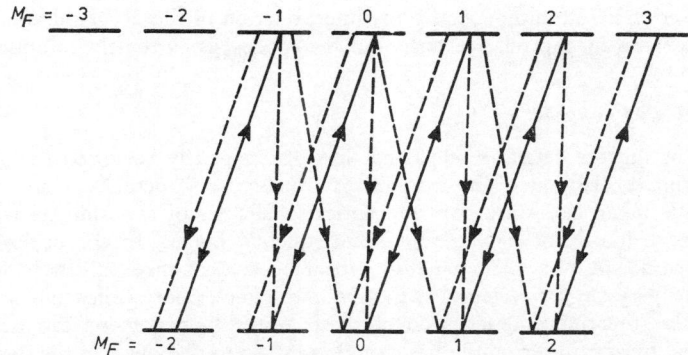

Figure 5. Optical pumping effects in a transition $F = 2 \to F = 3$ excited by σ^+ polarized light.

in atomic spectra, this is possible in particular cases, mainly thanks to the use of appropriately polarized light.

2.8.2. Real laser properties. It is certainly beyond the scope of this review to give even a brief technical survey on available lasers, more especially as the situation is continuously changing. Besides, if one excepts a few cases of accidental coincidences between fixed-frequency laser lines and atomic transitions, most of the experiments on resonant interactions between free atoms and laser radiation have been made possible thanks to tunable lasers, and more precisely to dye lasers. The cw types are, of course, best adapted since single-mode emission can now be obtained for many dyes from blue to red, but it must be noticed that a few single-mode pulsed lasers have been recently built (Pinard and Liberman 1977). In any case, the jet-stream technique leads to powers as high as 100 mW for cw dye lasers and therefore, even for atomic resonance lines which are the most difficult to saturate, saturation parameters larger than ten can be obtained. In some cases, such as Doppler-free two-photon spectroscopy (see §5), for example, these powers are sufficient to observe even non-resonant processes but, of course, the ultra-high powers ($10^9 \sim 10^{15}$ W) required to detect some strongly non-linear phenomena such as multiphoton ionization of noble gases, for example, can be provided by pulsed lasers only.

For cw single-mode lasers, monochromaticity is essentially limited by fluctuations; the laser line then exhibits a jitter which determines its spectral width. It has been possible, with special care of course, to construct fixed-frequency gas lasers with a short-term spectral linewidth of the order of 1 kHz. In this case, the jitter is mainly due to very rapid phase fluctuations (Haken 1970). For dye lasers, the situation is less favourable, and in spite of numerous theoretical approaches the origin of the jitter is not very well understood. Spectral linewidths of a few parts in 1 MHz have been obtained but, with commercial dye lasers, the jitter is typically of the order of a few MHz. Therefore, with this type of laser, the spectral width of the laser line is of the same order of magnitude as the natural linewidth of the atomic transition. However, if one excepts a trivial loss of resolution, no supplementary effect appears with respect to the monochromatic approximation. Obviously, this is not true in multimode laser operation, especially if high orders of non-linearity are involved in the phenomenon considered (see, for example, Lecompte *et al* 1974,

Decomps *et al* 1976), and generally the interpretation of the observed effects (which strongly depend on the relative phases of the modes) is extremely complicated.

2.9. Concluding remarks

Most of the theoretical models that have been briefly reviewed in this section were originally elaborated for magnetic resonance, and actually many resonance experiments using lasers are only the optical analogues of experiments which were carried out a long time ago in the radiofrequency range. In the optical domain, however, some specific phenomena do appear because of three effects which can be ignored for very large wavelengths: the Doppler effect, fluorescence and propagation effects. The theoretical description of these 'optical' phenomena can nevertheless be achieved by starting from the basic models introduced above. In the next sections such a description will be given for a variety of distinguishable cases.

3. Response of an atomic system to resonant laser excitation

The scope of this section is to describe selected experimental situations allowing the observation of the main phenomena that occur in the interaction between atoms and resonant laser radiation. Propagation effects, however, will be studied in a separate section and therefore are ignored here. According to the discussion of §2.1, one of the essential parameters characterizing an optical resonance is the inhomogeneous width of the transition. First, we consider experiments with atomic beams, transversely illuminated by continuous radiation. In this case, no inhomogeneous effect has to be taken into account, of course. Then, we discuss the opposite case, corresponding to a collection of atoms in a cell and we pay special attention to hole-burning techniques that allow one to carry out Doppler-free experiments. Finally, we discuss briefly the various ways of studying transient effects directly in the time domain.

3.1. Atomic beams transversely illuminated

This technique is particularly well-adapted to spectroscopic measurements and a great deal of experiments have now been performed to investigate fine and hyperfine structures, isotope shifts, etc, of many atomic transitions. For details, the reader is referred to a review written recently on this subject (Jacquinot 1976a). Moreover, this situation is the simplest one for the study of basic phenomena in the interaction between monochromatic light and atoms. In fact, it is quite similar to the one encountered in magnetic resonance, and most of these phenomena have already been observed many years ago in the radiofrequency range. They can now be observed in the optical domain thanks to the monochromaticity and power of laser radiation. All the phenomena depending on the light power are more or less directly related to the so-called dynamic Stark effect (Autler and Townes 1955) since the Bloch–Siegert shift is quite negligible in the visible range. New manifestations of this effect can appear however, in particular in relation to spontaneous emission whose spectral characteristics are modified under the influence of a strong driving field.

3.1.1. Power broadening. First, let us consider the simplest experiment which can

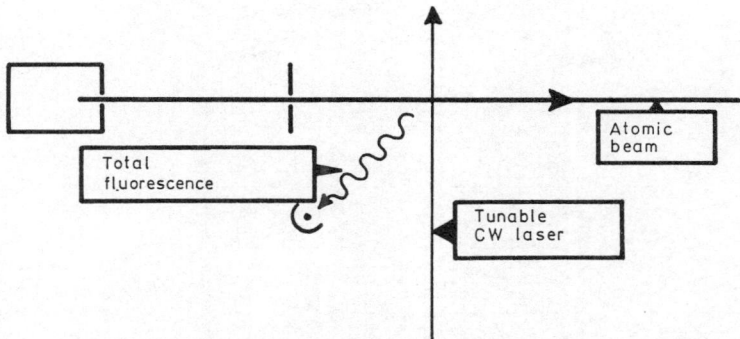

Figure 6. The simplest set-up for resonance experiments using an atomic beam transversely illuminated by tunable laser radiation.

be carried out with an atomic beam and a tunable cw single-mode laser (see figure 6). If the atom can be approximated by a two-level system, any method of monitoring the population of the upper level, be it optical (total fluorescence (Hartig and Walther 1973)) or non-optical (magnetic deviations (Duong et al 1973b), beam deflection (Jacquinot et al 1973), selective photoionization (Duong et al 1973a), field ionization (Ducas et al 1975, Pinard and Liberman 1977), etc) leads to a resonance curve with a Lorentzian profile centred on the frequency of the atomic transition, the width of this resonance being given, according to the theoretical results of §2 by:

$$\Gamma_R = 2\left(\Gamma_{eg}^2 + 4\frac{K_0^2}{\Gamma_2}\Gamma_{eg}\right)^{1/2} = (\gamma^2 + 8K_0^2)^{1/2} = \gamma(1+\chi^2)^{1/2}. \quad (3.1)$$

Since K_0 (or χ) is proportional to the square root of the energy density flux of the driving field, the width of the resonance is an increasing function of the light power. This power broadening, well-known in magnetic resonance, has been observed in the optical range in nearly all spectroscopic experiments using atomic beams and tunable cw single-mode lasers (see, for example, Lange et al 1973). In fact, in such experiments it appears as a limitation of the resolution and therefore one tries to eliminate it by working at sufficiently low light power.

In fact, as already noted in §2.8, the problem is often much more complicated for real systems, in particular when the upper level can decay to other levels. In this case, optical pumping effects appear, and for high powers of the light field, the continuous regime approximation is no longer justified. Under these conditions the resonance no longer has a Lorentzian profile since pumping efficiency depends on the instantaneous population of the upper level, that is to say, on Δ. Its width increases not only with the saturation parameter but also with the interaction time. More precisely, if Γ^{-1} and τ are, respectively, the lifetime of the upper level and the interaction time, for any value of Δ much smaller than $[K_0^2(\Gamma-\gamma)\tau]^{1/2}$, the detection signal is practically independent of Δ. Therefore, for very long interaction times and strong light powers, the resonance is so broad that it disappears completely. This phenomenon has been observed by various authors, especially on the resonance transition 3s–3p of sodium and it is illustrated in figure 7. These optical pumping effects are quite cumbersome in experiments for spectroscopic purposes since they modify intensity ratios, but we have shown in §2.8 that they

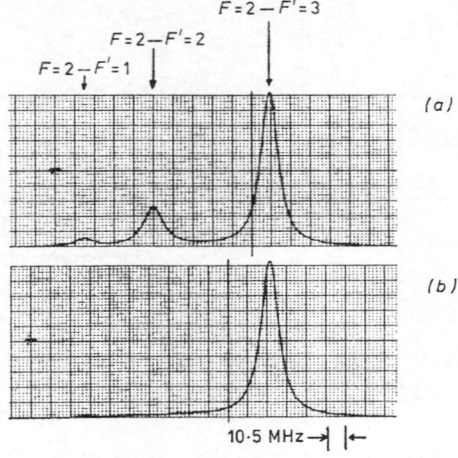

Figure 7. Optical pumping effects on the D_2 line of sodium. (a) Hyperfine structure recorded at low intensity and for short interaction times (negligible pumping). (b) Recorded signal after efficient pumping (from Grove et al 1977).

are very convenient for the creation of actual two-level systems in a real multilevel atom.

3.1.2. Two-step experiments, Autler–Townes splitting. The main characteristic of the spectrum of a two-level atom dressed by photons in a single mode is a doublet structure. Of course, no doublet appears in single-step resonance experiments if one records only a signal proportional either to the population of the upper level or to the population difference between the two levels. However, let us consider the following experiment that is shown schematically in figure 8: a three-level system, characterized by a ground state $|g\rangle$ and two non-degenerate excited states $|e\rangle$ and $|f\rangle$ with respective energies $\hbar\omega_g$, $\hbar\omega_e$ and $\hbar\omega_f$, is coupled with two driving fields with respective frequencies ω_1 and ω_2, quasi-resonant respectively for the transitions g↔e and e↔f. If the field quasi-resonant with the first step is very strong and has a fixed frequency, it is clear from figure 9 that in the weak-field limit for the second step two resonances appear respectively for

$$\Delta_2^\pm = \omega_2^\pm - \omega_f + \omega_e = -\tfrac{1}{2}[\Delta_1 \pm (\Delta_1^2 + 4K_1^2)^{1/2}]$$
$$= -\tfrac{1}{2}(\Delta_1 \pm \Omega_1) \tag{3.2}$$

where $\Delta_1 = \omega_1 - \omega_e + \omega_g$ and K_1 is the Rabi frequency at zero detuning for the first step. If the first step is exactly on resonance, the resonance shape on the second step is a symmetrical doublet, the separation between the two components of the doublet being given by K_1, which is proportional to the square root of the energy density flux in the first step. This doubling was observed for the first time many years ago by Autler and Townes (1955) on microwave transitions of the molecule COS.

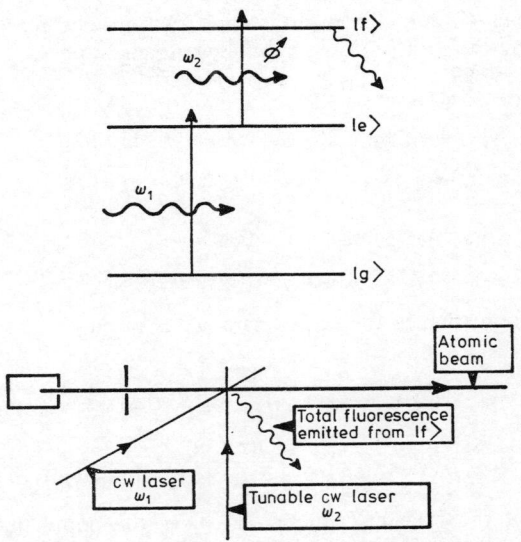

Figure 8. Typical set-up for two-step experiments with atomic beams.

Of course, in the optical range, relaxation phenomena play an important role. The simplest way to introduce them is to utilize density matrix formalism associated with a classical description of the fields (Feld and Javan 1969, Feldman and Feld 1970, Mollow 1972a). If the weak-field limit is valid for the second step, we can assume that the density matrix elements corresponding to the first step are not significantly affected by the second step transition, and therefore that they are still

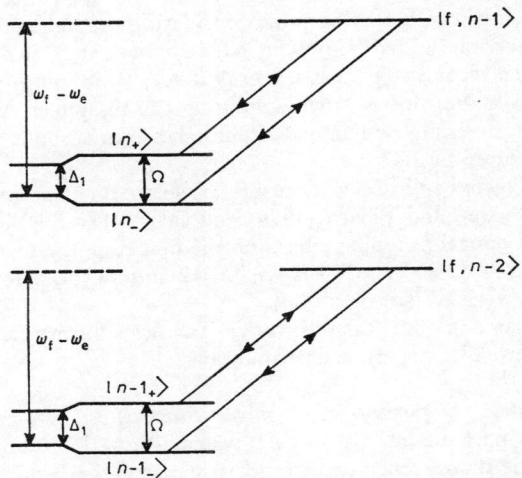

Figure 9. Interpretation of the Autler–Townes splitting in the dressed-atom picture.

given by equations (2.30). Then the stationary solutions for the density matrix elements involving the state $|f\rangle$ obey the following equations:

$$\gamma_f \rho_{ff} = iK_2(\alpha_{fe} - \alpha_{ef}) \tag{3.3(a)}$$

$$(i\Delta_1 - \tfrac{1}{2}\gamma_e - \tfrac{1}{2}\gamma_f)\alpha_{fe} + iK_1\alpha_{fg} - iK_2 \tfrac{1}{2}(w+1) = 0 \tag{3.3(b)}$$

$$(i\Delta_1 + i\Delta_2 - \tfrac{1}{2}\gamma_f)\alpha_{fg} + iK_1\alpha_{fe} - iK_2 \tfrac{1}{2}(u - iv) = 0 \tag{3.3(c)}$$

where

$$\alpha_{fe} = \alpha_{ef}^* = \rho_{fe} \exp[-i(\omega_2 t + \phi_2)]$$

$$\alpha_{fg} = \alpha_{gf}^* = \rho_{fg} \exp\{-i[(\omega_1 + \omega_2)t + \phi_1 + \phi_2]\}.$$

Therefore, the population of the upper state $|f\rangle$ is given by:

$$\rho_{ff} = \frac{K_2^2}{2\gamma_f} \left(\frac{(w+1)}{[-i\Delta_2 + \tfrac{1}{2}(\gamma_e + \gamma_f)] + K_1^2/(-i\Delta_1 - i\Delta_2 + \tfrac{1}{2}\gamma_f)} + \text{cc} \right)$$

$$+ \frac{K_2^2}{2\gamma_f} \left(\frac{K_1(v + iu)}{[-i\Delta_2 + \tfrac{1}{2}(\gamma_e + \gamma_f)](-i\Delta_1 - i\Delta_2 + \tfrac{1}{2}\gamma_f)} + \text{cc} \right). \tag{3.4}$$

The first term of the previous equation involves the population $\rho_{ee} = \tfrac{1}{2}(w+1)$ of the intermediate state and it corresponds actually to a two-step process resonant for $\Delta_2 \simeq 0$. On the contrary, the second term involves coherences only and it appears even if no population is created in the intermediate state. In the language of perturbation theory, this term corresponds to a two-photon process resonant for $\Delta_1 + \Delta_2 \simeq 0$. If Δ_1^2 is much larger than K_1^2, γ_e^2 and γ_f^2, two resonances appear in any case with the same total intensity but the two-photon resonance has a width much smaller than the two-step resonance and therefore the peak intensity of the two-photon process is much larger. If Δ_1 equals zero, two symmetrical resonances appear, provided that K_1 is larger than a critical value depending only on γ_e and γ_f; otherwise, only one resonance appears for $\Delta_2 = 0$. If γ_f is much smaller than γ_e (which is quite realistic if the transition g↔e is a resonant one) the critical value of K_1 corresponds to a saturation parameter for the first step roughly equal to $\tfrac{1}{8}$ (Feneuille and Schweighofer 1975), which is a rather small value considering the light power now provided by single-mode tunable lasers. However, let us remark that, if the weak-field limit is not valid for the second step, that is to say if K_2^2 is no longer small with respect to K_1^2, γ_e^2 and γ_f^2, the Autler–Townes splitting can disappear even for high values of K_1^2. This can be easily understood if one takes into account the supplementary broadening introduced by K_2^2.

The Autler–Townes splitting was recently observed on an atomic beam of sodium in one case by Picqué and Pinard (1976) and in another by Bjorkholm and Liao (1977). To avoid optical pumping effects on the first step, the considered transitions were 3s $^2S_{1/2}F=2 \to$ 3p $^2P_{3/2}F=3 \to$ 5s $^2S_{1/2}F=2$ and all the points discussed above were qualitatively verified (see figure 10).

We shall see later in §3.2 that the Autler–Townes splitting can be also observed in vapours for particular experimental conditions.

3.1.3. Power spectrum of fluorescence, dynamic Stark effect. Of course, the doublet structure of the spectrum of a two-level system dressed by single-mode photons appears also in the fluorescence spectrum. In particular, the fluorescence spectrum emitted to a third level $|f\rangle$ of the atom (rather far in energy from the ground state

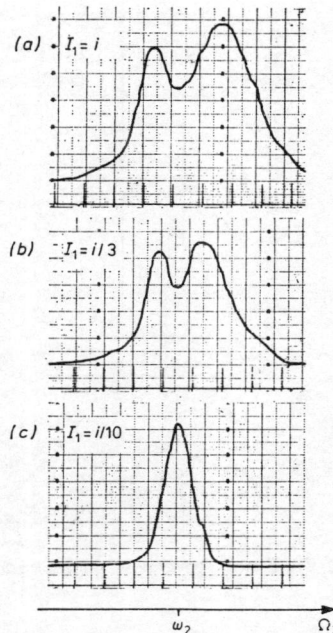

Figure 10. Autler–Townes splitting resulting from saturation of the transition $3\,^2S_{1/2}F = 2 \to 3\,^2P_{3/2}F = 3$ ($i \simeq 0.9$ W cm^{-2}) and observed on the transition $3\,^2P_{3/2}F = 3 \to 5\,^2S_{1/2}F = 2$ of sodium (from Picqué and Pinard 1976).

in order that the two-level approximation should remain valid) is a doublet. In the strong-field limit, the frequencies of the two components of this doublet are respectively (Notkin et al 1967):

$$\omega_e - \omega_f + \tfrac{1}{2}\Delta - (\tfrac{1}{4}\Delta^2 + K_0^2)^{1/2} \qquad \text{(excitation process)} \qquad (3.5(a))$$

$$\omega_e - \omega_f + \tfrac{1}{2}\Delta + (\tfrac{1}{4}\Delta^2 + K_0^2)^{1/2} \qquad \text{(Raman process)}. \qquad (3.5(b))$$

At exact resonance ($\Delta = 0$), the splitting remains only if K_0 is larger than a critical value depending only on the natural width of the considered transitions.

The shift of the Raman line as a function of the intensity of the driving field was studied more than ten years ago by using high-power lasers with fixed frequency. (For details, the reader is referred to the review paper of Papoular and Platz (1972).) Instead of looking at the fluorescence emitted to a third level, one can study the resonance fluorescence spectrum for the two levels coupled by the driving field. In this case, since spontaneous transitions take place between identical doublets, a symmetrical triplet is expected (Rautian and Sobel'man 1962), at least in the strong-field limit (see figure 11). However, in the quantum approach, a rigorous treatment of this problem appears very complicated at first sight because of cascade and interference effects. Indeed, the various paths starting from the doublet $|n \pm \rangle$ and arriving at $|n-N \pm \rangle$ with emission of N photons with definite frequencies but with a different order of emission cannot be physically distinguished and therefore interferences of $N!$ quantum amplitudes have to be considered. Since in experiments involving

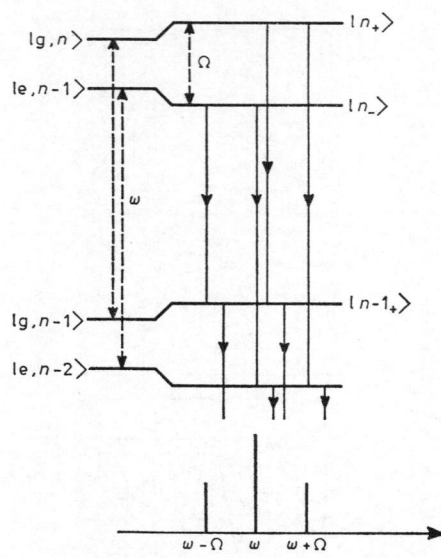

Figure 11. Interpretation of the fluorescence power spectrum in the dressed-atom picture.

thermal atomic beams the interaction time is often a hundred times longer than the radiative lifetime, N can be as high as 50 if the atomic transition is saturated, and it is obvious that such an approach is not very convenient (Cohen-Tannoudji 1975a).

In fact, to derive the spectral distribution of the resonance fluorescence light $\mathscr{I}(\omega')$, such a detailed analysis of the emission processes is not actually useful. $\mathscr{I}(\omega')$ can be directly related to the atomic density matrix elements and, more precisely, it has been shown in various references that $\mathscr{I}(\omega')$ is none other than the Fourier transform of the correlation function of the atomic dipole moment, D. Therefore, $\mathscr{I}(\omega')$ is proportional to

$$\int_0^\tau dt \int_0^\tau dt' \langle D_+(t)D_-(t')\rangle \exp\left[-i\omega'(t-t')\right] \tag{3.6}$$

where $D_+ = d|e\rangle\langle g|$ and $D_- = d|g\rangle\langle e|$, d being the matrix element $\langle e|D|g\rangle$ which is assumed to be real. With this result, it is no longer necessary to use a quantum approach to the driving field and one can start from the Bloch equations introduced in §2.5. This was done for the first time by Mollow (1969) and recently reconsidered by Cohen-Tannoudji (1975a, b) in the framework of the Bloch vector model.

The key points of these theoretical treatments are the following.

(i) By writing $D_\pm(t) = \langle D_\pm(t)\rangle + \delta D_\pm(t)$, where $\delta D_\pm(t)$ is the deviation from the average value, one obtains:

$$\langle D_+(t)D_-(t')\rangle = \langle D_+(t)\rangle\langle D_-(t')\rangle + \langle \delta D_+(t)\delta D_-(t')\rangle.$$

This separation between mean values and fluctuations clearly shows that the spectrum of the fluorescent light contains, on the one hand, an elastic component which is the light radiated by the average motion of the dipole, and, on the other hand, an inelastic component radiated by the fluctuation.

(ii) The evolution in time of the correlation function $\langle D_+(t)D_-(t')\rangle$ is the same for $\rho_{eg}(t)$ if $t > t'$ as for $\rho_{ge}(t)$ if $t < t'$. The physical meaning of this important result, which is a particular case of the so-called 'quantum regression theorem' (Lax 1968), is that the evolution of the dipole after a fluctuation is the same as the transient behaviour of the mean dipole moment starting from a non-steady-state initial condition.

From these results and those given in §2.6 the calculation of $\mathscr{I}(\omega')$, though quite tedious, is straightforward. First of all, one obtains for the ratio of the respective total intensities of the elastic and the inelastic parts:

$$I_{el}/I_{inel} = (\Gamma^2 + 4\delta^2)/8K_0^2. \tag{3.7}$$

In other words, for very high intensities of the external field $K_0^2 \gg \Gamma^2$, δ^2, most of the light is scattered inelastically. This can be easily understood if one remarks that, in this case, the transition being fully saturated, the mean value of the dipole moment reduces to zero. For such high intensities, three Lorentzian peaks appear in the inelastic spectrum: one central component around $\omega' = \omega$ and two symmetrical sidebands around $\omega' = \omega \pm \Omega$. This result is in agreement with the qualitative prediction of the dressed-atom model, but here one also obtains the peak intensities ($\frac{1}{3}:1:\frac{1}{3}$) and the half-widths ($\frac{3}{4}\Gamma:\frac{1}{2}\Gamma:\frac{3}{4}\Gamma$) of the various components. This triplet structure has been observed by various authors on the sodium D$_2$ line. The most recent experiments (Wu et al 1975, Hartig et al 1976, Grove et al 1977) show a perfect agreement between experimental data and theoretical predictions (see figure 12).

We have already noticed that a full quantum approach to this problem leads to complicated calculations. Such calculations have been carried out however and of course, when they are correct, they confirm the previous results (Oliver et al 1971, Carmichael and Walls 1975). A more basic contribution was recently given by Cohen-Tannoudji and Reynaud (1977a, b) who have shown that a dressed-atom approach allows one, for intense laser beams, to derive simple general results on

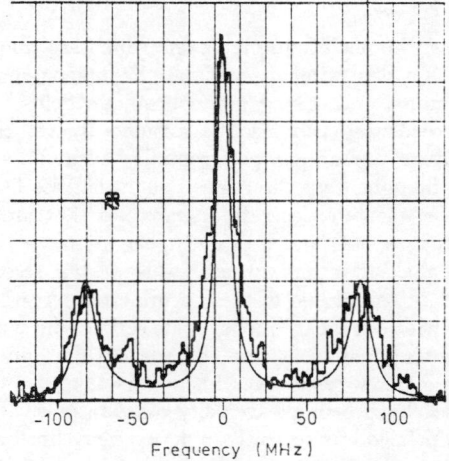

Figure 12. On-resonance fluorescence spectrum observed on the transition $3^2S_{1/2}F = 2 \rightarrow 3^2P_{3/2}F = 3$ of sodium. Theoretical spectrum shown as smooth curve (from Grove et al 1977).

resonance fluorescence or on the absorption rate of photons from a weak probe tunable laser beam (Mollow 1972b), even for multilevel atoms. A detailed description of this approach is beyond the scope of this review but one can indicate that the principle is to solve the master equation describing spontaneous emission from the eigenstates of the dressed atom by neglecting the drift of ρ_n which has been discussed in §2.5.

3.1.4. Modulated excitation. Another way to exhibit the doublet structure of the spectrum of a two-level system dressed by single-mode photons is to use modulated excitation. In particular, it has been theoretically shown (Armstrong and Feneuille 1975) that, if the two-level atom is excited in a resonant way by a single-mode laser weakly modulated at low frequency, the scattered light, modulated at the same frequency, has an amplitude of modulation showing a resonant behaviour when the frequency splitting of the doublets is equal to the modulation frequency. This can be physically understood in the following way. The modulated strong laser beam is equivalent to a triplet of coherent monochromatic waves. The strong central component produces the doublet structure whereas the two weak sidebands act as a probe which excites coherently the two sidebands of the resonance triplet if the modulation frequency has the correct value. The modulation of the scattered light then appears as the beat note between the two corresponding waves.

Furthermore, another theoretical analysis has recently shown (Feneuille *et al* 1976) that, if the two-level atom is excited in a resonant way by a strong single-mode laser beam fully modulated in amplitude at low frequency (or by two coherent laser beams), the dynamical Stark effect can be observed not only through the modulation rate or the phase shift of the scattered light but also through the variations of the mean total intensity with the modulation frequency. Thus its experimental observation should be particularly easy in this case but, unfortunately, no experiment using modulated narrow-band excitation has been carried out up to now.

3.2. Vapours

In previous sections, the use of atomic beams which are transversely illuminated appears to be essential for observing the various phenomena characterizing resonant interactions between atoms and strong monochromatic fields. In fact, this is not true since Doppler broadening can also be avoided by velocity selection. More precisely, the atoms of a vapour are preferentially excited by the laser radiation if their velocity along the light wave direction is such that the Doppler shift balances exactly the detuning between the laser frequency and the transition frequency for atoms at rest. Therefore, a hole and a peak appear in the velocity distributions of atoms respectively in the lower and upper levels of the resonant transition (see figure 13) (Bennett 1962); the profile of these symmetric perturbations is Lorentzian and the corresponding width in the frequency scale is the same as that of the resonance studied in §3.1.1 for a single atom at rest. Therefore, the only remaining problem for the observation of narrow resonances inside the Doppler contour is to find a method to probe these changes in the velocity distributions. This can be done in three different ways, which lead respectively to fluorescence line narrowing, absorption line narrowing and saturated absorption techniques.

3.2.1. Fluorescence line narrowing. The most straightforward way to probe the peak

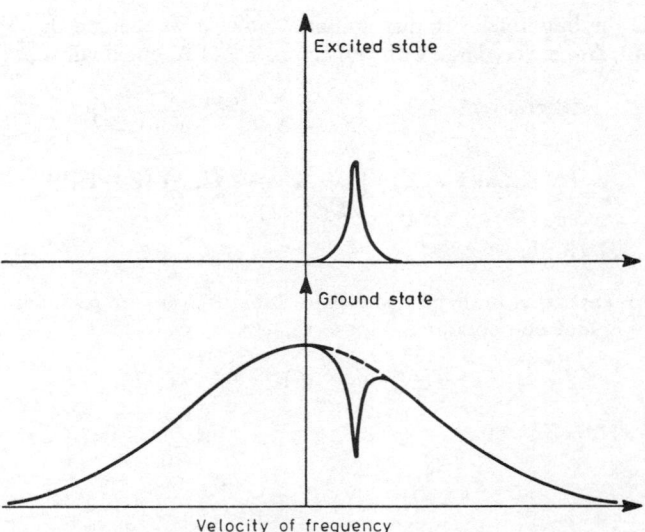

Figure 13. 'Hole-burning' in velocity distributions of atoms in a cell interacting with a quasi-resonant monochromatic field.

created in the velocity distribution for atoms in the upper level is to analyse the spectral distribution of the fluorescence emitted from this level either in the same direction as the incoming radiation or in the opposite direction. If a sufficiently small aperture angle is used, the profile of this spectral distribution depends only on the homogeneous widths of the excitation and fluorescence transitions, on the ratio of their frequencies, and on the power of the driving field (Feld and Javan 1969). For sufficiently low values of this power, the profile is Lorentzian and its width is power-independent. This situation is obviously the best for spectroscopic purposes and corresponds to the few experiments actually performed up to now (Schweitzer et al 1967, Cordover et al 1967). A question arises, however: is it possible to observe in a vapour the dynamical Stark splittings described in §3.1.3 for atoms at rest? First, let us consider the case in which one analyses the fluorescence emitted to a third level far from the lower level of the saturated transition. In the strong-field limit, according to equations (3.5), atoms emit light at frequency ω' only if their velocity v in the direction of the light beam obeys the following equation (see, for example, Chebotayev 1976):

$$\omega' - k'v = \omega_e - \omega_f + \tfrac{1}{2}(\Delta + kv) \pm [\tfrac{1}{4}(\Delta + kv)^2 + K_0^2]^{1/2} \quad (3.8(a))$$

or

$$[(\Delta + kv) - 2(\Delta' - k'v)]^2 - (\Delta + kv)^2 - 4K_0^2 = 0 \quad (3.8(b))$$

where $\Delta' = \omega' - \omega_e + \omega_f$; k/k' is positive ($k/k' = \omega/\omega'$) if the fluorescence light is observed in the same direction as the incoming radiation and is negative ($k/k' = -\omega/\omega'$) in the opposite case. If we denote by v_+ and v_- the two solutions of the previous equations, the spectral distribution of the fluorescence light is given by

$$\mathscr{I}(\omega') = \mathscr{I}_0 \left(\frac{2K_0^2}{4K_0^2 + (\Delta + kv_+)^2} g(v_+) + \frac{2K_0^2}{4K_0^2 + (\Delta + kv_-)^2} g(v_-) \right) \quad (3.9)$$

where $g(v)$ is the initial velocity distribution. Now, if we assume that the inhomogeneous width, Δ_D, is very large with respect to K and Δ, one finally obtains:

$$\mathscr{I}(\omega') = 2K_0^2 \mathscr{I}_0' \left(\frac{1}{4K_0^2 + \Delta_+^2} + \frac{1}{4K_0^2 + \Delta_-^2} \right) \quad (3.10)$$

where

$$\Delta_\pm = \frac{(\Delta' - Z\Delta)(1 - 2Z) \pm [(\Delta' - Z\Delta)^2 + 4K_0^2 Z(Z-1)]^{1/2}}{2Z(Z-1)} \quad (3.11)$$

$Z = K'/k$.

The two terms appearing in the previous expression of $\mathscr{I}(\omega')$ are resonant respectively for $\Delta' = Z\Delta \pm K$ and one obtains for these values:

$$\mathscr{I}(\Delta' = Z\Delta \pm K) \gtrsim \tfrac{1}{2}\mathscr{I}_0'$$

while

$$\mathscr{I}(\Delta' = Z\Delta) = 0 \quad \text{for} \quad 0 \leqslant Z \leqslant 1$$

$$= \mathscr{I}_0' \frac{4Z(Z-1)}{4Z(Z-1)+1} \quad \text{for} \quad \begin{array}{l} Z < 0 \\ Z > 1. \end{array}$$

Therefore, a splitting appears in the spectral distribution of the fluorescent light only if $-\epsilon < Z < 1 + \epsilon$, ϵ being a small quantity depending mainly on the Doppler width and on the saturation of the transition g↔e. So, the conditions on the two frequencies ω_{eg} and ω_{ef} to observe a splitting are very different according to the direction of observation of fluorescent light (Feld and Javan 1969). If one observes the fluorescent light in the opposite direction with respect to the incoming radiation, the condition is in fact $\omega_{eg} \gg \omega_{ef}$ and it does not correspond to realistic cases. On the contrary, if the fluorescent light is observed in the same direction as the incoming radiation, the condition is $\omega_{ef} \lesssim \omega_{eg}$ and many realistic three-level systems satisfying this condition can be found in atomic spectra. However, to our knowledge, this doublet structure has never been observed, probably because of the weakness of the power of the driving fields actually used in the few experiments performed up to now.

From the previous results, it is clear that the resonance fluorescence spectrum for the two levels coupled by the driving field exhibit a triplet structure only if the fluorescent light is observed in the same direction as the incoming radiation, but it is clear that, in this case, the difficulties in separating the scattered and the direct beams would prevent any significant observation.

3.2.2. Absorption line narrowing. Another obvious way to probe the peak and/or the hole respectively created in the velocity distributions for atoms in the upper and lower levels is to use a second laser beam for studying absorption (or induced emission) from one of the two coupled levels to a third one. Three situations can be encountered (see figure 14) and each of them has been studied in detail in various theoretical papers (for references, see Chebotayev 1976). If the probe beam is sufficiently weak, the perturbation profile in the excitation curve is exactly the same as the spectral distribution of the fluorescent light studied in the previous subsection, provided that the sign of Z is changed when the cascade scheme is considered. Therefore, for the two-step case discussed in §3.1.2, a splitting is still observed in vapours, provided that the saturating and the probe beams counter-propagate and that the

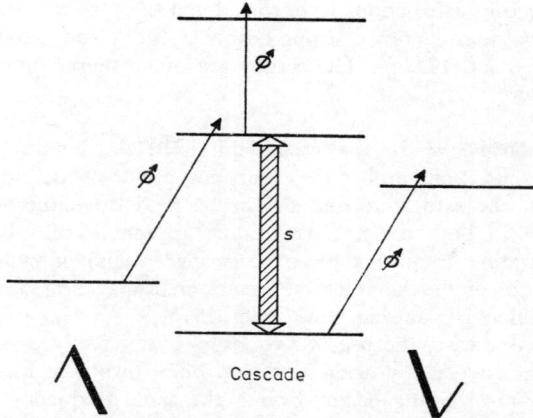

Figure 14. The three situations encountered in absorption line narrowing experiments.

frequency of the probe is smaller than the frequency of the saturating beam (Hänsch and Toschek 1970, Feneuille and Schweighofer 1975). These predictions have been recently verified in several experiments on noble-gas spectra (Shabert *et al* 1975, Cahuzac and Vetter 1976, Delsart and Keller 1976a, b). One of the corresponding results is illustrated in figure 15.

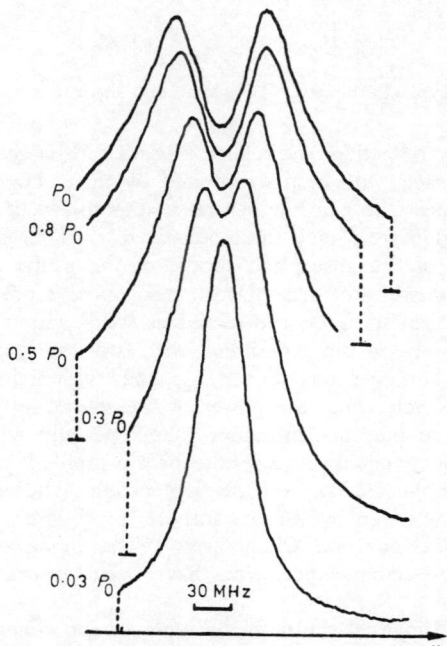

Figure 15. Observation of the Autler–Townes splitting on the cascade $1s_5\ (J=2) \rightarrow 2p_4\ (J=2) \rightarrow 2s_2\ (J=1)$ of neon (from Delsart and Keller 1976a).

Furthermore, recent calculations have considered the case of a three-level system interacting with two quasi-resonant strong beams (Whitley and Stroud 1976, Cohen-Tannoudji and Reynaud 1977c). The results are much more complicated and will not be discussed here.

3.2.3. Saturated absorption. In the weak-field limit, the results of the previous paragraph remain valid if one probes the absorption profile of the saturated transition itself. Here again, the saturating and the probe field can either co-propagate or counter-propagate; in both cases, narrow dips appear in the absorption profile either at the saturating frequency (co-propagating beams) or symmetrically with respect to the centre of the Doppler line (counter-propagating beams). However, only a few experiments (Cahuzac and Vetter 1975) of the latter type have been reported since, in this case, the use of two lasers is not really essential to observe narrow resonances. Indeed, by using a mirror, one can obtain from a single laser beam two counter-propagating beams having the same frequency (see figure 16).

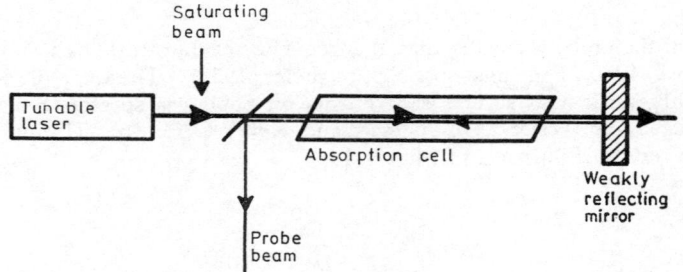

Figure 16. Typical arrangement for saturated absorption experiments.

If the mirror is weakly reflecting, the reflected beam can be considered as a probe. This probe beam interacts with atoms saturated by the stronger beam only if the laser frequency coincides with the line centre, that is to say, if the laser frequency is equal to the transition frequency for atoms at rest. Therefore, a resonant dip appears at the centre of the absorption profile of the probe wave (Letokhov and Chebotayev 1969). The width of these narrow resonances is not difficult to evaluate if the saturating field can itself be considered as weak with respect to relaxation constants. More precisely, in the experiment with two lasers, the width is close to the zero-field ($K_0 \to 0$) homogeneous width, Γ_{eg}, while with a single laser it becomes twice this value. However, when the power of the saturating beam increases, coherence effects begin to play an important role, especially when the decay rates of the two levels are nearly equal. As a result, the dip profile is no longer Lorentzian and depends in a complicated way on the field power. Theoretical expressions of these profiles have been given by various authors for counter-propagating and co-propagating waves (Baklanov and Chebotayev 1971a, b, Haroche and Hartmann 1972), but most of the actual experiments have been performed with low-power fields.

It is now usual to range all the previous types of experiments among saturated absorption methods, but in fact they allow one to observe narrow resonances even if the saturation parameter is very small. Thus, it should certainly be more correct to limit the use of this expression to a slightly different type of experiment in which atoms interact with a standing wave. Apparently, the situation is similar to the previous case of two counter-propagating waves with the same frequency, but in fact it is rather different since no distinction can be made between saturating and probe beams. Nevertheless, it is clear that the transmission for each of the two counter-propagating waves is influenced by the other only if they interact with the same atoms, that is to say if the frequency of the standing wave coincides with the centre of the line. If this condition is fulfilled, the interaction between atoms and the standing wave takes a non-linear character which can be detected in various ways.

This effect is in fact at the origin of the power dip appearing at the centre of gain profiles of gas lasers (Bennett 1962). The phenomenon was first investigated by Lamb (1964) who showed that, in the weak saturation limit, the profile of this dip is Lorentzian with a width equal to $2\Gamma_{eg}$, a result in agreement with the predictions of the simple hole-burning model.

Later, a variety of effects have been observed in connection with saturated absorption. One of the most sensitive techniques uses the modulation of one of the two counter-propagating waves, the saturation signal being detected by the modulation of the second wave, which appears only because of the non-linear character of the interaction. Other methods utilize fluorescence (Freed and Javan 1970), intermodulated fluorescence (Sorem and Schawlow 1972), birefringence properties (Wieman and Hänsch 1976, Delsart and Keller 1977), etc.

It is beyond the scope of this review to give a survey of the wide range of applications based on saturated absorption techniques. Let us say, however, that they now play a very important role in high-resolution spectroscopy, laser stabilization and metrology. For details, the reader is referred to the numerous reviews written on this subject (see, for example, Letokhov 1976, Evenson and Petersen 1976).

The effect of the light power on saturated absorption has also been extensively discussed in the literature. In the general case, the problem can be solved only by computer calculations. Here again, the main difficulties appear because of coherence effects and spatial inhomogeneity of the absorption of a standing wave. If these effects are ignored, analytical expressions of the shape of the Lamb dip can be found (Greenstein 1968, Shimoda and Uehara 1971a, b, Baklanov and Chebotayev 1972). The main results can be summarized as follows: there is a maximum depth for the dip which is obtained for $\chi = 1.42$; this corresponds to a relative decrease of absorption equal to 0.133. This is completely different from the results obtained under the probe beam approximation since in that case the absorption can decrease to zero at the centre of the line when increasing the saturating field power. Moreover, for large saturation, the shape of the dip is nearly Lorentzian and its width is close to $\Gamma_{eg}(1+\chi)^{1/2}$, which is precisely the width of the Bennett hole. The influence of coherence and spatial inhomogeneity do not strongly modify the previous results. They lead only to a small decrease ($\sim 20\%$) in the depth of the dip and a very small increase in its width (Feldman and Feld 1970).

3.3. Transient effects

In §§3.1 and 3.2, it has been assumed that the laser excitation is continuous and that the interaction time is very long with respect to all the other times characterizing the resonances. Under these conditions, only the stationary parts of the phenomena are actually observed. We briefly review the opposite case here: some experimental observations of transient effects which appear when atoms begin to interact with the field. Since we ignore the specific role of propagation which will be studied in §4, the only transient phenomenon predicted by theory (see §2) is the optical analogue of the well-known spin nutation in magnetic resonance. This optical nutation was demonstrated on Rb atoms by recording the total fluorescence (integrated over time) emitted from the state $5P_{1/2} m_J = \frac{1}{2}$ after resonant pulsed excitation (Gibbs 1973). The experiment was performed with a pulsed Hg II laser on a Rb atomic beam in a 75 kG magnetic field (see §2.8.1). The pulsed duration was adjusted by using a Pockels cell. Under these conditions, the recorded signal is clearly proportional to the population of the excited state and, according to the Rabi formula, must be a periodic function of the input pulse area:

$$\theta = 2 \int_{-\infty}^{t} K(t) \, dt.$$

An example of the measurements is shown in figure 17; the result is in perfect agreement with the predictions of the Rabi formula.

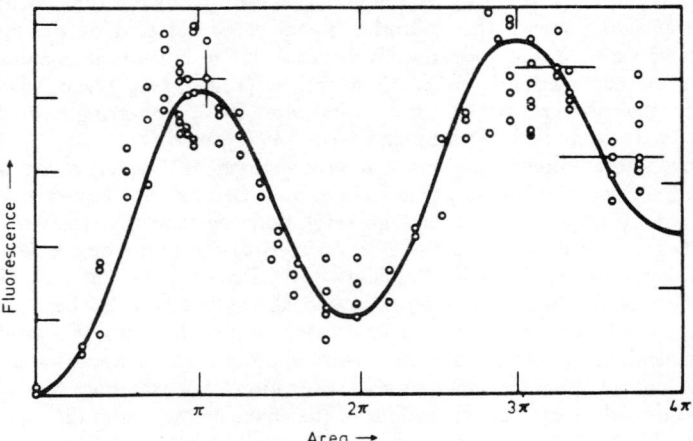

Figure 17. Experimental evidence of Rabi oscillations in the excited-state population of rubidium atoms excited by a pulsed Hg II laser (from Gibbs 1973).

Optical nutation can be observed in other types of experiment on two- or three-level systems. The most famous experiment was performed by Brewer and Shoemaker (1971) on some molecular transitions. In this experiment, the trick is to use continuous detuned excitation and to make it resonant very abruptly by switching on or off a static field which causes a sudden Stark shifting or splitting in the atomic resonance. Because of the velocity selection, discussed previously, this experiment can be done on vapours and in this case there is always a group of atoms in resonance for any value of the static field. Therefore, because of optical nutation, a modulation in time appears on the transmitted light every time the static field is switched on or off (see figure 18). The same arrangement allowed Brewer and Shoemaker (1972) to observe another type of transient effect very well known in magnetic resonance and called 'free-induction decay'. However, this effect belongs to a wide class of phenomena which do not characterize the interaction between atoms and fields but

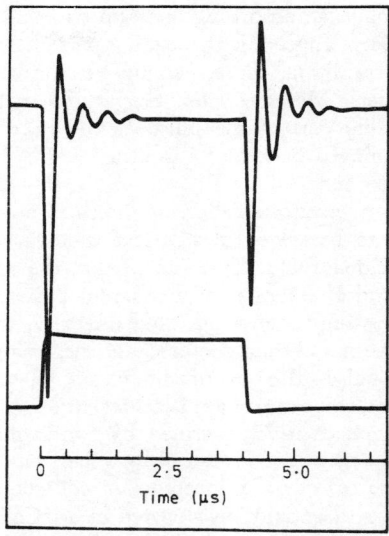

Figure 18. Damped optical Rabi oscillations observed by Stark-field switching (from Brewer and Shoemaker 1971).

the free evolution of atomic states after coherent excitation. Therefore, although considerable attention has been paid or is now being paid to some of these phenomena, such as photon echoes (see, for example, Allen and Eberly 1975), quantum beats (Haroche 1976) and Dicke super-radiance (Dicke 1954), their discussion is beyond the scope of this review.

4. Propagation effects

This section is devoted to a brief survey of propagation effects which have been ignored in all the previous sections. Therefore, while most of the resonant phenomena encountered up to now had been observed in magnetic resonance a long time before the advent of lasers, the effects described here are specific to the optical range and were unknown in magnetic resonance studies.

Another difference lies in the fact that only a classical description of the field can be actually used since, as has been already noticed in §2, a quantum description of the field is not well adapted to describe any effect connected with amplitude or phase variations. Therefore, the origin of new phenomena appearing in relation to the predictions of purely classical optics (Beer's law of linear absorption) must be sought solely in the non-linear character of Bloch equations. This non-linear character is particularly pronounced in the strong-field limit and for interaction times short in comparison with dipole and population decay times. Thus, new phenomena can be expected only in the propagation of short and intense light pulses.

4.1. Incoherent saturation effects on the propagation of pulses

According to the results of §§2 and 3, a sufficiently strong incoherent pulse can be transmitted through an absorbing medium with only very weak attenuation, owing to saturation effects. Indeed, if the leading part of the pulse alone is sufficient to saturate the atomic transition, the remaining part propagates without any interaction with the absorber. Moreover, the propagation velocity will appear to be reduced since, the leading part of the pulse having been absorbed, the centre of gravity of the pulse is moved backward. In other words, a delay appears in transmission through the absorber.

For incoherent pulses, a rate equation approach is obviously sufficient and the transmission problem can be solved in terms of time-varying transmittance (Gires and Combaud 1965, Selden 1967). One can predict theoretically that, under these conditions, distortion and attenuation of incoherent pulses propagating through an absorber will be a decreasing function of input intensity, while the pulse delay first increases up to a maximum and then decreases for increasing input intensity. Moreover, for high intensity pulses, the phenomena do not depend on the inhomogeneous character of the transition since saturation is achieved for any class of velocity. These predictions have been quantitatively verified by Smith and Allen (1973) on lead vapour illuminated by 12 ns pulses produced by a lead vapour laser.

Now, the question arises of the importance of coherence effects in propagation. This problem, which was essentially investigated by McCall and Hahn (1967, 1969), is briefly examined in the next two sections.

4.2. The quantum-mechanical 'area theorem'

Assuming that the electric field propagates along the z axis and that damping phenomena can be neglected, the Bloch equations for a two-level system at a point z are directly obtained from equations (2.12) by replacing ϕ by $-kz$. Now, if we assume that Δ is zero for atoms at rest and that $K(t, z) = 0$ for times t' larger than $t_0(z)$ (see figure 19), integration of the Bloch–Maxwell equations over time and

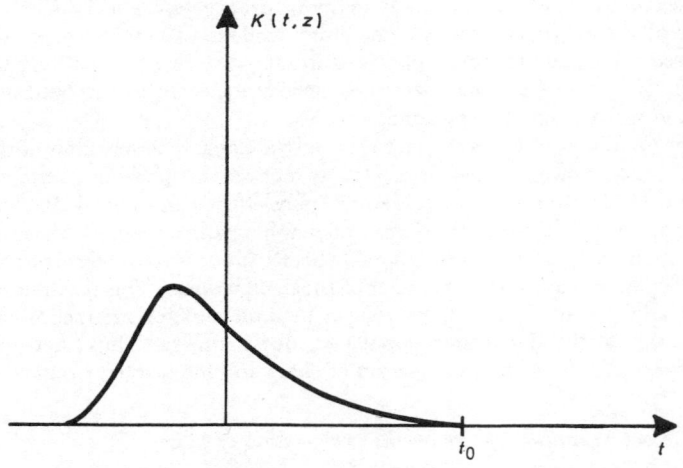

Figure 19. Description of a light pulse in the time domain.

atomic velocity distribution can be carried out exactly and lead for the envelope area defined by

$$\theta(t', z) = 2\int_{-\infty}^{t'} K(t, z)\,dt \qquad (4.1)$$

to the following differential equation:

$$\partial\theta/\partial z = \tfrac{1}{2}\alpha v_0(t_0, z) \qquad (4.2)$$

where α is none other than the classical reciprocal absorption length and $v_0(t_0, z)$ is the value of $v(t, z)$ for atoms at rest ($\Delta = 0$). Of course, $\theta(t', z)$ depends on $v_0(t_0, z) = v(t', z)$ for $t' > t_0$ only because damping phenomena have been completely ignored.

For atoms at rest, the Bloch equations reduce to the particularly simple system:

$$\dot{v}_0(t, z) = -2 K(t, z) w_0(t, z) \qquad (4.3(a))$$

$$\dot{w}_0(t, z) = 2 K(t, z) v_0(t, z) \qquad (4.3(b))$$

which is exactly soluble. More precisely,

$$v_0(t, z) = -\sin \theta(t, z) \qquad (4.4(a))$$

$$w_0(t, z) = -\cos \theta(t, z). \qquad (4.4(b))$$

In other words, the evolution in space of $\theta(t, z)$ is given by:

$$\partial\theta(t', z)/\partial z = -\tfrac{1}{2}\alpha \sin \theta(t_0, z). \qquad (4.5)$$

This is the famous quantum area theorem derived by McCall and Hahn in 1967. Naturally, in the weak-field limit, the pulse area is small at any point. Thus $\sin \theta \approx \theta$ and equation (4.2) reduces to the classical linear relation:

$$\partial\theta(t', z)/\partial z = -\tfrac{1}{2}\alpha\theta(t', z) \qquad (4.6)$$

but as soon as θ takes values larger than $\tfrac{1}{4}\pi$, large non-linearities appear since, in particular if $\theta = n\pi$, the pulse area is not modified in propagation whatever the value of n. However, integration of equation (4.5) show that pulse areas $n\pi$ are stable in propagation only if n is even (see figure 20). Consequently, whatever the input pulse area is during propagation into the absorber, the pulse is modified in such a way that its area tends towards an even multiple of π. So it appears to be possible to compress strong pulses by sending them through an absorbing medium and thus to amplify them if they are sufficiently short to make any energy-loss mechanisms impossible. This was verified for the first time in 1971 by Gibbs and Slusher (see figure 21).

4.3. Self-induced transparency

Knowing that the area of a $2n\pi$ pulse is stable in propagation, another question arises about shape stability of such a pulse. This problem was solved for a 2π pulse first by McCall and Hahn (1967). By numerical integration of the coupled Bloch–Maxwell equations, they show that after propagation through a few absorption lengths, the shape of a 2π pulse appears to be stable as long as the propagation time

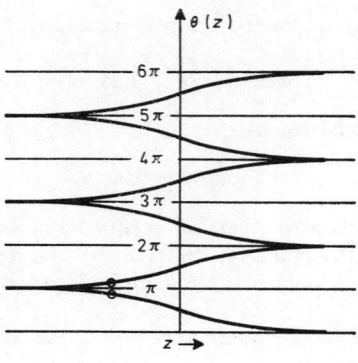

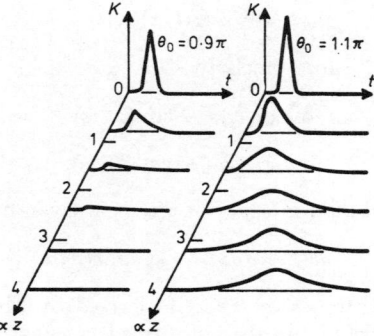

Figure 20. The McCall–Hahn quantum optical area theorem (from McCall and Hahn 1969).

remains small in comparison with the population and dipole decay times. This property was verified experimentally by the same authors (McCall and Hahn 1969) on a liquid-helium-cooled ruby rod with a Q-switched liquid-nitrogen-cooled ruby laser. They named the phenomenon self-induced transparency since the 2π character of a pulse makes the absorber transparent for this pulse. Of course, after the 2π pulse has become stable it has a well-defined shape. From the steady-state character of a pulse stable in shape and in area it is possible to prove that this shape is that of the hyperbolic secant:

$$K(t, z) = \frac{2}{\tau} \text{sech}\left(\frac{t - z/v}{\tau}\right) \qquad (4.7)$$

where τ is the pulse duration and v is the pulse velocity which is a complicated function of α, τ and Δ_D. However, if τ is much longer than Δ_D^{-1}, v is given by the particularly simple formula:

$$v = c \frac{1}{1 + \tfrac{1}{2}\alpha c \tau} \qquad (4.8)$$

which shows that the pulse velocity can be much smaller than c since in many pulse

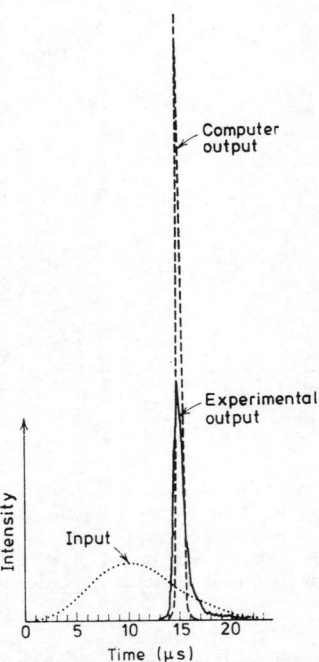

Figure 21. Experimental evidence of compression and peak amplification of an optical pulse with input area $3 \cdot 5\,\pi$ propagating in a passive absorber (from Gibbs and Slusher 1971).

propagation experiments $c\tau$ can be of the order of several metres and therefore much larger than α^{-1}.

It must be noted that the hyperbolic secant pulse is the only one which leads to a simple factorization of $v_\Delta(t, z)$, $u_\Delta(t, z)$ and $w_\Delta(t, z)$ for off-resonance atoms. The factorization is as follows:

$$v_\Delta(t, z) = v_0(t, z)\, F(\Delta)$$

where $F(\Delta)$ is given by

$$F(\Delta) = \frac{1}{1+(\tau\Delta)^2}. \tag{4.9}$$

Then it can be shown that the evolution equation for the pulse area at any position is

$$\frac{\partial^2 \theta}{\partial t^2} - \frac{1}{\tau^2} \sin\theta = 0 \tag{4.10}$$

which is none other than the pendulum equation. Moreover, the first physical explanation of self-induced transparency was given by McCall and Hahn on the analogy with the pendulum problem illustrated in figure 22. Numerous analogies can be found in various parts of physics, including classical physics. For example, the fact that 2π pulses can travel very long distances at very low velocities recalls the quite old problem of solitar waves in hydrodynamics solved by Korteweg and

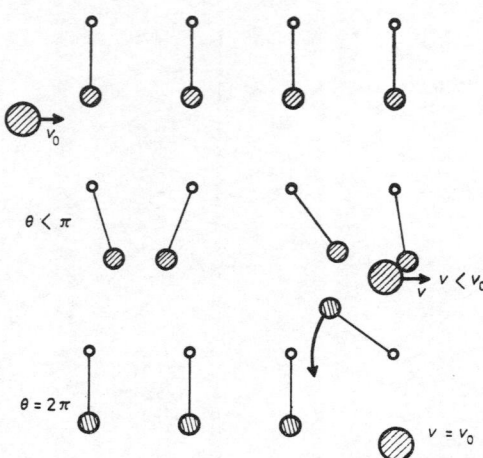

Figure 22. Analogy explaining self-induced transparency. If a slowly rolling ball hits a sequence of pendula, it slows down rapidly (linear absorption). On the other hand, with a ball rolling sufficiently fast, the pendula swing right round and hit the ball on the back side (stimulated emission) returning all the energy initially absorbed from the ball (self-induced transparency).

de Vries (1895). More generally, all the problems connected with stability conditions in non-linear regimes exhibit deep similarities and in recent years the general study of non-linear equations has become a very active subject concerning phenomena as diverse as dislocations in solids, superconductivity and elementary particle stability.

Of course, the question of shape stability also arises for the larger $2n\pi$ pulses which are also stable in area during propagation. In fact, it has been shown that, if one excepts infinite pulse-train solutions, 2π pulses are the only steady-state ones. The area of the larger $2n\pi$ pulses is stable but their shape is not. Non-shape-preserving pulses were theoretically investigated first by Lamb (1971), who showed that pulses with an area larger than 3π split into 2π pulses with different pulse durations and therefore pulse velocities. From this result the problem can be seen as a collision process between 2π pulses and this new point of view has opened a wide range of possibilities which unfortunately cannot be described in so brief a survey.

All the features predicted within the framework of non-linear interactions between a light pulse and an absorbing medium have been experimentally observed by Slusher and Gibbs (1972) in a series of very beautiful experiments. The light pulses came from a ^{202}Hg II laser ($\lambda = 7944 \cdot 66$ Å) and propagated in dilute ^{87}Rb vapour (10^{11}–10^{13} atoms cm^{-3}). The coincidence between the laser light and the atomic transition ($\lambda = 7947 \cdot 64$ Å) was achieved by Zeeman tuning. This system was, in fact, nearly ideal for self-induced transparency observations because the condition $1/\Delta_D \ll \tau \ll 1/\gamma$ was actually fulfilled. More precisely, the pulse duration was 7 ns while $1/\Delta_D$ was 0·8 ns and the natural lifetime of the upper atomic level is 28 ns. The corresponding experimental results provided the first unquestionable proof of the validity of the theory of McCall and Hahn since, in fact, the first experiments on ruby could have been interpreted, at least qualitatively, in terms of incoherent saturation which was discussed in §4.1. On the contrary, pulse breakup into separate 2π pulses and pulse amplification can be understood only after introducing coherence effects.

5. Multiphoton processes

Up to now, we have been mainly concerned with resonant interaction between atoms and laser radiation. Of course, the use of lasers and particularly of tunable lasers also plays a very important role in the study of non-resonant processes. One can even say that the most part of quantum optics is now devoted to these non-resonant processes which play an important role in tunable laser research especially in the VUV and IR ranges (Stoicheff and Wallace 1976, Mooradian 1976). However, since this review is limited to the interaction between laser radiation and free atoms, we have chosen to consider two types of multiphoton processes only. The first is the so-called Doppler-free multiphoton excitation which is closely related to three-level problems which have been discussed in §3.3. Thus it will be discussed only very briefly in spite of the important role it now plays in spectroscopy (Bloembergen and Levenson 1976, Grynberg and Cagnac 1977). The second one is multiphoton ionization. The corresponding field is very broad and a complete review would be necessary to cover it completely (Bakos 1974, Lambropoulos 1976). Therefore, we shall discuss only resonant multiphoton ionization which can be easily understood within the frame of an exactly soluble model and nevertheless realistic (Beers and Armstrong 1975).

5.1. Doppler-free multiphoton excitation

The principle of this method is extremely simple. If an atom interacts with several travelling light waves characterized by their wavevector k_i and if the geometrical arrangement of the corresponding light beams is such that $\sum_i k_i = 0$, the total momentum of the absorbed photons can be zero and if this condition is fulfilled it is clear that the energy absorbed by the atom is equal to $\sum_i \hbar \omega_i$ whatever its velocity is. In other words, the Doppler effect is actually suppressed for all the atoms.

The case of two photons was first considered both theoretically (Vasilenko et al 1970, Cagnac et al 1973) and experimentally (Cagnac et al 1974, Levenson and Bloembergen 1974). Then the previous condition is fulfilled for two counter-propagating waves with the same frequency. This can be achieved with a single laser and a mirror in the same way as in saturated absorption experiments, but now the resonances are detected by recording the total fluorescence emitted from the upper level of the two-photon transition (see figure 23). With such an arrangement, if the two counter-propagating waves have the same polarization, two photons belonging to the same wave can be absorbed and obviously this process is not Doppler-free. Therefore, the resonance profile appears as the sum of a Gaussian profile (with the Doppler width and a small amplitude) and of a Lorentzian profile (with the homogeneous width and a large amplitude) (see figure 24). However, by an adequate choice of the respective polarizations of the two counter-propagating waves, it is sometimes possible to make forbidden the transition with two co-propagating photons while the transition with two counter-propagating photons remains allowed. Under these conditions, the Gaussian background must

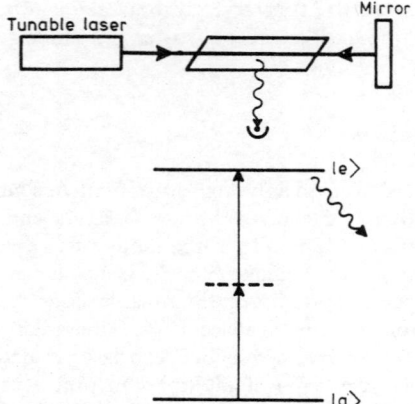

Figure 23. Typical arrangement for Doppler-free two-photon experiments.

disappear. This was originally suggested and experimentally demonstrated by Cagnac et al (1973, 1974).

Of course, two-photon atomic transition probabilities are extremely weak and it must be noticed that the demonstration of the two-photon absorption process in the optical range had to await the advent of lasers: it was actually realized in 1961, thirty years after the theoretical predictions of Goeppert-Mayer (1931). However, the light intensity required to observe the phenomenon is greatly reduced if there is a quasi-resonant intermediate level. Most of the actual experiments using tunable lasers have been carried out on atomic systems (mainly sodium and neon) satisfying this condition. In this case, the situation becomes quite similar to stepwise excitation which was discussed in §3.2.2, and an exact treatment of the problem can be made

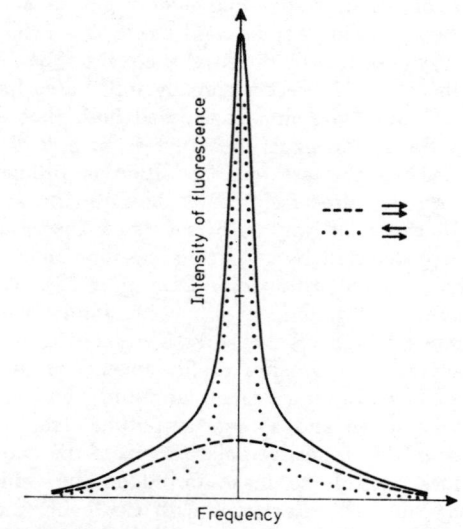

Figure 24. Two-photon signal for linearly polarized light.

(see, for example, Bloembergen and Levenson 1976). Important differences remain, however: first, in the two-photon process, all the atoms participate in the resonance signal, while for stepwise excitation the Doppler effect is reduced by velocity selection and thus the resonance condition is fulfilled for only one class of velocity; and secondly, in the two-photon process, power-dependent effects are usually very small and reduce to light shifts (Harvey et al 1975) while, as shown in §3.2.2, resonance profiles in stepwise excitation depend strongly on the light intensity since a dynamic Stark splitting can be observed. When the intermediate level has a structure, other differences can appear (Jacquinot 1976b) but their discussion is beyond the scope of this review. Let us say, however, that one can imagine a continuous transition from the two-step process to the two-photon one and this has been done by Bjorkholm and Liao (1974) and Liao and Bjorkholm (1975) in a series of experiments on the sodium transitions 3s→3p→5s, 4d.

To investigate transitions between states of opposite parity, at least a three-photon process must be used to suppress the Doppler effect. An experiment of this type, based on a hyper-Raman process in which two photons are absorbed and another one is emitted in a stimulated way (see figure 25), has been recently achieved by Grynberg et al (1976). In such a case there is no need for a quasi-resonant intermediate level to obtain reasonable values of the transition probability.

5.2. Resonant multiphoton ionization

Multiphoton ionization belongs typically to the class of phenomena which were unobservable before the advent of lasers, though they were theoretically predicted many years earlier (Goeppert-Mayer 1931). In fact, the first observation of this multiphoton process was achieved by Voronov and Delone (1965) on Xe atoms. At the present time more than 50 papers are published per year on this subject.

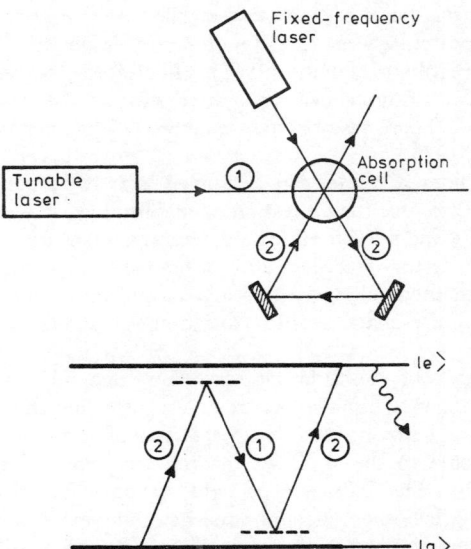

Figure 25. Doppler-free three-photon experiment (from Grynberg et al 1976).

In spite of the fact that very high intensities (typically of the order of $10^8 \sim 10^{10}$ W cm^{-2}) are required to observe multiphoton ionization, the corresponding experimental data can usually be understood within the frame of perturbation theory by keeping the lowest non-vanishing terms for the process. However, this is no longer true when an intermediate level is quasi-resonant with a certain number of photons. Indeed, the corresponding resonant phenomena are perturbed by various effects (shift and asymmetry of resonances, variations of the order of non-linearity) which depend on the light intensity in a complicated way. During the last few years, these 'super non-linear' effects have been observed experimentally, especially on noble gases and alkalis. Of course, it is possible to reproduce them theoretically by introducing higher-order terms of time-dependent perturbation theory (Gontier and Trahin 1973), but the corresponding calculations are extremely complicated even for the simplest atoms and, in any case, it is rather difficult to deduce general conclusions and to give a simple physical explanation of the observed phenomena. A different approach (Armstrong et al 1975) was recently proposed to solve these difficulties. The basic idea is to describe resonant multiphoton ionization in terms of effective interactions. More precisely, only two channels are introduced:

(i) A non-resonant n-photon process induced by an effective interaction which is obtained from perturbation theory to nth order.

(ii) A two-step process: first, a p-photon transition from the ground state to the quasi-resonant state, then a $(n-p)$-photon transition from this quasi-resonant state to a continuum state, the corresponding effective interactions being obtained from perturbation theory respectively to pth and $(n-p)$th orders. This model can be solved in a particularly simple way if higher-order terms can be ignored both in the non-resonant process and in one of the two-step process (Feneuille and Armstrong 1975, Armstrong et al 1975). In this case, the ionization cross section and the non-linearity can be written for continuous excitation as analytical functions depending on light intensity and on two atomic parameters only. In particular, if higher-order terms are large only for the transition between the resonant state and the continuum, a close analogy appears between multiphoton ionization and the theory of autoionizing states as described by Fano (1961). Therefore, the resonance has a Fano profile and the order of non-linearity plotted against the laser frequency has a dispersion-like shape. These results are in qualitative agreement with recent observations (Morellec et al 1976).

In fact, the considered model can be solved exactly in the general case (Beers and Armstrong 1975) to give the ionization probability near a resonance as a function of time, light intensity and a few atomic parameters; a quantum or a classical approach can be used to describe the laser field and, of course, these lead to the same results. In particular, this method allowed the study of pulsed excitation and emphasized the role of the light pulse characteristics (duration and shape) (Crance and Feneuille 1976, 1977).

The characteristics of multiphoton ionization depend not only on the field strengths and the frequency but also on the polarization and the correlation functions of high orders of the ionizing field. A great deal of theoretical and experimental work has been devoted to the study of the role of polarization (see, for example, Lambropoulos 1976). The influence of light statistics has also been investigated theoretically but very little theoretical or experimental work has been actually carried out for resonant processes (Armstrong et al 1976, Mostowski 1975).

6. Concluding remarks

The use of lasers to investigate in detail the various aspects of the interaction between light and matter is a new subject which is becoming more and more active. As long as lasers had fixed frequency, the volume of work increased rather slowly, but since the advent of tunable lasers the number of people involved in the corresponding studies, and therefore the number of articles published on this subject, approximately doubles each year. Moreover, in many cases, several research groups work on quite similar problems and therefore, as soon as a new question arises or a new technique appears, a high concentration of effort is immediately directed towards it. As a consequence of this very changing situation, new perspectives are constantly being opened while some studies, apparently very promising, disappear nearly completely after a few beautiful experiments. Under these conditions, it is quite difficult to see clearly what the main research directions of this subfield will be in the near future.

It seems that most of the basic phenomena described in this review are now well understood. Of course, one can still imagine more and more complicated experimental situations involving several atomic levels and numerous laser beams, but then the interpretation of the observed signals reduces to more or less tedious calculations without exhibiting really new aspects of the interaction between light and matter. However, it must not be assumed from this slightly pessimistic point of view that there are no promising trends in atomic physics with lasers. In particular, atomic physicists are just at the beginning of the exploration of the 'terrae incognitae' (Rydberg states, unstable isotopes, forbidden transitions, etc) recently accessible thanks to the new techniques elaborated during the last five years. Furthermore, a fascinating field now seems to be open by simultaneous studies of laser- and collision-induced processes (see, for example, Lidow *et al* 1976) which up to now have been studied separately. Of course, these statements are certainly questionable. They express more of a personal feeling of the author than a real certitude.

Acknowledgments

The author would like to thank Professor P Jacquinot and Dr S Liberman for their helpful comments during the preparation of this manuscript.

References

ALLEN L and EBERLY J H 1975 *Optical Resonance and Two-Level Atoms* (New York: Wiley-Interscience)
ARGYRES N P and KELLEY P L 1964 *Phys. Rev.* **134** A98–111
ARMSTRONG L JR, BEERS L B and FENEUILLE S 1975 *Phys. Rev.* A **12** 1903–10
ARMSTRONG L JR and FENEUILLE S 1973 *J. Phys. B: Atom. Molec. Phys.* **6** L182–7
—— 1975 *J. Phys. B: Atom. Molec. Phys.* **8** 546–51
ARMSTRONG L JR, LAMBROPOULOS P and RAHMAN N K 1976 *Phys. Rev. Lett.* **36** 952–5
AUTLER S H and TOWNES C H 1955 *Phys. Rev.* **100** 703–22
BAKLANOV E V and CHEBOTAYEV V P 1971a *Zh. Eksp. Teor. Fiz.* **60** 552–68
—— 1971b *Zh. Eksp. Teor. Fiz.* **61** 922–9
—— 1972 *Zh. Eksp. Teor. Fiz.* **62** 541–50
BAKOS J S 1974 *Adv. Electron. Electron Phys.* **36** 57–152

BEERS L B and ARMSTRONG L JR 1975 *Phys. Rev.* A **12** 24–54
BENNETT W R JR 1962 *Phys. Rev.* **126** 580–93
BJORKHOLM J E and LIAO P F 1974 *Phys. Rev. Lett.* **33** 128–31
—— 1977 *Opt. Commun.* to be published
BLOCH F 1946 *Phys. Rev.* **70** 460–74
BLOCH F and SIEGERT A 1940 *Phys. Rev.* **57** 522–7
BLOEMBERGEN N and LEVENSON M D 1976 *High-Resolution Laser Spectroscopy* ed K Shimoda (Berlin: Springer-Verlag) pp315–69
BREWER R G and SHOEMAKER R L 1971 *Phys. Rev. Lett.* **27** 631–4
—— 1972 *Phys. Rev.* A **6** 2001–7
CAGNAC B, GRYNBERG G and BIRABEN F 1973 *J. Physique* **34** 845–58
—— 1974 *Phys. Rev. Lett.* **32** 643–5
CAHUZAC P and VETTER R 1975 *Phys. Rev. Lett.* **34** 1070–3
—— 1976 *Phys. Rev.* A **14** 270–2
CARMICHAEL H J and WALLS D F 1975 *J. Phys. B: Atom. Molec. Phys.* **8** L77–81
CHEBOTAYEV V P 1976 *High-Resolution Laser Spectroscopy* ed K Shimoda (Berlin: Springer-Verlag) pp207–50
COHEN-TANNOUDJI C 1968 *Cargèse Lectures in Physics* vol 2, ed M Levy (New York: Gordon and Breach)
—— 1975a *Frontiers in Laser Spectroscopy* ed R Balian, S Haroche and S Liberman (Amsterdam: North-Holland) pp3–104
—— 1975b *Laser Spectroscopy* ed S Haroche, J C Pebay-Peyroula, W Hänsch and S E Harris (Berlin: Springer-Verlag) pp334–9
COHEN-TANNOUDJI C, DUPONT-ROC J AND FABRE C 1973 *J. Phys. B: Atom. Molec. Phys.* **6** L214–7
COHEN-TANNOUDJI C and HAROCHE S 1969 *J. Physique* **30** 125–44
COHEN-TANNOUDJI C and REYNAUD S 1977a *J. Phys. B: Atom. Molec. Phys.* **10** 345–63
—— 1977b *J. Phys. B: Atom. Molec. Phys.* **10** 365–83
—— 1977c *J. Phys. B: Atom. Molec. Phys.* to be published
CORDOVER R H, BONCZYK P A and JAVAN A 1967 *Phys. Rev. Lett.* **18** 730–2
CRANCE M and FENEUILLE S 1976 *J. Physique Lett.* **37** L333–7
—— 1977 *Phys. Rev.* A to be published
DECOMPS B, DUMONT M and DUCLOY M 1976 *Laser Spectroscopy of Atoms and Molecules* ed H Walther (Berlin: Springer-Verlag) pp283–347
DELSART C and KELLER J C 1976a *J. Phys. B: Atom. Molec. Phys.* **9** 2769–75
—— 1977 *Opt. Commun.* **20** 147–9
DICKE R H 1954 *Phys. Rev.* **93** 99–110
DUCAS T W, LITTMAN M G, FREEMAN R R and KLEPPNER D 1975 *Phys. Rev. Lett.* **35** 366–9
DUONG H T, JACQUINOT P, LIBERMAN S, PINARD J and VIALLE J L 1973a *C.R. Acad. Sci., Paris* B **276** 909–13
DUONG H T, JACQUINOT P, LIBERMAN S, PICQUÉ J L, PINARD J and VIALLE J L 1973b *Opt. Commun.* **7** 371–3
EVENSON K M and PETERSEN F R 1976 *Laser Spectroscopy* ed H Walther (Berlin: Springer-Verlag) pp352–68
FANO U 1961 *Phys. Rev.* **124** 1866–78
FELD M S and JAVAN A 1969 *Phys. Rev.* **177** 540–62
FELDMAN B J and FELD M S 1970 *Phys. Rev.* A **1** 1375–96
FENEUILLE S and ARMSTRONG L JR 1975 *J. Physique* **36** L235–7
FENEUILLE S and SCHWEIGHOFER M G 1975 *J. Physique* **36** 781–6
FENEUILLE S, SCHWEIGHOFER M G and OLIVER G 1976 *J. Phys. B: Atom. Molec. Phys.* **9** 2003–9
FREED C and JAVAN A 1970 *Appl. Phys. Lett.* **17** 53–6
GIBBS H M 1973 *Phys. Rev.* A **8** 446–55
GIBBS H M and SLUSHER R E 1970 *Phys. Rev. Lett.* **24** 638–41
—— 1971 *Appl. Phys. Lett.* **18** 505–7
GIRES F and COMBAUD F 1965 *J. Physique* **26** 325–30
GLAUBER R 1963 *Phys. Rev.* **131** 2766–88
GOEPPERT-MAYER M 1931 *Ann. Phys., Lpz.* **9** 273–94
GONTIER Y and TRAHIN M 1973 *Phys. Rev.* A **7** 2069–73

GREENSTEIN H 1968 *Phys. Rev.* **175** 438–52
GROVE R E, WU F Y and EZEKIEL S 1977 *Phys. Rev. A* **15** 227–33
GRYNBERG G, BIRABEN F, BASSINI M and CAGNAC B 1976 *Phys. Rev. Lett.* **37** 283–5
GRYNBERG G and CAGNAC B 1977 *Rep. Prog. Phys.* **40** 791–841
HAKEN H 1970 *Quantum Optics* ed S M Kay and A Maitland (New York: Academic) pp242–63
HANNAFORD P, PEGG D T and SERIES G W 1973 *J. Phys. B: Atom. Molec. Phys.* **6** L222–5
HÄNSCH T and TOSCHEK P Z 1970 *Z. Phys.* **236** 213–44
HAROCHE S 1976 *High-Resolution Laser Spectroscopy* ed K Shimoda (Berlin: Springer-Verlag) pp256–314
HAROCHE S and HARTMANN F 1972 *Phys. Rev. A* **6** 1280–300
HARTIG W, RASMUSSEN W, SCHIEDER R and WALTHER H 1976 *Z. Phys. A* **278** 205–10
HARTIG W and WALTHER H 1973 *Appl. Phys.* **1** 171–4
HARVEY K C, HAWKINS R T, MEISEL G and SCHAWLOW A L 1975 *Phys. Rev. Lett.* **34** 1073–6
HEITLER W 1954 *Quantum Theory of Radiation* (Oxford: Oxford University Press)
JACQUINOT P 1976a *High-Resolution Laser Spectroscopy* ed K Shimoda (Berlin: Springer-Verlag) pp52–91
—— 1976b *Very High Resolution Spectroscopy* ed R A Smith (London: Academic) pp1–12
JACQUINOT P, LIBERMAN S, PICQUÉ J L and PINARD J 1973 *Opt. Commun.* **8** 163–5
JAYNES E T 1973 *Coherence and Quantum Optics* ed L Mandel and E Wolf (New York: Plenum) pp35–81
KORTEWEG D J and DE VRIES G 1895 *Phil. Mag.* **39** 422–43
LAMB G L JR 1971 *Rev. Mod. Phys.* **43** 99–124
LAMB W E JR 1964 *Phys. Rev.* **134** A1429–50
LAMBROPOULOS P 1976 *Adv. Atom. Molec. Phys.* **12** 87–164
LANGE W, LUTHER J, NOTTECK B and SCHRÖDER H W 1973 *Opt. Commun.* **8** 157–9
LAX M 1968 *Phys. Rev.* **172** 350–61
LECOMPTE C, MAINFRAY G, MANUS C and SANCHEZ F 1974 *Phys. Rev. Lett.* **32** 265–8
LETOKHOV V S 1976 *High-Resolution Laser Spectroscopy* ed K Shimoda (Berlin: Springer-Verlag) pp95–173
LETOKHOV V S and CHEBOTAYEV V P 1969 *Pis'ma Zh. Eksp. Teor. Fiz.* **9** 364–7
LEVENSON M D and BLOEMBERGEN N 1974 *Phys. Rev. Lett.* **32** 645–8
LIAO P F and BJORKHOLM J E 1975 *Phys. Rev. Lett.* **34** 1–4
LIDOW D B, FALCONE R W, YOUNG J F and HARRIS S E 1976 *Phys. Rev. Lett.* **36** 462–4
MCCALL S L and HAHN E L 1967 *Phys. Rev. Lett.* **18** 908–11
—— 1969 *Phys. Rev.* **183** 457–85
MOLLOW B R 1969 *Phys. Rev.* **188** 1969–75
—— 1972a *Phys. Rev. A* **5** 1522–7
—— 1972b *Phys. Rev. A* **5** 2217–22
MOORADIAN A 1976 *Tunable Lasers and Applications* ed A Mooradian, T Jaeger and P Stokseth (Berlin: Springer-Verlag) pp60–9
MOORE C E 1949 *Atomic Energy Levels* (Washington, DC: National Bureau of Standards)
MORELLEC J, NORMAND B and PETITE G 1976 *Phys. Rev. A* **14** 300–12
MOSTOWSKI J 1975 *Proc. 2nd Int. Conf. on Interaction of Electrons with Strong Electromagnetic Field* (Budapest: Balatonfured) II 47–8
NOTKIN G E, RAUTIAN S G and FEOKTISTOV A A 1967 *Sov. Phys.-JETP* **25** 1112–21
OLIVER G, RESSAYRE E and TALLET A 1971 *Lett. Nuovo Cim.* **2** 777–83
PAPOULAR R and PLATZ P 1972 *Proc. 1st Int. Conf. on Interaction of Electrons with Strong Electromagnetic Fields* (Budapest: Balatonfured)
PICQUÉ J L and PINARD J 1976 *J. Phys. B: Atom. Molec. Phys.* **9** L77–81
PINARD J and LIBERMAN S 1977 *Opt. Commun.* **20** 344–6
RABI I I 1937 *Phys. Rev.* **51** 652–4
RAUTIAN S G and SOBEL'MAN I I 1962 *Sov. Phys.-JETP* **14** 328–31
SCHWEITZER W G JR, BIRKY M M and WHITE J A 1967 *J. Opt. Soc. Am.* **57** 1226–30
SELDEN A C 1967 *Br. J. Appl. Phys.* **18** 743–8
SERIES G W 1970 *Quantum Optics, Scottish Universities Summer School in Physics, 1969* ed S M Kay and A Maitland (London: Academic) pp395–482
SHABERT A, KEIL R and TOSCHEK P E 1975 *Appl. Phys.* **6** 181–4
SHIMODA K and UEHARA K 1971a *Jap. J. Appl. Phys.* **10** 623–9

—— 1971b *Jap. J. Appl. Phys.* **10** 460–7
SHIRLEY J H 1965 *Phys. Rev.* **138** 979–87
SLUSHER R E and GIBBS H M 1972 *Phys. Rev. A* **5** 1634–59
SMITH K W and ALLEN L 1973 *Opt. Commun.* **8** 166–70
SOREM M S and SCHAWLOW A L 1972 *Opt. Commun.* **5** 148–51
STENHOLM S 1973a *Phys. Rep. C* **6** 1–121
—— 1973b *J. Phys. B: Atom. Molec. Phys.* **6** L240–6
STOICHEFF B P and WALLACE S C 1976 *Tunable Lasers and Applications* ed A Mooradian, T Jaeger and P Stokseth (Berlin: Springer-Verlag) pp1–20
TORREY H C 1949 *Phys. Rev.* **76** 1059–68
VASILENKO L S, CHEBOTAYEV V P and SHISHAEV A V 1970 *Pis'ma Zh. Eksp. Teor. Fiz.* **12** 161–4
VORONOV G S and DELONE N B 1965 *Pis'ma Zh. Eksp. Teor. Fiz.* **1** 42–55
WALLS D F 1972 *Phys. Lett.* **42A** 217–8
WHITLEY R M and STROUD C R JR 1976 *Phys. Rev. A* **14** 1498–513
WIEMAN C and HÄNSCH W 1976 *Phys. Rev. Lett.* **36** 1170–2
WU F Y, GROVE R E and EZEKIEL S 1975 *Phys. Rev. Lett.* **35** 1426–9

Interaction of laser radiation with free atoms

Serge Feneuille

ADDENDUM written in December 1984

Since the review was written, the use of lasers in atomic physics has been continuously increasing. However, in most situations, the laser is used only for preparing the atoms in a well defined state, or for spectroscopic purposes. This is the case in the numerous studies concerning atomic Rydberg states (see for example, Feneuille and Jacquinot 1981) or photo-ionisation of excited atoms by synchrotron radiation (Le Gouet et al 1982). Therefore, the number of new phenomena observed in the interaction of laser radiation with free atoms is rather limited. Moreover, some of these phenomena can be interpreted in terms of collective phenomena only, and some others do not characterise properly the interaction between atoms and fields, but the free evolution of atoms after coherent excitation. Since they were not described in the review of 1977, they have not been considered here. Finally, only three new situations are briefly presented here.

The first one concerns collinear laser spectroscopy on neutral fast atomic beams (Kaufman 1976, Bonn et al 1979, Bendall et al 1984). This technique was initially elaborated on ion beams. It exploits a velocity-bunching phenomenon to produce extremely narrow optical lines and it is particularly well adapted for high resolution spectroscopic studies of on-line produced short-lived isotopes. However, for studying the interaction between laser radiation and free atoms, their main interest is to allow one to use the Doppler effect for simulating a completely harmonic three-level atom in two-photon spectroscopy (Salomaa and Stenholm 1975, 1976; Poulsen and Winstrup 1981). This system exhibits remarkable features such as shift vanishing for equal Rabi frequencies on the two steps.

The second phenomenon concerns the creation of multiply charged ions by multiphoton absorption. This situation has been observed in noble gas atoms irradiated by a strong laser field (10^{12} to 10^{15} W cm^{-2}) at 193, 530 and 1060 nm (L'Huillier et al 1983a, b, c, Luk et al 1983). The charge state distribution has been recently interpreted by using a simple model based on a statistical description of multiphoton ionisation (Crance 1984).

Finally, one must notice recent proposals of stable optical traps for neutral atoms (Gordon and Ashkin 1980, Dalibard et al 1983). This particular aspect of the interaction between laser radiation and free atoms was not considered in the review of 1977 although deflection of an atomic beam by a crossed resonant laser beam had been observed for a long time. This certainly was a mistake because cooling and trapping of neutral atoms by laser beams appears now as a very promising subject, even if the present results remain essentially theoretical.

Of course, there are many other aspects of the interaction between laser radiation and free atoms which have been studied during the last seven years. Therefore, the three phenomena introduced above must not be considered as forming an exhaustive list, but only as being representative. The same is true for the list of new references. In any case, they show that this particular subfield of atomic physics and quantum optics remains active and fruitful.

References

Bendali H, Duong H T, Saint-Jalm J M and Vialle J L 1984 J.Physique $\underline{45}$ 421-7
Bonn J, Klempt W and Neugart R 1979 Phys.Lett. B $\underline{82}$ 47-50
Crance M 1984 J.Phys.B: At.Mol.Phys. $\underline{17}$ 3503-10
Dalibard J, Raynaud S and Cohen-Tannoudji C 1983 Opt.Comm. $\underline{47}$ 395-9
Feneuille S and Jacquinot P 1981 Adv.Atom.Mol.Phys. $\underline{17}$ 99-166
Gordon J P and Ashkin A 1980 Phys.Rev. A $\underline{21}$ 1606-17
Kaufman S L 1976 Opt.Comm. $\underline{17}$ 309-12
Le Gouet J L, Picqué J L, Wuilleumier F, Bizau J M, Dhez P, Koch P and Ederer D L 1982 Phys.Rev.Lett. $\underline{48}$ 600-3
L'Huillier A, Lompré L A, Mainfray G and Manus C 1983a J.Physique $\underline{44}$ 1247-55
—— 1983b Phys.Rev. A $\underline{27}$ 2503-12
—— 1983c J.Phys.B: At.Mol.Phys. $\underline{16}$ 1363-81
Luk T S, Pummer H, Boyer K, Shahidi M, Egger H and Rhodes C K 1983 Phys. Rev.Lett. $\underline{51}$ 110-3
Poulsen O and Winstrup N I 1981 Phys.Rev.Lett. $\underline{47}$ 1522-25
Salomaa R and Stenholm S 1975 J.Phys.B: At.Mol.Phys. $\underline{8}$ 1795-805
—— 1976 Opt.Comm. $\underline{16}$ 292-4

Optical bistability and related devices

E Abraham and S D Smith

Department of Physics, Heriot-Watt University, Riccarton, Edinburgh EH14 4AS, UK

Abstract

Optical bistability is a general title for a number of static and dynamic phenomena that result from the interplay of optical non-linearity and feedback. The object of this review is to give a broad description of optical bistability: from practical applications, as an optical transistor or optical memory element, to its phase transition interpretation. The theory is divided into three parts. The first is a simple discussion that covers most of the basic experimental effects and concepts of practical importance. The second part applies to atomic systems where a semiclassical as well as a quantum-mechanical approach is possible. The third one discusses the mechanisms compatible with large non-linearities observed in InSb and GaAs as semiconductors constitute the most promising materials for applications. Current experimental progress in all-optical and hybrid devices is also discussed and finally a scaling example is presented to indicate the possibilities in high-speed all-optical signal processing.

1. Introduction

1.1. Aim of this review

The field of optical bistability (OB) has grown very rapidly over the past five years. Important contributions have been made to its various aspects which range from the demonstration of devices to theories involving non-equilibrium statistical mechanics. Some review articles have been written but these are rather specialised: it is the aim of the present paper to review comprehensively the full subject in a way that could be of general interest as well as relevant to new workers and to those already working on some aspect of it. We think that in order to accomplish this task a tutorial approach is necessary. To this end, and wherever convenient, we derive expressions from first principles, or else we give the appropriate references by way of textbooks or papers where these are available. The list of references is meant to be exhaustive, any omission being merely accidental. Bistability in lasers is not considered, as our only concern here is bistability in passive optical systems.

1.2. Historical remarks and general discussion

An optically bistable system is one which can exhibit two steady transmission states for the same input intensity. The simplest example of such a device is a Fabry–Perot resonator with the optical cavity filled with a material whose refractive index is intensity-dependent. This *non-linear interferometer* can have an output–input characteristic which shows differential gain, i.e. the output changes more than the input, as well as *hysteresis*. This device may be operated with two beams so that one of them can control the gain in the other, thus constituting a three-port 'optical transistor' which, because it operates by transferred optical thickness, is termed a 'transphasor' by Miller and Smith (1979). Utilising the bistability, optical memory and logic functions may also be performed.

The principles behind these 'optical circuit' elements were first put forward by Szöke *et al* (1969). In their paper they predicted that bistability would occur at exact resonance if a Fabry–Perot resonator was filled with a *saturable absorber* in which the absorption coefficient is a decreasing function of local intensity. At low light intensities, interference—which would be constructive at resonance—is prevented by the absorption. In the absence of interference the intracavity intensity I_c will be approximately related to the incident intensity I_I by

$$I_T \cong TI_c \cong I_I T^2$$

where T is the transmissivity of the resonator mirrors and I_T is the transmitted intensity. The described conditions will hold while $I_c < I_s$ where I_s is the *saturation intensity*. Constructive interference can occur at high input intensities if $I_c > I_s$ implying effectively $I_I = I_T$ and approximately $I_c = I_I/T$†. The conditions $I_s > I_I T$ (interference

† In the next section we justify these expressions.

prevented) and $I_s < I_I/T$ (interference allowed) can exist simultaneously if $I_I = I_s$, so bistable behaviour should be possible. Physically, more intensity can enter the cavity under conditions of constructive interference so that a lower external intensity can maintain the saturation once it is established. Szöke et al (1969) presented results for a resonator containing SF_6 gas irradiated by a CO_2 laser at 10.6 μm. Non-linear behaviour was shown but bistability was not demonstrated. It turns out that this *absorptive bistability* is rather hard to achieve. Further attempts were made to observe absorptive bistability. Austin and De Shazer (1971) studied the transmission of a single-mode ruby-laser pulse through a Fabry–Perot cavity containing cryptocyanine in methanol. They observed pulse narrowing and asymmetrising which they attributed to optical hysteresis. Spiller (1971) investigated the transmission of pulses from a ruby or Nd laser through etalons containing organic dyes. Although marked non-linearities were noticed, 'bleaching' of the medium was not possible and bistability not observed.

The first numerical study of absorptive bistability by McCall (1974) became the initial step towards the first observation of OB by Gibbs et al (1976) using Na vapour (cf § 6). The experiment was shown to depend on a dispersive mechanism about which possibility the experimenters were unaware at the time of the experiment. Almost simultaneously, Felber and Marburger (1976) gave the simplest explanation of dispersive OB (cf § 2).

In addition to device applications, OB also shows many fascinating theoretical aspects as it has the character of a phase transition in a system lying in a stationary but non-equilibrium state. This was originally pointed out by Bonifacio and Lugiato (1976) whose contributions triggered a lively interest in the subject by putting forward an analytically solvable model—the mean field model. As it turns out, a bistable system can be considered to be the passive counterpart of the laser.

Non-linearity and optical feedback form the basis of OB. The former is provided by the medium and the latter can either be provided by the mirrors of the cavity or by the use of some external electronic circuit. In the case when the feedback is all-optical the device is said to be '*intrinsic*'. If the output power is sensed and a voltage proportional to this signal applied to the crystal an 'artificial' non-linearity is created: such devices are termed *hybrid*. Hybrid devices were proposed by Kastal'skii (1973) and they were first constructed by Smith and Turner (1977) using an electro-optic crystal inside a resonator. The concept was advanced by Garmire et al (1978a, b, c) in which the resonator was replaced merely by using the feedback to a crystal placed between crossed polarisers, thus eliminating the resonator.

It will be shown that a large value for refractive non-linearity is the best approach for intrinsic OB systems; the introduction of semiconducting materials in 1979 by Gibbs et al (1979a, b, c, d) and Miller et al (1979) marked an important step forward following the discovery of very large non-linear refraction near the absorption edge of these materials. This in turn allows the construction of micron-sized optical resonators.

There are two special cases of observed *intrinsic* bistability that do not require a resonator. One is the non-linear interface (Smith et al 1979) first suggested by Kaplan (1976, 1977) in which feedback takes place via an intensity-dependent refractive index which in turn makes the angle for total internal reflection intensity-dependent. The other one (Bjorkholm et al 1981) is based on self-focusing. There are other novel systems but these are still at a theoretical stage (cf § 4).

1.3. Organisation of the review

In view of the many facets of OB we provide in § 2 the basic theory for those particularly interested in the experiments as a first reading. In § 3 the semi-classical theory of two-level atomic systems is discussed by means of the Maxwell–Bloch equations which have been the object of extensive investigation. We show steady-state and time-dependent phenomena both in Fabry–Perot and ring cavity configurations. This is followed by an introduction to the Fokker–Planck equation that we use for a phase transition analogy. This section finishes with a fully quantum-mechanical approach. In § 4 we take a glance at other theoretical approaches and novel bistable systems. Section 5 contains theoretical and experimental studies on non-linear optics of semiconductors, namely InSb and GaAs. In § 6 we present all the experiments on intrinsic devices whereas the hybrid ones appear in § 7. In § 8 we discuss switching speeds and energies as fast optical processing constitutes one of the most promising applications of OB. Finally, in § 9, we present a summary and conclusions.

2. Basic theory

The non-linear response of the medium manifests itself through the intensity dependence of otherwise intrinsic basic properties such as refraction and absorption. We now refer to the case in which OB is obtained with a purely refractive—often referred to as 'dispersive'—non-absorptive non-linearity (Felber and Marburger 1976) placed in a Fabry–Perot cavity (figure 1); this is followed by a discussion of the effect of linear absorption (Miller 1981).

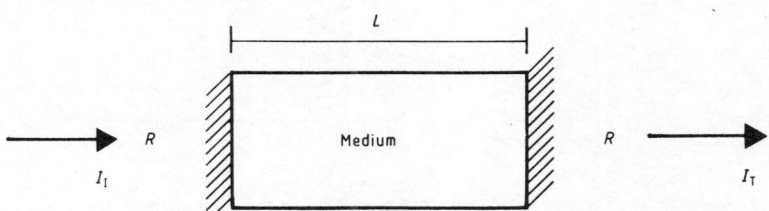

Figure 1. Fabry–Perot interferometer of length L and mirror reflectivity R, I_I is the incident intensity and I_T is the transmitted intensity.

Suppose we have a linear material of refractive index n_0 filling the cavity of figure 1. The transmission τ of the cavity, i.e. the ratio between the transmitted intensity I_T and the input intensity I_I, is given by† (see, for example, Born and Wolf 1965)

$$\tau \equiv \frac{I_T}{I_I} = \frac{1}{1+F\sin^2 \Phi/2} \tag{2.1}$$

where $F = 4R/T^2$ is the finesse, $R(T+R=1)$ is the mirror reflectivity, $\Phi = 2\pi n_0 L(\lambda/2)^{-1}$ is the round-trip phase shift, λ is the wavelength of the radiation and L is the length of the cavity. Whenever $\Phi = 2\pi N$ (N is integer) there is full transmission ($\tau = 1$) which can be achieved, e.g. by changing the optical path $n_0 L$. One way of doing this is by making the refractive index non-linear, i.e. we must replace n_0 by

† Plane wave theory and normal incidence assumed.

some $n(I_c)$. The simplest functional relationship is one that allows for a first-order effect in the intensity, namely

$$n(I_c) = n_0 + n_2 I_c \qquad (2.2)$$

where n_2 is a constant measured in $cm^2\ W^{-1}$; then

$$\Phi(I_c) = 4\pi n_0 \frac{L}{\lambda} + \frac{4\pi n_2 L}{\lambda} I_c$$

$$4\pi n_0 L/\lambda = 2\pi N - \theta \qquad (2.3)$$

where we have considered the cavity tuned away from resonance by θ rad. It is now possible to justify a hysteresis cycle (figure 2) as follows: when $I_I = 0$, $\Phi = 2\pi N - \theta$; as I_I is increased so is I_c and Φ begins to approach the resonance value $2\pi N$. At a critical value I_{th} of I_I there is a runaway effect as the increasing Φ increases I_c and vice versa. As a result, I_T switches from a low to a high value with $\Phi \geqslant 2\pi N$ as negative feedback is required for stability. On decreasing I_I the large value of I_c (cf § 1) keeps Φ close to $2\pi N$; a sudden decrease of I_T only occurs when $4\pi n_2 (L/\lambda)I_c - \theta = 0$ which corresponds to $I_I < I_{th}$. This hysteretic behaviour is the origin of bistability.

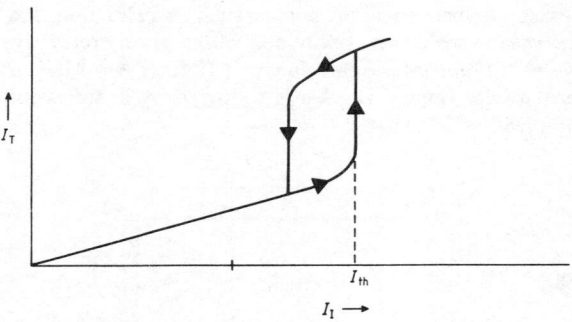

Figure 2. Characteristic curve of a non-linear optical cavity. As the input intensity I_I varies from zero to $I_I > I_{th}$ and back to zero, the transmitted intensity I_T can exhibit a hysteresis cycle.

The above qualitative discussion can be made quantitative by including a second expression for the transmission of a Fabry–Perot cavity also given by Born and Wolf (1965):

$$\tau = \frac{I_c}{I_I} \frac{(1-R)}{(1+R)} \qquad (2.4)$$

which relates external and internal fields. Only certain values of I_c, for a given I_I, satisfy self-consistency between (2.4) and (2.1) and they will emerge from a graphical solution (see below). In order to revert to our discussion of § 1 let us take T very low, e.g. 0.1. If we assume full constructive interference $\sin \Phi/2 \sim 0$ in (2.1), $\tau \sim 1$, and from (2.4) $I_c \sim I_I/T = 10 I_I$; if instead the interference is destructive, $\sin \Phi/2 \simeq 1$, $\tau \sim T^2$ (from (2.1)) and then from (2.4) $I_c \sim T I_I = 0.1 I_I$. Thus, depending on the interference conditions, the cavity can either store a large or a small amount of optical energy: when this is determined by a non-linear medium, hysteresis is possible.

For a graphical solution we plot the Airy function (2.1) and the straight line (2.4) in figure 3 using the parameters of Miller et al (1979) against $\Phi(I_c)$. We follow the intersection point as I_I is varied from zero to some maximum. At low intensities the point moves smoothly along the curve until it reaches A; here, a small change in I_I causes the point to jump to B. If I_I is decreased the intersection moves smoothly up to C where it jumps to D and a first bistable region is obtained. If, instead, I_I is increased further from B, successive bistable regions can appear (multistability).

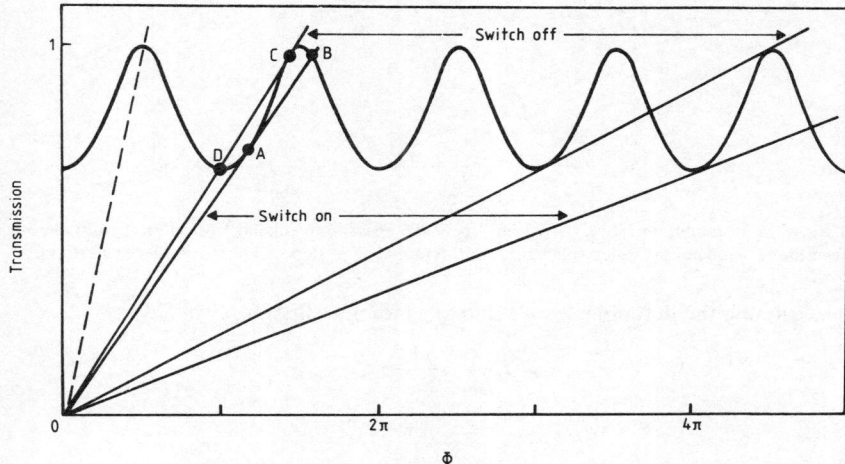

Figure 3. Intersections between relations (2.1) and (2.4) for $\theta = \pi$ and $F = 0.5$, demonstrating no bistability in first order. Only from the intersection, point A, progressively wider bistable regions begin to appear. Straight lines of shallower gradient correspond to higher input intensities I_I (from Miller et al 1979).

The first observation of multistability in an intrinsic device was achieved by Miller et al (1979, 1981b, c). Their original characteristic curve, shown in figure 4, can be made use of to show some of the applications of a bistable device. If the input (holding) intensity is fixed where the slope is greater than unity (bistability can be avoided by simple cavity detuning), a superimposed signal will experience differential gain or 'transistor' action. The flat regions imply that an increase in input intensity does not result in an increase of output intensity. This is the *limiter* action which can be used to cut off the unwanted top of an optical signal and improve measuring accuracy in the application of pulsed or locked lasers. Also a bistable region can be identified.

Following the discussion of § 1.2 we may ask what is the role played by non-linear absorption; we follow Miller (1981). If we take a saturable two-level system the real and imaginary parts of the susceptibility are, respectively (Yariv 1975),

$$\chi' \propto \frac{\Delta\omega}{1+\Delta\omega^2+I/I_s} \quad \text{(dispersion)}$$

$$\chi'' \propto \frac{1}{1+\Delta\omega^2+I/I_s} \quad \text{(absorption)}.$$

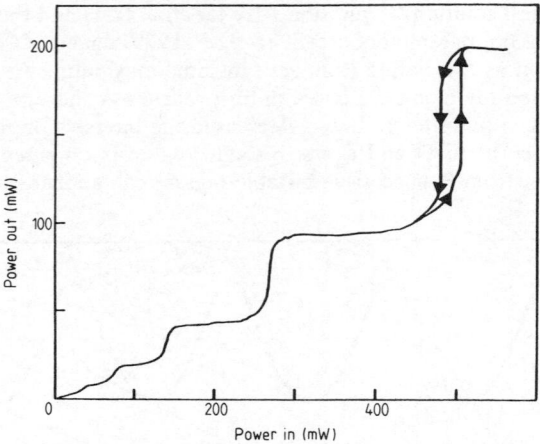

Figure 4. Multistability in InSb at 1895 cm^{-1} near the absorption edge at 5 K. Note the bistable region, power limiter mode of operation (plateau) and differential gain (slope > 1) (from Miller et al 1979).

Assuming the detuning $\Delta\omega \gg 1$ and expanding to first order in I/I_s:

$$\chi' \propto \frac{1}{\Delta\omega} - \frac{1}{\Delta\omega^3} I/I_s + \ldots$$

$$\chi'' \propto \frac{1}{\Delta\omega^2} - \frac{1}{\Delta\omega^4} I/I_s + \ldots$$

The change in non-linear absorption falls as $\Delta\omega^{-4}$ whereas the dispersion falls as $\Delta\omega^{-3}$. The transition becomes very difficult to saturate and large changes of intensity would be necessary to have any effect. Therefore refraction will be the predominant interferometric process as only small changes of refractive index are needed for switching. There still remains linear absorption, whose effect we now discuss.

For the design of devices we require to know the conditions when bistable behaviour arises. Furthermore we would like to know what is the minimum (critical) incident intensity I_{crit} for observing OB. A bistable region can be guaranteed if the finesse of the cavity is such that the Airy function is intersected at least three times by the straight line (2.4) (Felber and Marburger 1976). The non-linear term in (2.3) indicates that the longer the cavity the less intensity is necessary for these effects, but if the cavity is too long, then linear absorption becomes important and this results in a reduction of the effective finesse and bistability can be suppressed. Therefore I_{crit} should result from a compromise between non-linearity (n_2), linear absorption (α) and mirror reflectivity. In order to obtain first-order bistability we must have the slope (>0) at the inflection point greater than the slope of the straight line:

$$d\tau/dI_c > \tau/I_c. \tag{2.5}$$

Moreover (figure 5) there is also a critical value for the initial mistuning θ above which there is no bistability. The plane wave calculation of Miller (1981) shows that

$$I_{\text{crit}} \doteq \frac{1}{\beta} \frac{1}{\mu_0} \tag{2.6}$$

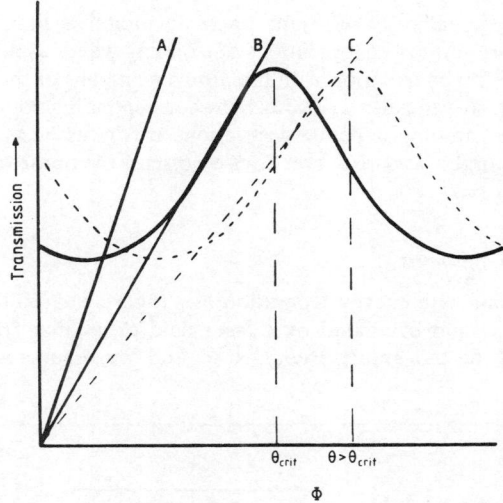

Figure 5. Diagram illustrating that (A) for $I_1 > I_{crit}$ the straight line only ever makes one intersection with the curve (i.e. no bistable region); (B) the critical condition $I_1 = I_{crit}$, $\theta = \theta_{crit}$; (C) bistability as there is multiple intersection of line and curve for $I_1 > I_{crit}$, $\theta > \theta_{crit}$ (from Miller 1981).

where $\beta = 3n_2/\lambda\alpha$, $\mu_0 \cong 0.65\pi(1 - R + A)^{-1}$ for an optimised high-finesse cavity, where $A = 1 - e^{-\alpha L}$ is the absorption per pass; β is the *figure of merit* of the material and μ_0 of the cavity. The critical initial detuning $\theta_{crit} = (2.72/\pi)(1 - R + A)$. The expression for β shows that the larger the value of α the larger n_2 must be for the same I_{crit}. As an example, for InSb at 77 K and $\lambda^{-1} = 1852 \text{ cm}^{-1}$, $n_2 = 3 \times 10^{-3} \text{ cm}^2 \text{ W}^{-1}$, $\alpha = 80 \text{ cm}^{-1}$; then $\beta = 0.14 \text{ cm}^2 \text{ W}^{-1}$. If $R = 0.95$ and $A = 0.05$ ($A = 1 - R$ is the optimisation condition) then $I_{crit} = 0.4 \text{ W cm}^{-2}$.

This simple discussion covers most of the basic experimental effects and concepts of practical importance to date.

3. Theory of optical bistability for atomic systems

A model that has been very successful in treating the interaction of light with atoms consists of considering the latter as two-level quantum-mechanical objects while considering the electromagnetic field as classical. In a full quantum-mechanical theory this is equivalent to taking the expectation value of products of operators as the product of expectation values—thereby neglecting correlations of the fluctuations—in the equations of motion. This *semi-classical* approximation is inadequate, e.g. to describe the initiation of superfluorescence (Bonifacio and Lugiato 1975a, b) which is caused by fluctuations, but very applicable in a wide range of optical phenomena like the coherent propagation of pulses (see Allen and Eberly 1975). Optical bistability is one such phenomenon in which the semiclassical approximation proves successful.

In this section we develop the theory in a way we consider pedagogical but which does not necessarily follow the historical development of the subject. We assume a system of N ($N \gg 1$) two-level atoms placed in a Fabry–Perot cavity. In some cases—to which we shall refer specifically—a more tractable model is that of a ring cavity because it does not present the complications of standing waves as in the former case. We study the interaction of radiation with the atoms by means of the Maxwell–Bloch equations which we derive and also write down the corresponding boundary conditions. Both the steady-state and time-dependent situations are discussed as well as the mean field approximation. In the last two subsections we discuss the phase transition analogy and quantum fluctuations.

3.1. Maxwell–Bloch equations

The N two-level atoms with energy separation $\hbar\omega_0$ (figure 6) are placed in a Fabry–Perot cavity (figure 1) and irradiated by a laser field of angular frequency ω. The resonance frequency of the empty cavity is ω_c and we assume that $|\omega - \omega_c| \ll \omega$, $|\omega - \omega_0| \ll \omega$.

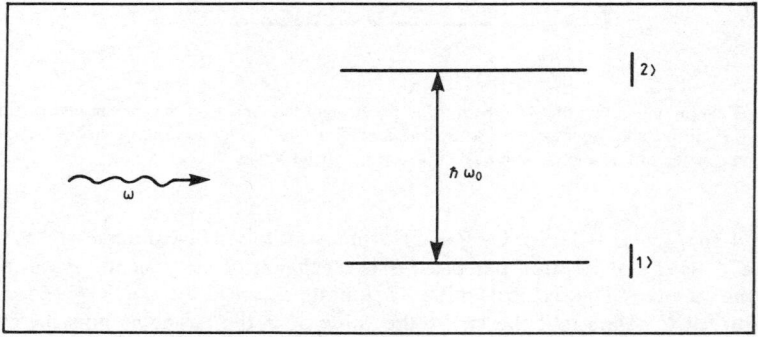

Figure 6. Two-level atom model; ω is the circular frequency of the input radiation.

Let us consider first the interaction of a single atom with a classical electromagnetic field by means of the density matrix formalism (also used in § 5). The density operator is defined as $\hat{\rho} = |\psi\rangle\langle\psi|$ with matrix elements $\rho_{ij} = \langle i|\psi\rangle\langle\psi|j\rangle$ where $|\psi\rangle$ is the state of the atom. Its equation of motion (see, for example, Sargent et al 1974) is given by

$$\frac{\partial \hat{\rho}(\bar{x},t)}{\partial t} = -\tfrac{1}{2}(\hat{\Gamma}\hat{\rho} + \hat{\rho}\hat{\Gamma}) - i\hbar^{-1}[\hat{H},\hat{\rho}] \tag{3.1}$$

($\hat{}$ is an operator and $\bar{x}$ is the atom's position) where $\hat{\Gamma}$ is a decay operator with matrix elements

$$\Gamma_{ij} = \gamma_i \delta_{ij} \qquad (i,j = 1, 2) \tag{3.2}$$

and $\hat{H}$ is the Hamiltonian in the electric-dipole approximation:

$$\hat{H} = \hat{H}_0 - \hat{\boldsymbol{\mu}} \cdot \boldsymbol{E}$$

where $\hat{H}_0$ is the unperturbed Hamiltonian of the atom, $\hat{\boldsymbol{\mu}}$ is the electric-dipole operator:

$$\boldsymbol{\mu}_{ij} = \boldsymbol{\mu}(1 - \delta_{ij}) \tag{3.3}$$

$$\langle \hat{\boldsymbol{\mu}} \rangle = \mathrm{Tr}(\hat{\rho}\hat{\boldsymbol{\mu}}) = \boldsymbol{\mu}(\rho_{12} + \rho_{21}). \tag{3.4}$$

Combining (3.2)–(3.4) in (3.1):

$$\frac{\partial \rho_{21}}{\partial t} = -\left(i\omega_0 + \frac{1}{T_2}\right)\rho_{21} + i\hbar^{-1}\boldsymbol{\mu} \cdot \boldsymbol{E}R_3$$

$$\frac{\partial R_3}{\partial t} = 2i\hbar^{-1}\boldsymbol{\mu} \cdot \boldsymbol{E}(\rho_{21} - \rho_{21}^*) - \frac{R_3 + 1}{T_1} \quad (3.5)$$

$$\rho_{12} = \rho_{21}^* \qquad R_3 = \rho_{22} - \rho_{11}$$

where R_3 is the population inversion ($R_3 = -1$ when $E = 0$) and $T_1 \equiv \gamma_1^{-1}$ and $T_2 \equiv \gamma_2^{-1}$ are the population and dipole relaxation times, respectively. As $\boldsymbol{E}$ is real we split it into its positive and negative frequency parts:

$$\boldsymbol{E} = \boldsymbol{E}^{(+)}(\boldsymbol{x}, t) \exp(-i\omega t) + \boldsymbol{E}^{(-)}(\boldsymbol{x}, t) \exp(i\omega t)$$
$$\boldsymbol{E}^{(+)} = \boldsymbol{E}^{(-)*} \quad (3.6)$$

and transform (3.5) into an interaction picture

$$\rho_{12} = R^{(-)}(\boldsymbol{x}, t) \exp(i\omega t)$$
$$\rho_{21} = R^{(+)}(\boldsymbol{x}, t) \exp(-i\omega t)$$
$$R_3 = D(\boldsymbol{x}, t) \quad (3.7)$$
$$R^{(-)} = R^{*(+)}.$$

Substituting (3.6) and (3.7) in (3.5) and neglecting rapidly oscillating terms of the form $\exp(2i\omega t)$ (rotating wave approximation)

$$\frac{\partial R^{(\pm)}}{\partial t} = -[i(\omega - \omega_0) + T_2]R^{(\pm)} + i\mu E^{(\pm)}D$$

$$\frac{\partial D}{\partial t} = -\frac{D+1}{T_1} + 2i\mu \hbar^{-1}(E^{(-)}R^{(+)} - E^{(+)}R^{(-)}). \quad (3.8)$$

In (3.8) the field drives the atoms but the re-radiation is neglected. This latter no longer holds in an extended system in which the dynamics of each atom is affected by the radiation of the rest. In other words, a macroscopic polarisation develops which acts as a source for the field. As a result, we have to include Maxwell's equation†

$$\frac{\partial^2 E(z,t)}{\partial z^2} - \frac{1}{c^2}\frac{\partial^2 E(z,t)}{\partial t^2} = \frac{4\pi}{c^2}\frac{\partial^2 P(z,t)}{\partial t^2} \quad (3.9)$$

where $P(z, t)$ is the macroscopic polarisation to account for propagation effects; from (3.4)

$$P(z, t) = n\mu(\rho_{12} + \rho_{21}) = n\mu(R^{(-)}\exp(i\omega t) + \text{CC}) \quad (3.10)$$

where n is the number density of atoms; equations (3.8) and (3.9) must be solved self-consistently (Lamb 1964). Since the atoms inhabit a Fabry-Perot cavity, the following expansions can be assumed ($k = \omega/c$):

$$E^{(+)}(z, t) = \sum_{m=1}^{\infty} E_m(z, t) \exp[i(2m-1)kz] + \sum_{m=-1}^{-\infty} E_m(z, t) \exp[i(2m+1)kz] \quad (3.11)$$

† From now on the plane wave approximation is made and linear polarisation of the electric field is assumed.

$$R^{(+)}(z,t) = \sum_{m=1}^{\infty} P_m(z,t) \exp[i(2m-1)kz] + \sum_{m=-1}^{-\infty} P_m(z,t) \exp[i(2m+1)kz] \quad (3.12)$$

$$D(z,t) = D_0(z,t) + \sum_{m=1}^{\infty} D_m(z,t) \exp(i2mkz) + \text{cc} \quad (3.13)$$

which allow for standing waves. The expansion of $E^{(+)}$ and $R^{(+)}$ in odd multiples of k and in even multiples of k for D can be deduced by utilising an all-order perturbation theory of a strong signal laser (Sargent *et al* 1974). Agrawal and Lax (1979) have shown that field amplitudes for $|m| > 1$ are negligible and the expansion of (3.11) and (3.12) can be truncated at $m = \pm 1$:

$$E^{(+)}(z,t) = E_F(z,t)\exp(ikz) + E_B(z,t)\exp(-ikz) \quad (3.14)$$

$$P^{(+)}(z,t) = P_F(z,t)\exp(ikz) + P_B(z,t)\exp(-ikz) \quad (3.15)$$

where we make the identifications:

$$\begin{aligned} E_F(z,t) &\equiv E_1(z,t) \\ E_B(z,t) &\equiv E_{-1}(z,t) \\ P_F(z,t) &\equiv P_1(z,t) \\ P_B(z,t) &\equiv P_{-1}(z,t) \end{aligned} \quad (3.16)$$

in which F(B) stands for 'forward' ('backward') and in (3.15) the amplitudes are slowly varying (known as the slowly varying envelope approximate, SVEA):

$$k^{-1}\left|\frac{\partial E_F}{\partial z}\right| \ll |E_F| \qquad \omega^{-1}\left|\frac{\partial E_F}{\partial t}\right| \ll |E_F| \qquad \text{etc.} \quad (3.17)$$

The expansion (3.13) does not have to be truncated (e.g. Carmichael 1980) if we are in the steady state (time derivatives = 0), but in order to treat time-dependent phenomena we stop the expansion at $|m| = 1$:

$$D(z,t) = D_0(z,t) + D_1(z,t)\exp(2ikz) + \text{cc}. \quad (3.18)$$

If the atoms have a thermal motion, the Doppler effect can be accounted for by a Gaussian distribution of velocities:

$$G(u)\,du = \frac{1}{\sqrt{\pi}\sigma}\exp[-(u/\sigma)^2]\,du \quad (3.19)$$

where $u = (\omega_0/c)v_z/T_2^{-1}$, v_z is the z component of the velocity of the atom and $\sigma = T_2/T_2^*$, where T_2^* is the inhomogeneous dipole dephasing time. Substituting (3.18), (3.14) and (3.15) plus using (3.17) in (3.8) and (3.9) leads to the *Maxwell–Bloch equations*:

$$\begin{aligned} \frac{\partial P_F}{\partial t} &= -[T_2^{-1} - i(\Delta + u')]P_F + i\mu\hbar^{-1}(E_F^* D_0 + E_B^* D_1) \\ \frac{\partial P_B}{\partial t} &= -[T_2^{-1} - i(\Delta - u')]P_B + i\mu\hbar^{-1}(E_B^* D_0 + E_F^* D_1) \\ \frac{\partial D_0}{\partial t} &= \tfrac{1}{2} T_1^{-1}(D_0 + 1) + \frac{i\mu\hbar^{-1}}{2}(E_F^* P_F + E_B^* P_B - \text{cc}) \\ \frac{\partial D_1}{\partial t} &= -T_1^{-1} D_1 + \frac{i}{4}\mu\hbar^{-1}[(1-i)E_F^* P_B + (1+i)E_B^* P_F - \text{cc}] \end{aligned} \quad (3.20)$$

$$c^{-1}\frac{\partial E_{F(B)}}{\partial t} \pm \frac{\partial E_{F(B)}}{\partial z} = ig\int_{-\infty}^{\infty} P_{F(B)}^* G(u')\,du'$$

where $g = 4\pi\mu\omega_0 nc^{-1}$ and the different signs of $u' = T_2^{-1}u$ stem from the atom experiencing one field blue-shifted and the other one red-shifted. Note in (3.20) that D_1 couples E_F and E_B with P_B and P_F, respectively. The boundary conditions can be obtained by demanding conservation of the tangential components of the fields at the mirror (e.g. Abraham 1979, Abraham et al 1980):

$$E_F(L, t) = R^{1/2} E_B(L, t) \exp(-i\theta) + (1-R)^{1/2} E_I(t)$$
$$E_R(t) = -R^{1/2} E_I(t) + (1-R)^{1/2} E_B(L, t) \exp(i\theta) \quad (3.21)$$
$$E_B(0, t) = R^{1/2} E_F(0, t) \qquad E_T(t) = (1-R)^{1/2} E_F(0, t)$$

where E_R is the reflected field, E_T is the transmitted field, E_I is the incident field and $\theta = (\omega_c - \omega)/\nu$ where $\nu = c/2L$ is the free-spectral range of the cavity.

3.2. Steady state

The derivation of (3.20) is a fairly general one in that we include atomic detuning ($\Delta \neq 0$) and cavity mistuning ($\theta \neq 0$ in (3.21)). This is the case of dispersive OB since the appearance of a refractive index effect entails $\Delta \neq 0$. In order to study output–input characteristics of this system when irradiated by a CW laser, we seek stationary solutions of (3.20) by dropping all the time derivatives. In this way all the material medium equations become an algebraic system and the P and D can be expressed in terms of the fields†. We then substitute the P in the field equations which still are to be integrated over space. The fields can be normalised as

$$f \equiv \mu \hbar^{-1} (T_1 T_2)^{1/2} E_F$$
$$b \equiv \mu \hbar^{-1} (T_1 T_2)^{1/2} E_B$$
$$x \equiv \sqrt{3} \mu \hbar^{-1} (T_1 T_2)^{1/2} (1-R)^{-1/2} E_T \quad (3.22)$$
$$y \equiv \sqrt{3} \mu \hbar^{-1} (T_1 T_2)^{1/2} (1-R)^{-1/2} E_I$$

and the steady-state equations read

$$\frac{df}{dz'} = i \frac{4C(1-R)}{1+\sqrt{R}} \int_{-\infty}^{\infty} du\, G(u) P_F^*(f, f^*, b, b^*; \tilde{\Delta}, u)$$
$$\frac{db}{dz'} = -i \frac{4C(1-R)}{1+\sqrt{R}} \int_{-\infty}^{\infty} du\, G(u) P_B^*(f, f^*, b, b^*; \tilde{\Delta}, u) \quad (3.23)$$

with $z' = z/L$, $\tilde{\Delta} = \Delta T_2$ and $C = \frac{1}{2}(1+\sqrt{R})\mu g/\hbar T_2^{-1}$ is the bistability cooperation parameter (Bonifacio and Lugiato 1976) which reduces to $C = \alpha L/2(1-R)$ for $R \leq 1$ where $\alpha = \mu g \hbar^{-1} T_2$ is the linear absorption coefficient.

Equations (3.23) are integrated numerically (Abraham and Hassan 1980) with the statistical average over u carried out for each step of integration along z'. The characteristic curves of $|x|^2 (\propto I_T)$ against $|y|^2 (\propto I_I)$ are shown in figure 7 where we compare a homogeneously broadened line ($\sigma = 0$) with one with $\sigma = 1.5$ (all the other parameters unaltered). The different curves are obtained by fixing $C = 14$, $R = 0.9$ and $\tilde{\Delta} = 3$ and changing θ. In figure 7(a) we can immediately recognise some of the

† Here the situation is greatly simplified by the truncation on D. Releasing this restriction a correct distribution of fields is obtained for both high and low input intensities. In the low input intensity limit both approaches coincide as pointed out by, for example, Carmichael (1980) and Goll and Haken (1980).

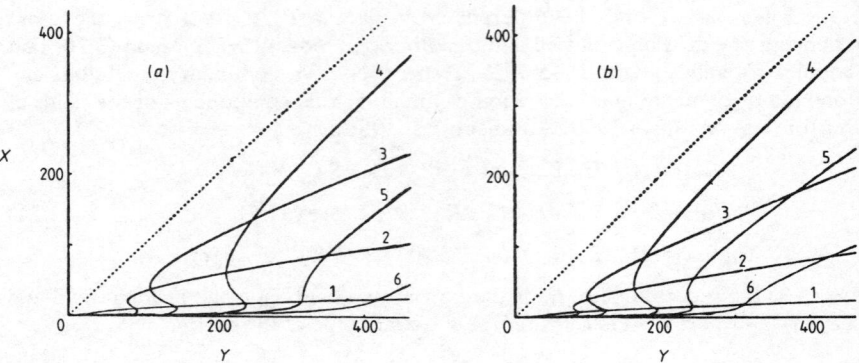

Figure 7. Effect of inhomogeneous (Doppler) broadening. (a) Normalised output intensity $X = |x|^2$ (see text) against the normalised input intensity $Y = |y|^2$ for $C = 14$, $R = 0.9$, $\tilde{\Delta} = 3$ and $\sigma = 0$. The number on the curves refer to different cavity detunings $\theta(i = 1-6) = 0.2\pi$, 0.08π, 0.04π, 0, -0.02π and -0.04π. (b) Same as (a) but for $\sigma = 1.5$ (from Abraham and Hassan 1980).

typical modes of operation (cf § 2). The curve labelled '3' shows bistability, '5' shows differential gain, and '1' limiter action. The general effect of inhomogeneous broadening (figure 7(b)) is to obliterate the bistable region and a larger value of C is necessary to regain bistability. When $|\tilde{\Delta}| \gg \sigma$, the motion of the atoms does not strongly affect their 'individual' mistuning and inhomogenous broadening has little effect. If we considered a Lorentzian broadening the effect would be even larger (Abraham and Hassan 1980) than that of a Gaussian owing to the long tails of the Lorentzian profile.

If the values of αL and $\tilde{\Delta}$ are substantially increased, there is a drastic change in the characteristic curves and a region of multistability is obtained (figure 8). This multistable behaviour stemming from the Maxwell-Bloch equations was first predicted by Ikeda (1979) for a ring cavity and a homogeneously broadened line. The first calculation for a Fabry-Perot cavity showing multistability was done by Abraham and Hassan (1980) who also included inhomogeneous broadening in the Maxwell-Bloch equations. The stability of these branches in the Fabry-Perot cavity are yet to be studied but for a ring cavity Ikeda (1979) predicted new instabilities (cf § 3.3).

The integration of the Maxwell-Bloch equations in the steady state was first done by McCall (1974) for the purely absorptive case ($\tilde{\Delta} = \theta = \sigma = 0$). Later, Gibbs et al

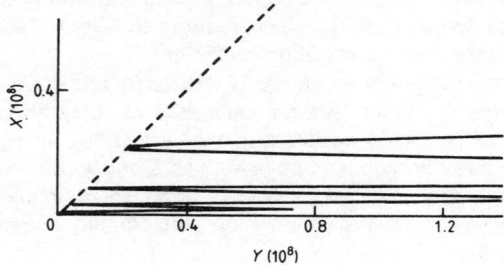

Figure 8. Multiple-valued branch for a Fabry-Perot cavity; X and Y as in figure 7. From Rb data (Grischkowsky 1978) $C = 92\,000$, $R = 0.7$, $\tilde{\Delta} = 2000$, $\theta = 0.24\pi$, $\sigma = 30$ (from Abraham and Hassan 1980).

(1976) presented similar curves to those of figure 7(a) which substantiated qualitatively their experimental results. Hermann (1979) and Roy and Zubairy (1980a, b) analytically solved these equations for ($\tilde{\Delta} = \sigma = \theta = 0$) in a Fabry–Perot cavity and obtained agreement with the numerical results of Meystre (1978) and Abraham et al (1979). Agrawal and Carmichael (1979) interpreted dispersive OB ($\sigma = 0$) in a Fabry–Perot cavity as an elementary cusp catastrophe; Carmichael and Agrawal (1980) have also studied the effect of inhomogeneous broadening. Others authors (Roy and Zubairy 1980a, b, Carmichael and Hermann 1980, Gronchi and Lugiato 1980) have given exact analytical treatments of dispersive OB with spatial effects in a ring cavity. There is a convergence in the conclusions of all of these authors and the ones we outlined above.

In our above discussion it has emerged that for $\sigma \gg 1$ a large value of C, for example high atomic density, is necessary to regain bistability. In the Doppler case the effect can be done away with by taking advantage of the two oppositely running waves in the cavity. This idea was first suggested by Arecchi and Politi (1978) who showed that Doppler-free bistability was possible. Hermann and Thompson (1980) extended these results to consider spatial effects and they show that these cause different switching processes in the Raman and double-absorption cases. Agrawal and Flytzanis (1980) investigated different frequency two-photon (double-beam) OB; they considered the degenerate case and the effect of optical Stark shift. The effect of the latter in two-photon OB was also investigated by Agarwal (1980). Hermann (1981) showed that level degeneracy in two-photon OB makes possible a form of tristable behaviour. Many of these authors invoke the mean field approximation which we now describe.

3.3. The mean field approximation

This model was introduced by Bonifacio and Lugiato and it has been the object of extensive research by an increasing number of authors. For the sake of simplicity we take $\tilde{\Delta} = \sigma = \theta = 0$ in (3.20) (absorptive OB); the equations then read:

$$\frac{\partial P_F}{\partial t} = -T_2^{-1} P_F + \mu \hbar^{-1}(E_F D_0 + E_B D_1)$$

$$\frac{\partial P_B}{\partial t} = -T_2^{-1} P_B + \mu \hbar^{-1}(E_B D_0 + E_F D_1)$$

$$\frac{\partial D_0}{\partial t} = -T_1^{-1}(D_0 + 1) - \mu \hbar^{-1}(E_F P_F + E_B P_B)$$

$$\frac{\partial D_1}{\partial t} = -T_1^{-1} D_1 - \tfrac{1}{2}\mu \hbar^{-1}(E_F P_B + E_B P_F)$$

$$\frac{1}{c}\frac{\partial E_F}{\partial t} + \frac{\partial E_F}{\partial z} = g P_F$$

$$\frac{1}{c}\frac{\partial E_B}{\partial t} - \frac{\partial E_B}{\partial z} = g P_B$$

(3.24)

in which all the quantities are real. From this system Bonifacio and Lugiato (1978a) derived their mean field equations as follows: we take a spatial average of (3.24) over the length of the cavity, i.e. for some variable $A(z, t)$

$$\overline{A(z, t)} = (1/L) \int_0^L A(z, t) \, dz = A(t) \tag{3.25}$$

where the bar indicates spatial average; moreover, it is assumed that

$$\overline{A(z, t)B(z, t)} = \overline{A(z, t)} \, \overline{B(z, t)} = A(t)B(t) \tag{3.26}$$

where $A(z, t)B(z, t)$ can be, for example, $E_F D_0$, etc, and also the ansatz

$$\bar{P}_F = \bar{P}_B = P$$
$$\bar{E}_F = \bar{E}_B = E \tag{3.27}$$

is made. With (3.25)–(3.27) plus the boundary conditions (3.21), equations (3.24) become

$$\dot{P}(t) = \mu \hbar^{-1} ED - T_2^{-1} P$$
$$\dot{D}(t) = -\mu \hbar^{-1} EP - T_1^{-1}(D+1) \tag{3.28}$$
$$\dot{E}(t) = -gP - K(E - E_I/\sqrt{T})$$

with $D(t) = \bar{D}_0 + \bar{D}_1$; $t_c \equiv K^{-1} = 2L/cT$ is the cavity build-up time. Using the normalisation (3.22) in the steady state ($\dot{E} = \dot{P} = \dot{D} = 0$) one obtains for the transmissivity

$$\tau = \frac{x}{y} = \frac{1}{1 + 2C(1 + x^2)^{-1}}$$

or equivalently

$$y = x + 2Cx(1 + x^2)^{-1}. \tag{3.29}$$

In (3.29) the only parameter that counts is C which explains its use in (3.23); its interpretation will become apparent in § 3.5. Plots of (3.29) are shown in figure 9 for different values of C, and it is clear that the bistability condition is $C > 4$; the critical value $C = 4$ is obtained by demanding $dy/dx = d y^2/dx^2 = 0$. The lower and upper thresholds of the bistable regions for $C \gg 4$ are given by

$$y_m = x_m + \frac{2Cx_m}{1 + x_m^2} \simeq \sqrt{8C}$$

$$x_m = (C - 1 + \sqrt{C^2 - 4C})^{1/2} \simeq \sqrt{2C}$$

$$y_M = x_M + \frac{2Cx_M}{1 + x_M^2} \simeq C + 1$$

$$x_M = \left(\frac{2C + 1}{C - 1 + \sqrt{C^2 - 4C}} \right)^{1/2} \simeq 1.$$

The stability of the branches of figure 9 can be determined as follows. In (3.28) we write

$$D(t) = D_\infty + \delta D(t)$$
$$P(t) = P_\infty + \delta P(t) \tag{3.30}$$
$$E(t) = E_\infty + \delta E(t)$$

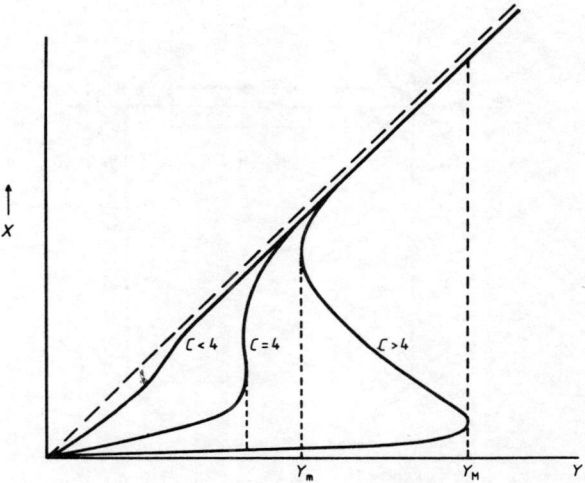

Figure 9. The mean field 'state equation' (3.29) shows that a bistable region exists when $C > 4$. Y_m and Y_M are the lower and upper thresholds, respectively.

where the subscript ∞ means steady state and $|\delta E(t)| \ll E_\infty$, etc, and a linear stability analysis is performed (second-order terms are neglected):

$$\delta \dot{P}(t) = -T_2^{-1}\delta P + \mu \hbar^{-1}E_\infty \delta D + \mu \hbar^{-1}D_\infty \delta E$$
$$\delta \dot{D}(t) = -\mu \hbar^{-1}E_\infty \delta P - \mu \hbar^{-1}P_\infty \delta E - T_1^{-1}\delta D \quad (3.31)$$
$$\delta \dot{E}(t) = -g\delta P - K\delta E$$

whose solution is

$$\begin{bmatrix} \delta P(t) \\ \delta D(t) \\ \delta E(t) \end{bmatrix} = \exp(\lambda t) \begin{bmatrix} \delta P(0) \\ \delta D(0) \\ \delta E(0) \end{bmatrix} \quad (3.32)$$

where $\delta P(0)$, $\delta D(0)$, $\delta E(0)$ are the initial perturbations. The condition of stability in (3.32) is that $\text{Re}\lambda < 0$. Bonifacio and Lugiato (1978a) showed that $\text{Re}\lambda < 0$ whenever $dy/dx > 0$. Thus, the middle branch of figure 9 is unstable and we have *bistability*.

Bonifacio and Lugiato (1978b) have investigated the conditions under which the mean field approximation is valid by solving exactly the Maxwell–Bloch equations for a ring cavity (figure 10) which read

$$\frac{\partial P_F}{\partial t} = -(T_2^{-1} - i\Delta)P_F + i\mu \hbar^{-1}E_F D_0$$

$$\frac{\partial D_0}{\partial t} = -T_1^{-1}(D_0 + 1) + i\mu \hbar^{-1}(E_F^* P_F - E_F P_F^*) \quad (3.33)$$

$$\frac{1}{c}\frac{\partial E_F}{\partial t} + \frac{\partial E_F}{\partial z} = igP_F^*$$

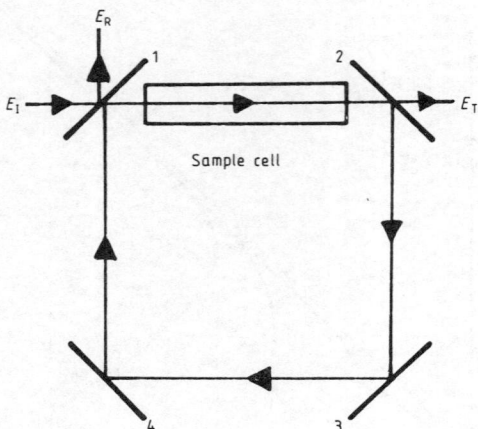

Figure 10. Ring cavity configuration.

showing that only the forward mode is assumed to exist. The boundary conditions are

$$E_F(0, t) = (1-R)^{1/2} E_I(t) + RE(L, t-t_0) \exp(i\theta)$$
$$E_T(t) = (1-R)^{1/2} E(L, t) \tag{3.34}$$

where $\theta = (\omega - \omega_c)\tilde{L}/c$, $\tilde{L} = 2(L+l)$ and $t_0 = (2l+L)/c$. In the steady state and for the absorptive case Bonifacio and Lugiato found a state equation in implicit form that reduced to (3.29) in the double limit $\alpha L \to 0$, $(1-R) \to 0$ with $C = $ constant. Physically this means that low atomic absorption ensures slight variation of fields inside the cavity; low mirror transmissivity makes (3.27) plausible. In other words, the spatial average is effectively an average over constants.

The first treatment of dispersive OB in a ring cavity was given independently by Bonifacio and Lugiato (1978c) and Hassan et al (1978); both groups of authors include inhomogeneous broadening with the effects described above. In both papers the problem is formulated quantum-mechanically but finally the mean field approximation is made with the following state equation (homogeneous broadening):

$$|y|^2 = |x|^2 \left[\left(1 + \frac{2C}{1+\tilde{\Delta}^2+|x|^2}\right)^2 + \left(\tilde{\theta} - \frac{2C\tilde{\Delta}}{1+\tilde{\Delta}^2+|x|^2}\right)^2 \right] \tag{3.35}$$

in which $\tilde{\theta} = \theta/(1-R)$. The effect of non-linear absorption is contained in the first brackets whereas non-linear dispersion is in the second. In the dispersive case the mean field approximation is valid if a further condition is added, namely $\theta \to 0$ with $\theta/(1-R) = $ constant. This means that as the cavity linewidth approaches zero, θ should do the same, otherwise there is zero transmission.

Abraham et al (1980) found the same state equation (3.35) for a Fabry–Perot cavity. In figure 11 we show the comparison between the exact numerical calculation and the mean field; in the right limits the agreement is excellent.

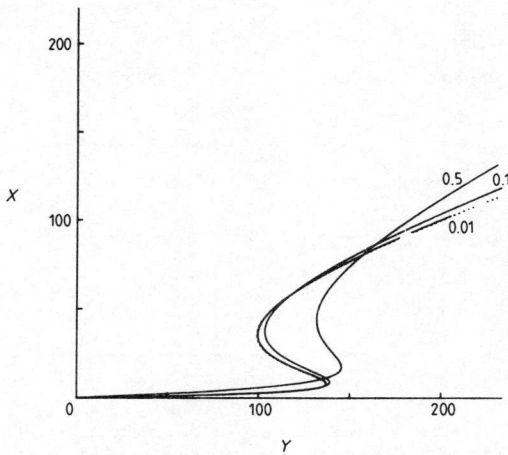

Figure 11. Comparison with the mean field equation (3.35) (broken curve) and numerical results (full curves) for $C = 14$, $\tilde{\Delta} = 3$ and different (R, θ): $(0.99; 0.0045\pi)$, $(0.9; 0.045\pi)$ and $(0.5; 0.225\pi)$. Note that $\tilde{\theta} \equiv \theta(1-R)^{-1}$ is the same in all cases and that the curve for $R = 0.99$ appears superimposed on the mean field curve (from Abraham et al 1980).

The conditions for a bistable region (Bonifacio et al 1979a) in the dispersive case ($\sigma = 0$) are

$$2C > \tilde{\Delta}\tilde{\theta} - 1 \qquad (3.36(a))$$

$$(2C - \tilde{\Delta}\tilde{\theta} + 1)^2(C + 4\tilde{\Delta}\tilde{\theta} - 4) > 27C(\tilde{\Delta} + \tilde{\theta})^2 \qquad (3.36(b))$$

where again $C > 4$ is the condition for bistability.

In the spatial average of the mean field approximation standing waves do not come into play. This is certainly the case when the atoms diffuse a distance $\lambda/4$ in a time much shorter than T_1. McCall and Gibbs (1980) showed that standing waves, even when $\alpha L \to 0$, $T \to 0$, make the condition for bistability somewhat more restrictive in that $C > 4.96$. The agreement between numerical calculations of Abraham et al (1980) and the mean field lies in the truncation adopted. Agrawal and Carmichael (1980) derived a different state equation—in the mean field limit—taking standing waves fully into account.

3.4. Time-dependent phenomena

One of the aims with OB devices is to make them part of an optical circuit where pulses or fluctuations may give rise to a variety of time-dependent phenomena. Indeed these phenomena *per se* are very interesting. The first study of transient phenomena with the Maxwell–Bloch system was done by Bonifacio and Meystre (1978) who studied the response of an OB device to stepwise excitation in the mean field approximation and also taking propagation effects into account. Two extreme cases in absorptive OB were considered: $T_1, T_2 \gg t_c$ ('bad quality' cavity) and $T_1, T_2 \ll t_c$ ('good quality' cavity). Their simulations were confirmed by Abraham et al (1979) who also extended their study of the full Maxwell–Bloch equations to generation of hysteresis cycles with pulses. In figure 12 the response to stepwise excitation (Abraham 1979) is shown.

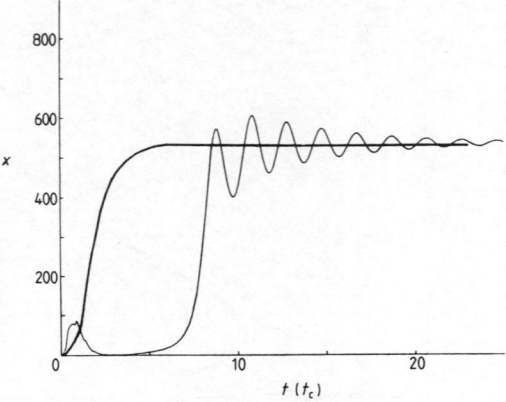

Figure 12. Plot of $|x(t)|^2$ when $y(t)$ is a step function of amplitude 30 units switched in at $t = 0$ for $\tilde{\Delta} = \sigma = \theta = 0$, $T_1 = 2T_2$, $C = 20$, $R = 0.9$. The smooth curve corresponds to $t_c = 10T_1$ ('good cavity' limit) whereas the oscillatory (frequency ∝ Rabi frequency) one corresponds to $t_c = 0.1T_1$ ('bad cavity' limit). Time is measured in units of t_c (from Abraham and Bullough 1980).

In the $T_1, T_2 \gg t_c$ limit we see the appearance of oscillations once $|x|^2$ 'jumps' to the upper branch. Bonifacio and Lugiato (1976, 1978a) showed that the eigenvalues (3.32) in the upper branch were complex with $\text{Re}\lambda < 0$ and $\text{Im}\lambda = \Omega_I = (1-R)^{-1/2}\mu\hbar^{-1}E_I$, where the last quantity is proportional to the Rabi frequency and has the meaning of minimum time for atomic inversion. In this case atomic dynamics is predominant. When $T_1, T_2 \ll t_c$, the cavity dominates and we have essentially an exponential type of response. Numerical simulations show (Abraham 1979) that Rabi oscillations persist in the dispersive case and also when inhomogeneous broadening is present (Abraham and Hassan 1980). A quasi-steady-state response to pulses for generation of hysteresis cycles is possible if $\tau_p > t_c > T_1, T_2$ where τ_p is the pulse duration (figure 13). Another interesting time-dependent simulation with the Maxwell–Bloch system is shown in figure 14 where we see 'transistor' action.

The change of sign of (3.32) when we change from a stable to an unstable branch means that $\lambda \to 0$ as we approach the thresholds y_m and y_M. As $\lambda \to 0$ any fluctuation is damped more and more slowly: this is called *critical slowing down*. Bonifacio and Meystre (1979) found this response numerically in a mean field context; their results were confirmed analytically by Benza and Lugiato (1979a, b).

When propagation effects are taken into account the stability analysis can be different. With the right choice of parameters positive slope regions may become unstable. This was first pointed out by McCall (1978) who predicted and observed it with a hybrid device. Bonifacio and Lugiato (1978d) analysed the Maxwell–Bloch equations (3.33) for a homogeneously broadened system and $\Delta = \varphi = 0$. After linearising the equations around the steady state and integrating over space, a general solution of the form

$$\delta E(z, t) = \sum_{n=-\infty}^{\infty} \epsilon_n(z, t) \exp(\lambda_n t) + \text{cc} \quad (3.37)$$

with

$$\lambda_n = -i\Delta\omega_n - \text{loss} + \text{gain}$$

$$\Delta\omega_n = 2\pi nc/\tilde{L}$$

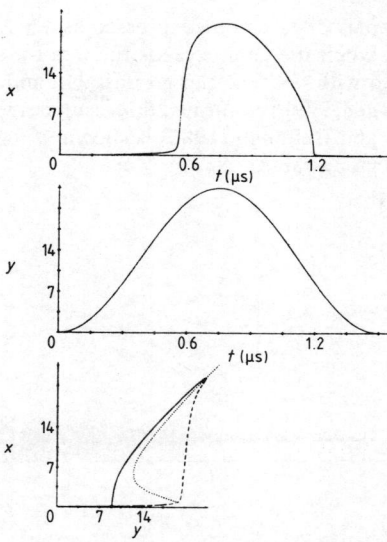

Figure 13. Dynamic hysteresis cycle. Plot of $x(t)$ and $x(y)$ for a pulsed input field $y(t)$ for $R = 0.9$, $C = 20$, $\tilde{\Delta} = 0$, $\theta = 0$. In the $x(y)$ plots the dotted curve is the corresponding steady-state characteristic curve (output against input); the broken curve corresponds to y increasing and the full curve for y decreasing (from Abraham et al 1979).

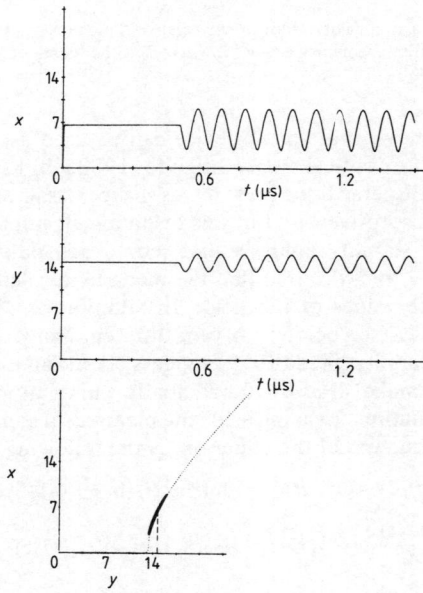

Figure 14. Transistor action for a modulated CW signal $y(t)$ (from Abraham et al 1979).

is obtained. As the real part of λ_n can be expressed as a loss term minus a gain term (Bonifacio et al 1979a), when the gain exceeds the loss for some $n \neq 0$, states in the high transmission branch with $x < C/2$ can go unstable and a new mode (or modes) are excited. Bonifacio et al (1979b) confirmed this result numerically. An independent simulation by Abraham and Bullough (1980) is shown in figure 15 where the modes $n = \pm 1$ were allowed to go unstable.

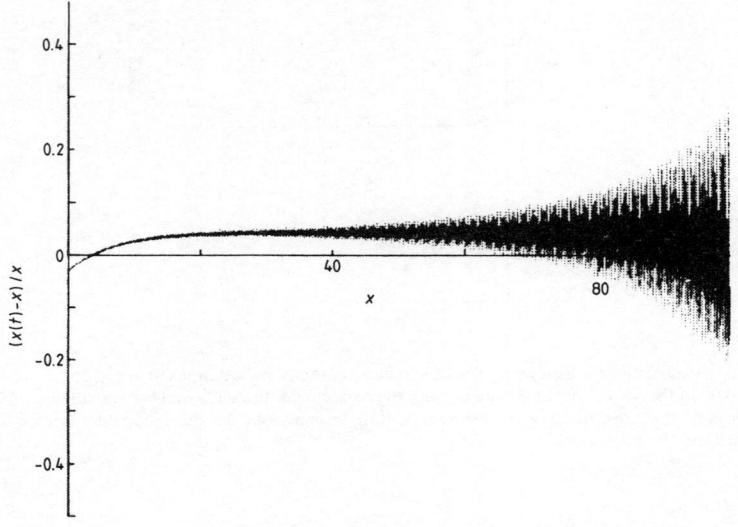

Figure 15. Self-pulsing in a ring cavity. Plot of the relative deviation $(x(t)-x)/x$ from the steady-state output field x. The modulation frequency is $\sim c/5L$ where L is the length of the sample in figure 10 (from Abraham and Bullough 1980).

This *self-pulsing* shows that an OB device can be used as a converter of DC light into a pulsating output or optical clock. Lugiato (1980a, b) has analytically predicted this behaviour in a dispersive ring cavity as have Casagrande et al (1980) in a Fabry–Perot cavity (absorptive OB), but this result awaits numerical confirmation.

The above linear stability analysis that led to self-pulsing was based on the assumption that $\alpha L \ll 1$ and $T \ll 1$ so that the mean field calculation can be appealed to for the steady-state values of the field. In addition, $\Delta = 0$ and no variable was eliminated *adiabatically* (see below). A very different analysis emerges when $\alpha L \gg 1$ and $\tilde{\Delta} \gg 1$ since multistability (Ikeda 1979) appears. Ikeda analysed (3.33) and assumed $T_1 \gg T_2$ in order to make $\partial P_F/\partial t = 0$ (adiabatic elimination of P_F) and formally integrated the field equation. Leaving aside the meaning of some terms and normalisation constants†, the structure of the equations is the following:

$$E_F(0, t) = (1 - R)^{1/2} E_I(t) + R E_F(0, t - t_R) \psi_1(D(t)) \tag{3.38}$$

$$\frac{dD(t)}{dt} = -\gamma_1 D(t) - |E_F(0, t - t_R)|^2 \psi_2(D(t)) \tag{3.39}$$

† We are only interested in the stability analysis; for details see Ikeda (1979); $t_R = 2(L + l)/c$ is the round-trip time.

where ψ_1 and ψ_2 are functions of $D(t)$. In the steady state the combination of (3.38) and (3.39) shows multistability (figure 16(a)): the stability analysis of these shows a new striking behaviour.

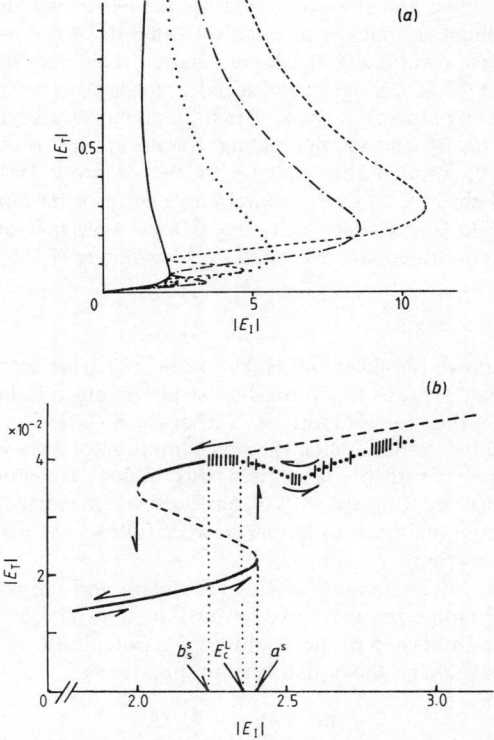

Figure 16. (a) Relation between the transmitted and incident fields for $C = 40$, $R = 0.95$, $\theta = 0$ and different $\tilde{\Delta}$ for a ring cavity configuration; from left to right $\tilde{\Delta} = 0$, $\pi/2$, π and 1.5π respectively. Note the multistable regions. (b) Linear stability analysis of (a) shows stable (full curve), chaotic (dotted curve) and periodic (vertical dashes) regions demarcated by b_s^s, E_c^L and a^s (from Ikeda 1979).

Suppose that E_I = constant and $t_R \gg T_1$ so that $D(t)$ in (3.39) can be adiabatically eliminated and substituted in (3.38). Equation (3.38) takes the functional form

$$E_F(0, t) = F(E_F(0, t - t_R); E_I) \tag{3.40}$$

which shows that the field at time t depends on the field at an earlier time $t - t_R$. This is equivalent to a two-dimensional—the field is complex—vectorial difference equation:

$$\bar{V}_{n+1} = \bar{F}(\bar{V}_n, a) \tag{3.41}$$

where the associations $t \equiv n + 1$ and $t - t_R \equiv n$ are made. This equation is linearised in the same way as, for example, (3.32) around the steady state but stability demands $|\lambda| < 1$ as opposed to Re $\lambda < 0$. If $|\lambda| < 1$ successive iterates of an initial perturbation are attenuated after each iteration. This stability analysis shows (figure 16(b)) for the first bistable region that the lower branch appears stable in the usual sense; the middle branch is unstable also in the usual sense. It is the upper branch that shows a new type of instability. The first segment is stable up to $|E_I| = b_s^s$ after which there is a bifurcation and the system begins to oscillate with a period $2t_R$; as $|E_I|$ is increased new bifurcations appear and at each bifurcation the period is doubled as $4t_R$, $8t_R$, etc. This period doubling continues until a critical value $|E_I| = E_c^L$ is reached after which the motion becomes completely aperiodic (chaos). This period doubling leading to chaos appears in physical, biological and social sciences (May 1976, Rabinovich 1978). One of the interesting aspects of chaos is that it appears as a stochastic process—with a broadband spectrum—and yet fluctuating 'forces' are not present in the equation. Ikeda et al (1980) applied this analysis to the Maxwell–Debye equations. The extension to the Fabry–Perot cavity containing a third-order non-linearity was done by Firth (1981) who found analytically these Ikeda-type instabilities. These results were numerically confirmed and extended by Abraham et al (1982).

3.5. Phase transition analogy

It has long been known (Graham and Haken 1968, 1970, DeGiorgio and Scully 1970) that laser action is analogous to a *second-order* temperature-induced phase transition in a ferromagnet in thermal equilibrium. Although the laser is a system in a far from thermal equilibrium situation, when the inversion is associated with the temperature and the magnetisation with the laser field, this analogy is firmly established. More recently, Bonifacio and Lugiato (1976) have shown that OB is a *first-order* phase transition in a non-equilibrium situation: in what follows we shall try to explain how this analogy comes about.

As much of our understanding of this subsection and the next one relies on the Fokker–Planck equation, we shall give an introduction (Haken 1978) explaining its meaning. Suppose we have a particle subject to a potential $U(x)$ and a friction force proportional to its velocity, the equation of motion reads

$$m\ddot{x} + \gamma \dot{x} = -\partial U/\partial x. \tag{3.42}$$

If the particle is overdamped ($\gamma \gg m$) then

$$dx/dt = -\partial U/\partial x \equiv f(x). \tag{3.43}$$

The solution of the deterministic equation (3.43) is a *unique* trajectory $x(t)$ for given initial conditions. This amounts to saying that the probability of finding the particle at the position $x(t)$ at time t is unity. This corresponds to a probability distribution given by a δ function:

$$\mathcal{P}(x, t) = \delta(x - x(t)) \tag{3.44}$$

which in a $\mathcal{P}$-x-t coordinate space would look like figure 17. When a fluctuating force $F(t)$ is added to (3.43) as in the case of the Brownian motion, then

$$dx/dt = f(x) + F(t) \tag{3.45}$$

and we can no longer speak of a definite trajectory. As $F(t)$ is stochastic we have to think of the probability dp of finding the particle between x and $x + dx$ at time t:

$$dp = \mathcal{P}(x, t) \, dx \tag{3.46}$$

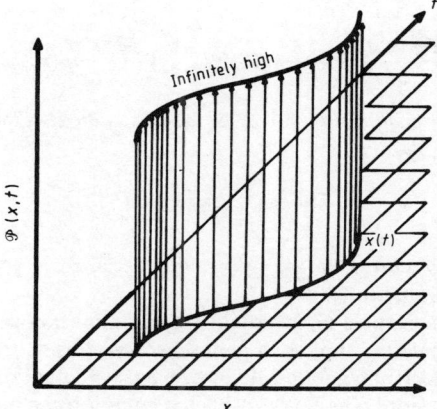

Figure 17. Infinitely peaked 'probability' distribution corresponding to $\mathcal{P}(x) = \delta(x - x(t))$; no random forces present (from Haken 1978).

where $\mathcal{P}(x, t)$ will now have a finite width. Moreover, we are interested in having an equation of motion for $\mathcal{P}(x, t)$ in order to determine its time evolution which qualitatively we expect to be as in figure 18; each curve is contained in a plane parallel to the $\mathcal{P}$-x plane. The random force $F(t)$ is assumed to have the following properties:

$$\langle F(t) \rangle = 0 \qquad (3.47)$$

$$\langle F(t)F(t') \rangle = 2\mathcal{D}\delta(t - t') \qquad (3.48)$$

where $\langle \rangle$ denotes statistical average; equation (3.48) expresses that forces at different times are uncorrelated. The equation of motion for $\mathcal{P}(x, t)$ is the *Fokker–Planck*

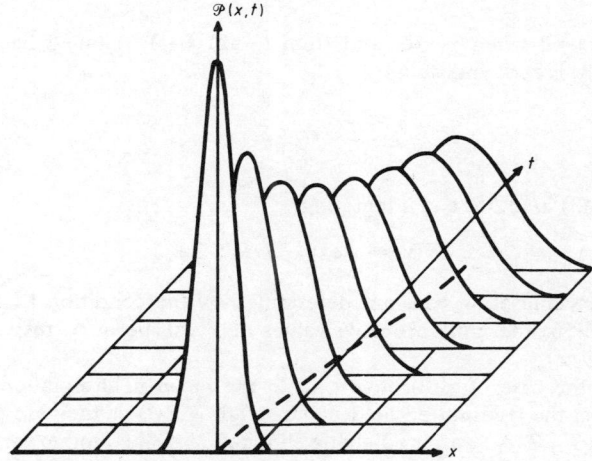

Figure 18. Probability distribution $\mathcal{P}(x, t)$ when random forces are present; note the finite width and height. The broken curve corresponds to the most probable path (from Haken 1978).

equation:

$$\frac{\partial \mathcal{P}(x,t)}{\partial t} = -\frac{\partial}{\partial x}[f(x)\mathcal{P}(x,t)] + D\frac{\partial^2 \mathcal{P}}{\partial x^2}(x,t) \tag{3.49}$$

where $f(x)$ is the *drift coefficient* and $\mathcal{D}$ is the *diffusion coefficient*. The mean path of the particle is

$$\bar{x}(t) = \int_{-\infty}^{\infty} \mathrm{d}x\, x\mathcal{P}(x,t) \tag{3.50}$$

and it is easy to show that $\mathrm{d}\bar{x}(t)/\mathrm{d}t = \overline{f(x)}$; the variance is equal to $\overline{x^2} - \bar{x}^2$ of the distribution is proportional to $\mathcal{D}$.

Our discussion of a particle undergoing Brownian motion and its connection with the Fokker–Planck equation applies in general to systems subject to random forces. One such case is the electromagnetic field in which the fluctuations come, for example, from spontaneous emission. The Fokker–Planck equation can be recast as a continuity equation:

$$\frac{\partial \mathcal{P}(x,t)}{\partial t} + \frac{\partial}{\partial x} j(x,t) = 0 \tag{3.51}$$

$$j = f(x)\mathcal{P}(x,t) - D\frac{\partial \mathcal{P}}{\partial x}(x,t) \tag{3.52}$$

with j being probability current. As we are interested in stationary solutions, $\partial \mathcal{P}/\partial t = 0$, and from (3.51)

$$\frac{\partial}{\partial x} j = 0. \tag{3.53}$$

Therefore

$$j = \text{constant}. \tag{3.54}$$

We expect $\mathcal{P}(x) \to 0$ when $x \to \infty$, and from (3.52) $j \to 0$ (detailed balance) so the constant of (3.54) is zero and we have

$$D\frac{\partial \mathcal{P}}{\partial x} = f(x)\mathcal{P}(x). \tag{3.55}$$

Substituting (3.43) in (3.55) and integrating:

$$\mathcal{P}(x) = \mathcal{N} \exp(-U(x)/\mathcal{D}) \tag{3.56}$$

where $\mathcal{N}$ is a normalisation constant determined by the condition $\int_{-\infty}^{\infty} \mathcal{P}(x)\,\mathrm{d}x = 1$. According to (3.56) the most probable values of x will be in correspondence with the minima of $U(x)$.

We turn to the case of absorptive OB. In the mean field equations (3.28) it is possible to insert the transmitted field $x(t)$ instead of $E(t)$ as they are proportional. Then, assuming $T_1, T_2 \ll t_c$, after adiabatic elimination of the atomic variables we get

$$t_c \frac{\mathrm{d}x}{\mathrm{d}t} = y - x - \frac{2Cx}{1+x^2}. \tag{3.57}$$

If a potential $U(x)$ is defined as

$$U(x) = -yx + \frac{x^2}{2} + C \ln(1+x^2) + \frac{y^2}{2} \quad (3.58)$$

then (3.57) can be written down as

$$t_c \frac{dx}{dt} = -\frac{\partial U}{\partial x}. \quad (3.59)$$

Adding to (3.58) a δ-correlated fluctuating force $F(t)$ as in (3.45) we can obtain the Fokker–Planck equation of this problem with a stationary solution identical to (3.56). Although this approach is somewhat coarse-grained† it is conceptually illustrative.

In the case of a ferromagnet the probability density for the fluctuations of the magnetisation M is

$$\mathcal{P}(M) = \mathcal{N}_0 \exp(-\mathcal{F}(M)/k_B T) \quad (3.60)$$

where $\mathcal{F}(M)$ is the free energy and $\mathcal{N}_0$ is a normalisation constant. The striking similarity between (3.60) and (3.56) makes us, as in the laser (DeGiorgio and Scully 1970), tentatively identify (3.58) as the free energy of our problem. The minima of (3.58) are obtained from the condition

$$\frac{\partial U}{\partial x} = 0 = -y + x + \frac{2Cx}{1+x^2} \quad (3.61)$$

and (3.61) is the equation of state (3.29); in the case of a ferromagnet the minima of $F(M)$ correspond to the equation of state (thermodynamic equilibrium).

In figure 19(a) we plot (3.58) for $C = 20$ and different values of y. The values of x at which $U(x)$ is minimum are the order parameters η. When $y < y_m$—compare with figure 19(b)—there is only one minimum, when $y_m < y < y_M$ two minima, and when $y > y_m$ one minimum again. By associating y with pressure in a fluid (Agarwal *et al* 1978a, b) and C with temperature, we find a region where two order parameters η'_1 and η'_2—that correspond to two different specific volumes—'coexist' as in the case of the liquid and vapour phases (which is a first-order phase transition). In general, when $U(x)$ has more than one minimum we have a first-order phase transition.

3.6. Quantum-mechanical approach

The Maxwell–Bloch description discussed in the previous subsections has been quite adequate for a considerable number of phenomena in OB. Yet by its semiclassical nature it fails to take into account the effect of fluctuations which can change the nature of the phase transition (Willis 1977) and also have practical implications in miniaturised systems. On the other hand, fluctuations bear a direct relationship with the spectrum of both the transmitted and fluorescent light. A semiclassical approach is unable to distinguish which of the stable branches that emerge from a linear stability analysis is the metastable one and which the truly stable one. This difficulty is particularly acute in regions of the input field where the system may develop large fluctuations and therefore tunnel from a metastable state (relative minimum of $U(x)$ in figure 19(a)) to the stable one (absolute minimum of $U(x)$ in figure 19(a)). It is

---

† Only a fully quantum-mechanical approach leads to the correct Fokker–Planck equation (Bonifacio *et al* 1978, Willis 1978).

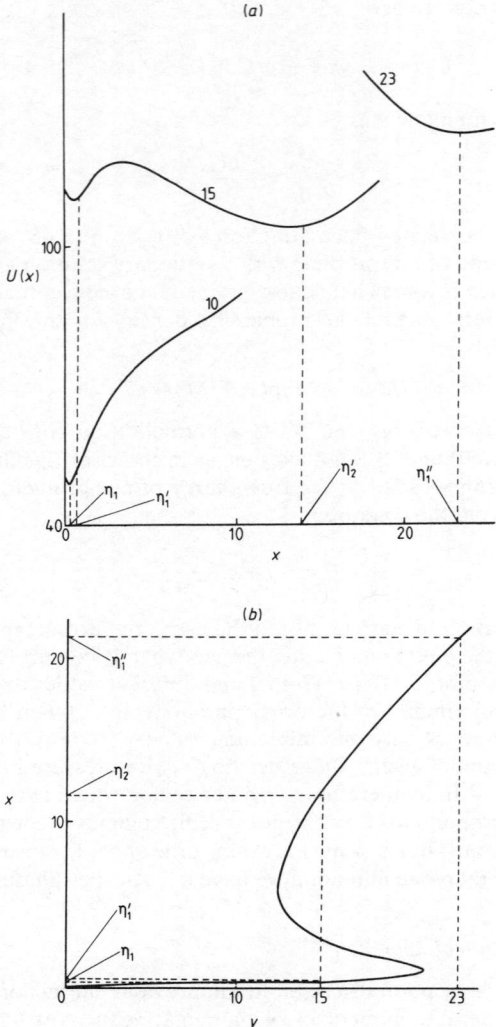

Figure 19. (a) Plot of $U(x)$—equation (3.58)—for $C = 20$ and $y = 10, 15$ and 23, respectively. The values of x that minimise $U(x)$ are the order parameters η. (b) Plot of the equation of state (3.29) that derives from $U(x)$.

clear then that a quantum-mechanical model is necessary. A rigorous treatment of the problem would require the use of a Hamiltonian that takes account of the finite size of the system and therefore propagation effects. This multimode approach is even difficult at a semiclassical level and a full operator theory is, at the moment, intractable. One way to approach this problem which various authors have independently adopted is the master equation approach (see, for example, Louisell 1973). A master equation is an equation for the density operator and consequently is a quantum-statistical description of the interaction of matter and radiation. As Willis (1977) pointed out, which master equation is adopted depends very much on the experimental conditions. Here we shall outline two extreme cases: (a) the field variables can be adiabatically eliminated (T_1, $T_2 \gg t_c$) and (b) the atomic variables can be adiabatically eliminated (T_1, $T_2 \ll t_c$). In both cases $\Delta = \theta = 0$ but we shall also mention other approaches that take account of dispersion.

As a general starting point—valid for both cases—we write down the master equation for the one-mode laser model for a ring cavity (Weidlich and Haake 1965, Scully and Lamb 1967, Haken 1970) without a pump term and including an injected signal. As previously, a system of N two-level atoms contained in a resonant cavity of length L and volume V is considered. The dynamics of each individual atom constitutes a pseudospin described by Pauli spin operators that satisfy angular momentum commutation relations:

$$[\hat{\sigma}^+, \hat{\sigma}^-] = 2\hat{\sigma}_z \tag{3.62}$$

$$[\hat{\sigma}_z, \hat{\sigma}^\pm] = \pm \hat{\sigma}^\pm \tag{3.63}$$

where $\hat{\sigma}^+ = \hat{\sigma}_x + \mathrm{i}\hat{\sigma}_y$ is the raising operator ($\sigma^+|1\rangle = |2\rangle$, cf figure 6), $\hat{\sigma}^-$ is the lowering operator ($\hat{\sigma}^-|2\rangle = |1\rangle$) and, in matrix representation,

$$\hat{\sigma}_x = \begin{bmatrix} 0 & 1 \\ 1 & 0 \end{bmatrix} \qquad \hat{\sigma}_y = \begin{bmatrix} 0 & \mathrm{i} \\ -\mathrm{i} & 0 \end{bmatrix} \qquad \hat{\sigma}_z = \begin{bmatrix} -1 & 0 \\ 0 & 1 \end{bmatrix} \tag{3.64}$$

in which $\hat{\sigma}_z$ represents the atomic inversion. The collection of N atoms is described by the total population inversion operator:

$$\hat{R}_3 = \sum_{j=1}^{N} \hat{\sigma}_{z,j} \tag{3.65}$$

and by the collective dipole operators:

$$\hat{R}^\pm = \sum_{j=1}^{N} \hat{\sigma}_j^\pm \exp(\pm \mathrm{i} \mathbf{k} \cdot \mathbf{x}_j) \tag{3.66}$$

where $\mathbf{k}$ is the wavevector of the incident radiation of amplitude E_I and angular frequency ω and $\mathbf{x}_j$ is the position of the jth atom. The $\hat{R}$ of (3.65) and (3.66) also satisfy angular commutation relations. The internal field that develops inside the cavity is described by photon creation ($\hat{a}^+$) and destruction ($\hat{a}$) operators that satisfy boson commutation relations:

$$[\hat{a}, \hat{a}^+] = 1 \tag{3.67}$$

$$[\hat{a}, \hat{a}] = [\hat{a}^+, \hat{a}^+] = 0. \tag{3.68}$$

The density operator $\hat{\rho}_{AF}$ for the coupled system of atoms plus internal field obeys the master equation

$$\frac{\mathrm{d}}{\mathrm{d}t}\hat{\rho}_{AF} = -\mathrm{i}\hat{L}_{AF}\hat{\rho}_{AF} + \hat{\Lambda}_F \hat{\rho}_{AF} + \hat{\Lambda}_A \hat{\rho}_{AF} \tag{3.69}$$

which is in the interaction picture—as all the operators—in order to eliminate high-frequency terms, and where

$$\begin{aligned} \hat{L}_{AF}\hat{\rho}_{AF} &= \hbar^{-1}[\hat{H}_{AF}, \hat{\rho}_{AF}] \\ \hat{H}_{AF} &= \mathrm{i}g(\hat{a}^+\hat{R}^- - \hat{a}\hat{R}^+) \end{aligned} \tag{3.70}$$

$$\hat{\Lambda}_F \hat{\rho}_{AF} = K\{[(\hat{a}-\alpha), \hat{\rho}_{AF}(\hat{a}^+ - \alpha^*)] + [(\hat{a}-\alpha)\hat{\rho}_{AF}, (\hat{a}^+ - \alpha^*)]\} \quad (3.71)$$

$$\hat{\Lambda}_A \hat{\rho}_{AF} = \sum_{j=1}^{N} \left(\frac{1}{2T_1}[\hat{\sigma}_j^-, \hat{\rho}_{AF}\hat{\sigma}_j^+] + [\hat{\sigma}_j^- \hat{\rho}_{AF}, \hat{\sigma}_j^+]\right)$$

$$+ \left(\frac{1}{T_2} - \frac{1}{2T_1}\right)\{[\hat{\sigma}_{z,j}, \hat{\rho}_{AF}\hat{\sigma}_{z,j}] + [\hat{\sigma}_{z,j}\hat{\rho}_{AF}, \hat{\sigma}_{z,j}]\}. \quad (3.72)$$

In (3.70) $g = (2\pi\hbar^{-1}\omega V^{-1}\mu^2)^{1/2}$ is the coupling constant, $K = cT/L$ and $\alpha = (V/8\pi\hbar\omega)^{1/2}E_I(1-R)^{-1/2} > 0$. The internal field is given by

$$E = (8\pi\hbar\omega/V)^{1/2}\langle\hat{a}\rangle \quad (3.73)$$

and the transmitted field

$$E_T = (1-R)^{1/2}E. \quad (3.74)$$

The term (3.70) accounts for the interaction between atoms and field; (3.71) accounts for the dynamics of the field in the cavity and includes the incident field, and (3.72) arises from atomic decay. If we want to find equations of motion for the expectation values of operators, we follow the prescription

$$\frac{d}{dt}\langle\hat{A}\rangle = \text{Tr}_{AF}\hat{A}\frac{d\hat{\rho}_{AF}}{dt} \quad \text{etc} \quad (3.75)$$

$$\langle\hat{A}\rangle = \text{Tr}_{AF}\hat{A}\hat{\rho}_{AF} \quad (3.76)$$

where Tr_{AF} is the trace over both atomic and field operators. In the semiclassical approximation (i.e. $\langle\hat{A}\hat{B}\rangle = \langle\hat{A}\rangle\langle\hat{B}\rangle$) we regain the mean field equations (3.28). For matter operators we use $\hat{\rho}_A = \text{Tr}_F\hat{\rho}_{AF}$ and for the field $\hat{\rho}_F = \text{Tr}_A\hat{\rho}_{AF}$.

When the field variables are eliminated adiabatically ($K \gg T_1^{-1}, T_2^{-1}$) we obtain the following reduced master equation (Narducci *et al* 1978a, b, Agarwal *et al* 1978a, b)

$$\frac{d\hat{\rho}_A}{dt} = -i\Omega_I[\hat{R}^+ + \hat{R}^-, \hat{\rho}_A] + \hat{\Lambda}_R\hat{\rho}_A + \hat{\Lambda}_A\hat{\rho}_A \quad (3.77)$$

where Ω_I is the Rabi frequency of the input field, $\hat{\Lambda}_A\hat{\rho}_A$ is given by (3.72), with $T_2 = 2T_1$ for simplicity, and

$$\hat{\Lambda}_R\hat{\rho}_A = \frac{2g^2}{K}(\hat{R}^-\hat{\rho}_A\hat{R}^+ - \tfrac{1}{2}\hat{\rho}_A\hat{R}^+\hat{R}^- - \tfrac{1}{2}\hat{R}^+\hat{R}^-\hat{\rho}_A). \quad (3.78)$$

The first term on the RHS of (3.77) represents the coherent interaction between atoms and the incident field, the second represents the collective atomic decay process into the resonant cavity mode, whereas the last one stands for incoherent atomic relaxation: it is the competition between these three processes that accounts for bistable behaviour. It is possible to show (Narducci *et al* 1978b, Walls *et al* 1978) that by dropping the third term in (3.77) a different behaviour is obtained which can be interpreted as a second-order phase transition: the transmitted light changes continuously with the input and there is no hysteresis. With all the terms kept the mean field state equation can be derived and Rabi oscillations and critical slowing down show up in the numerical simulations (Narducci *et al* 1978a).

The calculation of the spectrum of the transmitted light given in the steady state by

$$I(\omega) \propto \frac{1}{\pi}\text{Re}\int_0^\infty dt \exp[-i(\omega-\omega_0)t]\langle\hat{S}(t)\hat{S}(0)\rangle \quad (3.79)$$

shows a single peak that narrows as y approaches the upper threshold y_M from the origin; when $y > y_M$ the spectrum splits into three peaks (dynamical Stark effect) and becomes more resolved as y is increased. When y is decreased the three peaks continuously merge into a single peak and when $y < y_m$ there is a sudden broadening (figure 20). This hysteresis of the spectrum was originally calculated by Bonifacio and Lugiato (1976) and showed that the lower branch is *cooperative* whereas the upper branch shows *single-atom* behaviour.

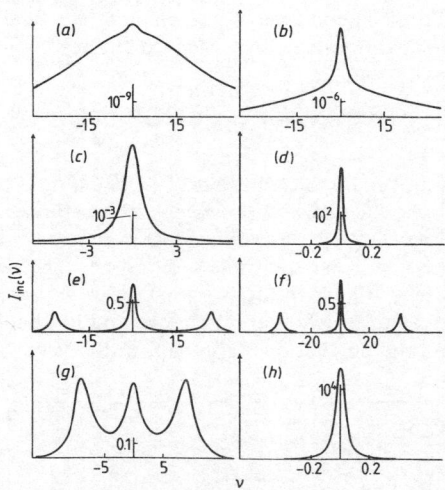

Figure 20. Hysteresis cycle of the incoherent part of the spectrum $I_{inc}(\nu)$ of transmitted field in arbitrary units; $\nu = (\omega - \omega_0)T_2$, $T_2 = 2T_1 \gg t_c$, $C = 20$ and different values of y: (a) 4.1, (b) 14.2, (c) 20.6, (d) 21.0, (e) 22.0, (f) 31.3, (g) 13.4 and (h) 12.5. In this case $y_m \sim 12.6$ and $y_M \sim 21.0$. Note the line narrowing from (a)–(d) in the lower branch and the sudden appearance of the dynamical Stark effect when a 'jump' to the upper branch occurs at (e). Further increase in y resolves the spectrum more whereas a decrease causes the peaks to merge (from Lugiato 1979).

In the limit $K \ll T_1^{-1}$, T_1^{-1} a new reduced master equation is obtained (Bonifacio et al 1978, Willis 1978); we follow Bonifacio et al:

$$\frac{d\hat{\rho}_F}{dt} = \hat{\Lambda}_F \hat{\rho}_F + g^2 N T_2 \{[\hat{a}^+, \hat{w}\hat{a}] + [\hat{a}, (\hat{\rho}_F - \hat{w})\hat{a}^+] + \text{HC}\} \quad (3.80)$$

$$\frac{d\hat{w}}{dt} = -T_1^{-1}\hat{w} + g^2 T_2[\hat{a}(\hat{\rho}_F - \hat{w})\hat{a}^+ - \hat{a}\hat{a}^+\hat{w} + \text{HC}] \quad (3.81)$$

where we set $d\hat{w}/dt = 0$ as it is eliminated adiabatically, being $\hat{w} = \text{Tr}_A(\hat{R}^+\hat{R}^-\hat{\rho}_{AF})$. For the derivations of (3.80) and (3.81) the relations $\text{Tr}_A(\hat{R}^{\pm}\hat{\Lambda}_A\hat{\rho}_{AF}) = -T_2^{-1}\text{Tr}_A(\hat{\rho}_{AF})$ and $\text{Tr}_A(\hat{R}^+\hat{R}^-\hat{\Lambda}_A\hat{\rho}_{AF}) = -T_1^{-1}\text{Tr}_A(\hat{\rho}_{AF})$ were used. In addition, atom–atom correlations that involve more than one atom are neglected, namely $\text{Tr}_A(\hat{\rho}_{AF}\hat{\sigma}_i^+\hat{\sigma}_j^-) = 0$ for $i \neq j$. At this stage $\hat{\rho}_F(t)$ is expressed in the Glauber representation $P(\beta, t)$†

$$\hat{\rho}_F(t) = \int d^2\beta |\beta\rangle\langle\beta| P(\beta, \beta^*, t) \quad (3.82)$$

† For a simple introduction to this representation and coherent states $|\beta\rangle$ see, for example, Sargent et al (1974, chap 15).

$$\hat{w}(t) = \int d^2\beta |\beta\rangle\langle\beta| w(\beta, \beta^*, t) \qquad (3.83)$$

where $d^2\beta = d(\mathrm{Re}\beta)d(\mathrm{Im}\beta)$. In (3.82) $P(\beta, \beta^*, t)$ represents the quasiprobability of finding the field with an amplitude β. Using polar coordinates, $\beta = r\exp(i\varphi)$, it is possible to find a Fokker–Planck equation for $P(x, \varphi, t)$ where $x = r/N_s^{1/2}$ is the transmitted field and $N_s = (T_1 T_2 4g^2)^{-1}$ is the saturation photon number.

Bonifacio et al show that when $g \equiv 2C/N_s \ll 1$ phase fluctuations are negligible and do not affect amplitude fluctuations so $\varphi \simeq 0$ and the final expression in the stationary case reads (see also Bonifacio and Lugiato 1978e)

$$\mathcal{P}(x) = \mathcal{N} \exp\left(-4N_s\right)\left(\frac{x^2}{2} + \ln x + 2C \ln(y-x)\right). \qquad (3.84)$$

The maxima of (3.84) are given by the solutions of (3.29). For $C > 4$ and $y_m < y < y_M$ (3.84) has two peaks (bimodal distribution) which characterise a first-order phase transition (cf § 3.5). A plot of $\langle x \rangle = \int x \mathcal{P}(x) dx$ against y coincides with the semiclassical solution everywhere except in a narrow transition region. The centre of this region corresponds to the value of y for which the two peaks have equal areas. At this point the system is 'undecided' as to which peak to choose and large fluctuations arise. In figure 21 is shown $\langle x \rangle$ and the fluctuations $\langle x^2 \rangle - \langle x \rangle^2$ in terms of y.

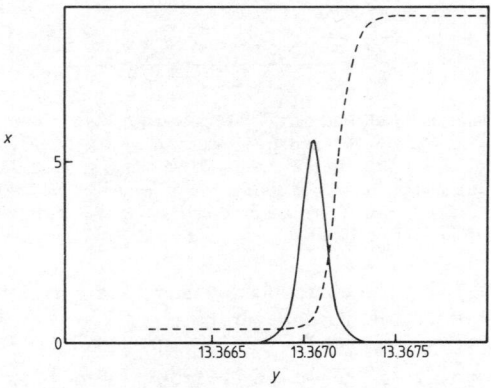

Figure 21. Mean value and relative fluctuation of the normalised transmitted field x. Note the large increase in fluctuations near the 'transition' point; N_s is the photon saturation number (from Bonifacio and Lugiato 1978a). ---, $\langle x \rangle$, ———, $(\langle x^2 \rangle - \langle x \rangle^2)/\langle x \rangle^2$. $N_s = 10^3$, $C = 20$.

Willis and Day (1979) have generalised their master equation to the dispersive case. They found that the diffusion coefficient of their Fokker–Planck equation is appreciably affected by the atomic detuning and reduces considerably the effect of fluctuations.

Different authors have treated the effect of fluctuations by different methods. Gragg et al (1979) gave a stochastic master equation description of absorptive OB and compared it with the microscopic theory discussed above. They adopted a macroscopic birth–death description (Haken 1978) to construct a master equation for the output intensity distribution, which, in the steady state, is bimodal, i.e. two peaks in the distribution in the region of optical bistability. The comparison of this technique for obtaining a macroscopic description of the fluctuations with that of Bonifacio *et al* (1978) shows considerable disagreement between them. The birth–death approach gives a larger standard deviation for the output intensity and the phase transition occurs at a different value of the incident field. The same technique was also applied by Bulsara *et al* (1978) to show that white noise can induce a phase transition in a bistable system in the region $C \le 4$. Good agreement is obtained with the macroscopic theory for the noise variance $\sigma^2 = 0.005$.

Brand and Schenzle (1978) obtained a soluble stochastic model for a first-order-type non-equilibrium transition. They study a non-symmetrical potential $U(x)$ for which the corresponding time-dependent Fokker–Planck equation is solved exactly. In this way they construct a soluble model for diffusion in a bistable potential. Schenzle and Brand (1978, 1979) have also studied fluctuations in OB in the limit $K \ll T_2^{-1}$, T_1^{-1}. They take the Langevin equations—the mean field equations plus fluctuating terms—and discuss separately the influence of these noise sources. They obtain an exact solution of the two-dimensional stationary Fokker–Planck equation that describes both amplitude and phase fluctuations in the case where the fluctuations of the incident field play a dominant role. They have also shown the significance of fluctuations for understanding the transient response of an OB device.

Hanggi *et al* (1980) have investigated the spectrum and dynamic-response function of the transmitted light in absorptive OB. They use a Fokker–Planck model with a non-linear diffusion coefficient and consider the 'good quality' cavity limit. The spectrum and response to a small additional injected coherent signal is studied while tunnelling between metastable and stable states takes place. Time-dependent fluctuations in absorptive OB both in the good- and bad-quality cavity limits have been studied by Zardecki (1980, 1981). For the first case this author rederives the Fokker–Planck equation for the quasiprobability distribution for the transmitted field and shows that on a semiclassical level different variants of the Fokker–Planck equation are possible. In order to calculate average moments $\langle x^n \rangle = \int x^n P(x, t) \, dx$ a Monte Carlo scheme is used. This consists of constructing random trajectories that result from the Langevin equation—see above—which is statistically equivalent to the Fokker–Planck equation. In the second case this author obtains a Fokker–Planck equation for the joint probability distribution of the atomic polarisation and inversion. Again a Monte Carlo scheme is used to evaluate numerically the adiabatic values of the output intensity and intensity fluctuations as functions of time.

Lugiato *et al* (1980) investigated the transient in absorptive OB in the good-quality cavity limit. They studied the time evolution of the Fokker–Planck equation of the quasiprobability distribution for the transmitted field amplitude. They show that the evolution of the system consists initially of a local relaxation around the minima of a defined potential, and later, in a much longer time scale, tunnelling takes place from one minimum to the other. For the first stage they give an analytical description of the mean value of the transmitted field and this is tested with an exact numerical integration. The long-time evolution stage was investigated by Bonifacio *et al* (1981). Kondo *et al* (1980) have studied the escape probability from metastable to a stable state in the mean field limit for the absorptive case; they have also given the stationary distribution of photons.

At some stage of the calculations of Bonifacio *et al* (1978) the condition of detailed balance is used but no simultaneous solutions are possible, which indicate that this condition is not satisfied. Graham and Schenzle (1981) develop methods for describing states lacking detailed balance. Systems exhibiting dispersive OB are examples of such cases. They derive a Fokker–Planck equation for the distribution of the electromagnetic field and show that in the limit of small fluctuations the former is reduced to a Hamilton–Jacobi equation for a non-equilibrium thermodynamic potential. The latter acts like a free energy for a first-order phase transition in a far from equilibrium system lacking detailed balance. They solved the Hamilton–Jacobi equation under different approximations.

4. Theory of alternative approaches to bistable systems

The combination of non-linearity and optical feedback may take different forms out of which different theoretical approaches have emerged. Bischofberger and Shen (1978, 1979a, b) have given a theory to explain their experiments (cf § 6) on Kerr media in a Fabry–Perot. It is a plane wave theory (including linear absorption) using the time-dependent Debye equation with a lowest-order non-linear driving term; a remarkable agreement with experiments was obtained. Marburger and Felber (1978) have extended their theory of a loss-less non-linear Fabry–Perot resonator. The non-linearity is cubic and in the plane wave (steady-state) case they obtain an exact solution in terms of elliptic functions. The theory was also extended to include resonators with spherical mirrors and finite beams where self-focusing plays a role. In the spherical mirror geometry the powers required for bistable operation are reduced vis-à-vis the plane wave. Ballagh et al (1981) in their study of Gaussian beam models in two-level atomic systems found a softer onset of saturation at line centre for the Gaussian case. This led to a raising in the critical values of C and y (cf § 3.3) for absorptive OB, whereas for the dispersive case ($\Delta \gg 1$) no substantial difference was found between models. The treatment of these authors is a mean field one. Drummond (1981) studied OB in a radially varying mode including dispersive effects and inhomogeneous broadening. In general, a larger value of C is necessary for the onset of bistability compared with the plane wave theory. Mattar (1981) developed computational methods for treating transverse and longitudinal reshaping in extended systems irradiated by single beams. In particular, OB can be studied with the appropriate boundary conditions. Firth and Wright (1982) have incorporated transverse effects in a Fabry–Perot cavity containing a third-order dispersive non-linearity by expanding the internal fields in orthonormal functions. Their main results are numerical with some analytical results of a general nature. They found that transverse coupling greatly reduces, and may eliminate, the optical switching range, but new hysteresis effects in the spatial character of the output field can occur.

Bistability in saturable optical cavities with distributed Bragg reflectors was investigated by Okuda et al (1976, 1978) and Okuda and Onaka (1977a, b, 1978). Hysteresis and non-linear effects were analysed in terms of the absorption coefficient, the length of the absorber and the coupling coefficient of the corrugated waveguide. A Kerr medium was also considered and the resonator had a multi-valued transmission characteristic. The response of the system to pulses was also studied. A similar approach was considered by Winful et al (1979) who predicted OB for systems with distributed feedback structures. Here the concentrated feedback mechanism of the end mirrors is replaced by a distributed one through a periodic variation of the linear refractive index n_0.

Bistability at a non-linear interface was predicted by Kaplan (1976, 1977). We shall explain the basic idea; the experiment will be referred to in § 6. Assume that the non-linear medium is such that

$$n = n_0 - \Delta + n_2 I_1 \qquad (\Delta > 0)$$

with n_0 being the refractive index of the linear medium.

When the incident light is of low intensity and the ψ is less than the critical angle ψ_c $(2\Delta/n_0)^{1/2}$ then total internal reflection (TIR) (see figure 22) occurs and no transmission results. There is, however, an evanescent field, which as a result of $n_2 > 0$,

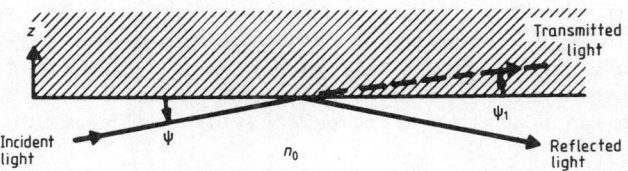

Figure 22. Light incident on the interface between linear and non-linear (shaded region) media (from Smith *et al* 1979). $n = n_0 - \Delta + n_2 I$.

increases the refractive index and reduces the critical angle. This in turn increases the evanescent field and a positive feedback is established. We can expect that at some threshold a sudden switch from TIR to a state where both transmitted and reflected fields exist can occur. Because the transmitted field increases the refractive index even further, it can be expected that a considerable reduction—below the above threshold—of the incident intensity is necessary to regain TIR. A hysteresis cycle in the reflection is exhibited (i.e. optical hysteresis). Kaplan (1981a, b, c) has also proposed a device based on a hybrid interface effect whereby the non-linear medium is driven electro-optically. Kaplan (1981a) has also studied the case when $n_2 < 0$. Wysin *et al* (1981) examined bistable operation in reflection with simultaneous excitation of the surface-plasmon mode at the interface with a non-linear Kerr medium. These authors found that the threshold power is one order of magnitude lower than in the case of grazing incidence geometry.

Other novel bistable systems have been proposed. Lugovoi (1977) investigated the case of an optical resonator filled with a Raman or Mandel'ṣtam–Brillouin medium for a single-mode (exciting and the first Stokes) radiation components in the cavity. Lugovoi (1979, 1981) has shown that optical Cerenkov-type parametric oscillators exhibit phenomena like bistability, hysteresis and self-pulsing.

Winful and Marburger (1980) showed that bistability may be observed in the phase conjugate signal or in the 'spontaneous' oscillation possible in degenerate four-wave mixing (DFWM) processes. Agrawal *et al* (1981) and Agrawal and Flytzanis (1981) have also suggested that through DFWM the reflectivity of the phase-conjugated signal exhibits bistability when the pump is varied continuously. They have suggested materials and applications. Walls and Zoller (1980) and Walls *et al* (1981) have shown that a medium of three-level atoms exhibits OB by utilising the population trapping that occurs in the coherent superposition of atomic sublevels. This system does not require saturation and is Doppler-free. Finally, Kitano *et al* (1981) show that a Fabry–Perot cavity filled with atoms with Zeeman sublevels in the ground state should exhibit optical *tristability*.

5. Non-linear refraction of semiconductors

In this section we concentrate on the physics of the refractive non-linearity in semiconducting materials. The pioneering experiments on bistability in semiconductors were done on InSb and GaAs and the unusually large non-linearities observed deserve an analysis before we describe the optical bistability experiments. Consequently this section will be a mixture of theory and experiments aimed at understanding the non-linear mechanism.

Semiconducting materials such as those mentioned have been known to exhibit a comparatively large passive non-resonant $\chi^{(3)}$ (Flytzanis 1975) around 10^{-8}–10^{-11} ESU, but the discovery of strong non-linear refraction in the region just below the optical band gap in both materials gives susceptibilities many orders of magnitude higher with $\chi^{(3)}$ in the region 10^{-2}–1 ESU. In both cases the giant effect can be explained by saturation mechanisms although the details vary: the non-linearity in GaAs relying on the existence of excitons whilst that in InSb is ascribed to interband transitions.

The advantages (Smith et al 1980) of using semiconductors in optically bistable systems are that (i) the fabrication technology is highly developed; (ii) semiconductor devices are small—in the range of 4–100 μm—and have short calculated cavity build-up times (1.2 ps for GaAs and 20 ps for InSb already demonstrated); (iii) the observed non-linearities are large ($n_2 \sim 4 \times 10^{-4}$ cm^2 kW^{-1} in GaAs and 1.0 cm^2 kW^{-1} in InSb at 77 K with $\chi^{(3)} \sim 1$ ESU) so that power densities as low as 10 W cm^{-2} or 8 mW incident power can be used; (iv) while switch-off times may be limited by material relaxation times, switch-on times, in principle, are not; this has been demonstrated to some extent in GaAs (Gibbs et al 1979a, b, c, d). Also, material relaxation times can be engineered to a considerable extent by, for example, doping, particularly for the non-linearity in InSb which is present also in impure material. These factors combined suggest that small, low-power, low-energy, fast switching, integrable devices may ultimately be possible with semiconductors.

5.1. Non-linearity in InSb

The first approach we adopt is a macroscopic one which was originally motivated by various striking non-linear effects observed by Miller et al (1978). In figure 23 we show these effects where it is posssible to see how a Gaussian beam, having passed through the sample, develops a power-dependent spatial profile. This was attributed to non-linear refraction. The macroscopic approach we outline here is due to Weaire et al (1979) and, in spite of its conceptual simplicity, proved powerful enough to explain the observations to a great extent.

The following assumptions for the theory are made.

(a) The condition

$$\frac{4n_2 L^2}{w_0^2 n_0} I_0 \ll 1 \tag{5.1}$$

where L is the thickness of the sample, w_0 is the (e^{-2} intensity) beam radius and I_0 is the peak beam intensity.

(b) The sample is assumed to be at the waist of the incident beam as was the case of the measurements, but this condition can be released.

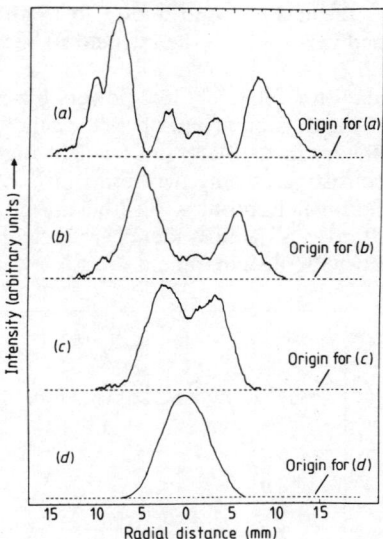

Figure 23. Diametric beam profiles in the far field (18.5 cm) behind the sample. Incident spot 300 μm diameter, laser wavenumber 1765 cm^{-1}, temperature 77 K. (a) 41 mW incident power, (b) 25 mW, (c) 9.4 W, (d) sample out (similar to sample in, lower power) (from Miller et al 1978).

With these assumptions the problem we are left with is that of the propagation in the z direction of a beam that emerges from the crystal at $z = 0$ with an amplitude

$$E(r, 0) = E(0, 0) \exp\left(\frac{-r^2}{w_0^2} + i\xi \exp\left(-2r^2/w_0^2\right)\right) \quad (5.2)$$

where the phase-shift parameter ξ is related to n_2 by

$$\xi = -\frac{\omega}{c} n_2 I_0 \frac{1 - \exp(-\alpha L)}{\alpha} \quad (5.3)$$

where the parameters have the meaning of § 3.

The crucial step for solving the problem is to expand (5.2) as follows:

$$E(r, 0) = E(0, 0) \sum_{m=0}^{\infty} \frac{(i\xi)^m}{m!} \exp(-r^2/\omega_m^2) \quad (5.4)$$

where the radius of each of the individual Gaussian terms is given by $\omega_m^2 = \omega_0^2 (2m+1)^{-1}$. In this way the Gaussian beam is expressed as a sum of beams of decreasing radius whose propagation is described by the theory of Kogelnik and Li (1966). At a distance z

$$E(r, z) = E(0, 0) \sum_0^\infty \frac{(i\xi)^m}{m!} \left(1 + \frac{z^2}{d_m^2}\right)^{-1/2} \exp\left(-\frac{r^2}{\omega_m^2(z)} - \frac{ikr^2}{2R_m(z)} - iP_m(z)\right). \quad (5.5)$$

In (5.5) $\omega_m(z) = \omega_m(0)(1 + z^2/d_m^2)^{1/2}$, $R_m(z) = z(1 + d_m^2/z^2)$, $P_m(z) = -\tan^{-1}(z/d_m)$, where $d_m = k\omega_m^2(0)/2$ is the diffraction length. The inequality (5.1) results from the comparison of sample thickness with such a length *within the sample*, defined for the value of m that makes the largest contribution to (5.5). Note that the sum over Gaussian beams is turned inside-out in the far field to become a sum over Gaussian beams of increasing radii.

Weaire *et al* took data for relatively low powers for which the structure in the beam profile is most easily understoood and where the validity criterion (5.1) is satisfied and compared it with theory in the near and far field (figure 24). The initial beam can be considered as consisting of only two terms in (5.4) and it is then clear how (5.5) gives the near- and far-field profiles. The best fit has $\xi = 3.5$ (self-defocusing). Self-focusing ($\xi = -3.5$) gives an almost identical far-field profile but the near field disagreed quite obviously with the experiment.

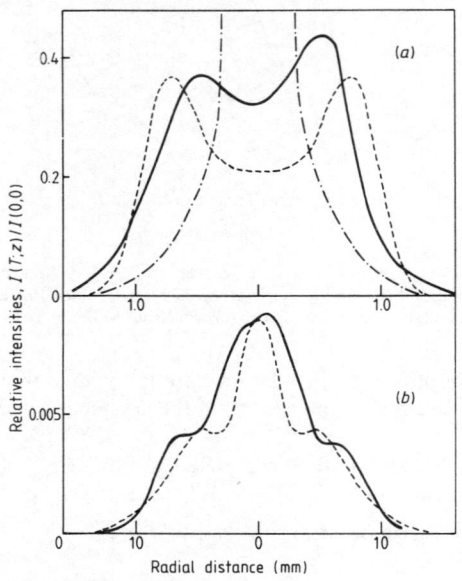

Figure 24. Experimental (full curves) and theoretical (broken curves) intensity profiles (*a*) in the near field at 7 cm from the sample and (*b*) in the far field at 189 cm. Data for 130 mW beam of 1.67 mm spot diameter on laser line at 1886 cm^{-1}; InSb sample and temperature as for figure 23. Theoretical profiles are shown for the self-defocusing condition ($\xi = 3.5$, broken curves) in the near and far field and for the self-focusing condition ($\xi = -3.5$, chain curve) in the near field to emphasise that the experimental results originate from a defocusing. In the far field, the focusing and defocusing results are practically identical. The theoretical and experimental plots are normalised to give the same power levels (from Weaire *et al* 1979).

Using (5.3) and noting that the sample has a measured $\alpha \sim 0.9$ cm^{-1} in the regime of interest, a value of $n_2 \simeq -6 \times 10^{-2}$ cm^2 kW^{-1} is inferred for radiation wavenumber 1886 cm^{-1} at 5 K, corresponding to *self-defocusing*. This latter has important consequences for the microscopic mechanism since in InSb (for photon energies below the band gap) the refractive index increases as the photon energy approaches the band-gap energy while the band-gap energy reduces with temperature. Thus, the refractive index should increase with temperature. For a Gaussian beam this should lead to self-focusing if thermal effects are the dominant mechanism. The experimentally found defocusing is therefore not consistent with a thermal origin.

The corresponding $\chi^{(3)}$ associated with n_2 is given by

$$n_2 = \frac{4\pi^2}{n_0^2 c} \operatorname{Re} \chi^{(3)} \tag{5.6}$$

where $n_0 = 4$ for InSb; the calculated $\chi^{(3)} \sim 10^{-2}$ ESU, which is several orders of magnitude larger than the quoted values of our introduction. This interesting huge non-linearity led to Miller et al (1980b) to develop a theory to understand the microscopic mechanism behind such large values. These authors discuss four contending processes associated with refraction: (i) the 'dynamic Burstein–Moss effect', (ii) the induced free-carrier plasma, (iii) direct saturation and (iv) saturation of exciton absorption. Of these, for steady-state experiments (i) is shown to dominate in InSb and to give magnitudes for n_2 consistent with experiments.

The electronic polarisation in the case of non-linear refraction can be expressed in terms of powers of the field strength $E(\omega)$ as follows:

$$P^{(3)}(\omega) = \chi^{(3)}(\omega, \omega, -\omega, \omega) E(\omega) E(-\omega) E(\omega) \tag{5.7}$$

where the monochromatic incident field is defined as

$$E(t) = E(\omega) \exp(-i\omega t) + E(-\omega) \exp(i\omega t). \tag{5.8}$$

For sufficiently weak fields, perturbation theory (Franken and Ward 1963) gives the standard result:

$$P^{(3)}(\omega) = \sum_{ijkl} \frac{-1}{\hbar^3 m^4 \omega^4} \frac{\tilde{P}_{il}\tilde{P}_{lk} \cdot E(\omega)}{\omega_{il} - \omega - \Gamma_{ij}} \frac{\tilde{P}_{kj} \cdot \tilde{E}(-\omega)}{\omega_{ki} - i\Gamma_{ki}} \frac{\tilde{P}_{ji} \cdot \tilde{E}(\omega)}{\omega_{ji} - \omega - i\Gamma_{ji}} \tag{5.9}$$

+ five terms from time ordering.

The sum is over all electronic states i, j, k, l per unit volume; $\tilde{P}_{ji}$ is the momentum matrix element and $\hbar\omega_{ji}$ is the energy difference between states j and i. $\Gamma_{ji}(\equiv T_2^{-1})$ is the inverse of the dephasing time of the excitation. For frequencies well below the band-gap resonance, e.g. $\omega < \omega_G/2$, theoretical values of $\chi^{(3)}$ calculated according to (5.9) are in acceptable agreement with experiment and in the range 10^{-11}–10^{-8} ESU for Si, GaAs, InSb, etc (Flytzanis 1975).

We turn to the case when the photon frequency is just below the band gap which, for InSb at 5 K, occurs at around 1900 cm^{-1}. The measured linear absorption α varies from 0.2 to 0.1 cm^{-1} in the range 1850 cm^{-1} to 1886 cm^{-1} and for laser intensities 10^2–10^3 W cm^{-2}. If we assume that at least some of this 'band-tail' absorption generates free carriers the rate is at most $\alpha I_1/\hbar\omega$, i.e. between 6×10^{20} to 3×10^{22} carrier cm^{-3} s^{-1}. For a recombination time $\tau_R \sim 10^2$ ns, a steady free-carrier density $N \sim 1 \times 10^{14}$ to 6×10^{15} cm^3 is obtained with

$$N = \alpha I_1 \tau_R / \hbar\omega. \tag{5.10}$$

These are two direct contributions from such intensity-dependent free-carrier densities to the non-linear refraction.

(i) In InSb the band gap is at zero wavevector. The occupation of the lowest wavevector condition states blocks interband transitions close to the band gap. In linear optics this is known as the Burstein–Moss shift† of the energy gap with

---

† This 'shift' is merely the filling of states at the 'bottom' of the bands, thereby blocking the absorptive transitions between these states and shifting the wavelength of the onset of optical absorption.

appropriate donor or acceptor concentration. A simultaneous change in refractive index is also expected as the blocked transitions can no longer contribute to refraction. Using as a model conduction and heavy hole valence bands at 0 K, the change in refractive index due to blocking of states beneath a Fermi wavevector k_F in the conduction band is

$$\Delta n = \frac{-2}{3\pi \hbar n_0} \left(\frac{ep^2}{\hbar \omega}\right)^2 \left(\frac{2m}{\hbar}\right)^{3/2} (\omega_G - \omega)^{1/2} (\eta_F - \tan^{-1} \eta_F) \quad (5.11)$$

where $p = i\hbar m^{-1} \langle \varphi_1 | P_x | \varphi_2 \rangle$ is the momentum matrix element between Bloch state functions φ_1 and φ_2 (Kane 1957) and $\eta_F = (\hbar k_F^2/2m)^{1/2}(\omega_G - \omega)^{-1}$ is the dimensionless Fermi wavevector.

Assuming negligible initial carrier concentration (5.11) is dominated by a term proportional to I_1 in the case of photoexcitation as $k_F \propto N^{1/3} \propto I_1^{1/3}$ (cf (5.10)) and $(\eta_F - \tan^{-1} \eta_F) \to \eta_F^3$ for small η_F. Therefore

$$n_2(\mathrm{BM}) = \frac{-2\pi}{3n_0} \left(\frac{ep}{\hbar \omega}\right)^2 \frac{\alpha \tau_R}{\hbar(\omega_G - \omega)\hbar \omega}. \quad (5.12)$$

(ii) The refractive index of a free-carrier plasma is given by

$$n^2 = (\epsilon_\infty - 4\pi N e^2 / m^* \omega^2) \quad (5.13)$$

where ϵ_∞ is the dielectric constant and m^* is the carrier effective mass. An excitation like (5.10) produces a non-linear refraction dominated by

$$n_2(\mathrm{plasma}) = -2\pi e^2 \alpha \tau_R / n_0 m^* \hbar \omega^3. \quad (5.14)$$

Such a mechanism has been proposed by Jain and Klein (1979) in their interpretation of degenerate four-wave mixing in Si and other semiconductors. Noting that $1/m^* \sim 4P^2/3\hbar^3 \omega_G$ for direct-gap semiconductors (Hilsum and Rose-Innes 1961), we have approximately

$$\frac{n_2(\mathrm{BM})}{n_2(\mathrm{plasma})} = \frac{\omega_G}{4(\omega_G - \omega)} \quad (5.15)$$

which shows that the Burstein–Moss refraction contribution dominates over the plasma contribution if the incident photons have energy greater than 3/4 of the band gap. The plasma non-linearity is, of course, quite general and has been used by Staupendhal and Schindler (1982) (cf § 6.3) via two- and three-photon generation in Te to demonstrate bistability at 10.6 μm. A value of n_2 of the order of 10^{-6} cm^2 kW^{-1} was shown.

(iii) Direct saturation is closely related to (i) but approaches the problem from a different limit, i.e. where the laser pulse is short so that 'state filling' occurs. In order to model excitation with $\omega < \omega_G$ a direct analogy to two-level systems broadened by a dephasing time T_2 is made. The semiconductor interband excitation is described as a set of independent two-level systems, one for each allowed k state, noting that radiative transitions couple only specific pairs of states (states of the same k value in different bands). The process is controlled by two time constants: the energy relaxation time T_1 within which population decays and T_2 which measures the uncertainty width of individual energy states.

The problem is treated by means of the density-matrix theory of § 3 and applying (3.4) to the momentum operator with the association $\mu \to (ie/m\omega)\tilde{P}_{ik}$ we have for $P(\omega)$

$$P(\omega) = \sum_{j,k} \frac{ie}{m\omega} (\tilde{P}_{jk}\rho_{kj}(\omega) + \rho_{jk}(\omega)\tilde{P}_{kj}) \tag{5.16}$$

where in the Schrödinger representation $\rho_{kj}(\omega)$ indicates that the frequency dependence is $\exp(-i\omega t)$. Saturation is accounted for by including non-radiative relaxation rates τ_{ij}. For uncoupled two-level systems the population redistribution can be accounted for exactly (Yariv 1975), thus:

$$P(\omega) = \chi(\omega, E)E(\omega) \tag{5.17}$$

with

$$\chi(\omega, E) = \frac{e^2}{\hbar^2 m^2 \omega^2} \sum_k \frac{|P_{kj}|^2}{\omega_{jk} - \omega - i\Gamma_{jk}} \frac{(\omega_{jk} - \omega)^2 + \Gamma_{jk}^2}{(\omega_{jk} - \omega)^2 + \Gamma_{jk}^2 + 4\Omega^2 \tau_{jk}\Gamma_{jk}} \tag{5.18}$$

where $\Omega = (e/\hbar\omega m)|\tilde{P}_{jk}\bar{E}(\omega)|$ is the Rabi flop frequency. In order to obtain the third-order susceptibility we have to expand (5.18) to first order in Ω^2, and the coefficient of this term is

$$\chi^{(3)}(\omega) = -\left(\frac{e}{m\omega}\right)^4 \frac{1}{\hbar^3} \sum_k \frac{|P_{jk}|^4}{\omega_{jk} - \omega - i\Gamma_{jk}} \frac{4\tau_{jk}\Gamma_{jk}}{(\omega_{jk} - \omega)^2 + \Gamma_{jk}^2} \tag{5.19}$$

and n_2 can be calculated from (5.6). The presence of τ_{jk} rather than some large denominator—as seen in general frequency mixing expressions—is responsible for the high value of $\chi^{(3)}$. In addition the right-hand energy denominator will resonate for $\omega_{jk} \simeq \omega$.

The discrete sum over k in (5.18) can be reduced to an integration over the wavevector k and sum over bands. For simplicity $\Gamma_{jk}^{-1} = T_2$, $\tau_{jk} = \tau$, i.e. independent of k. One gets

$$\chi^{(3)}(\omega) = -\hbar^{-3}\left(\frac{ep}{\hbar\omega}\right)^3 \frac{8}{15\pi^2} \int_0^\infty dk\, k^2 \frac{\tau}{T_2} \left|\omega_{cv}(k) - \omega - \frac{i}{T_2}\right|^{-2} \left(\omega_{cv}(k) - \omega - \frac{i}{T_2}\right)^{-1} \tag{5.20}$$

where $\omega_{cv} = \omega_G + \hbar k^2/2m_r$ is the heavy hole to conduction band gap at wavevector k, and m_r is the reduced effective mass. For frequencies *below* the band gap $(\omega_G - \omega) \gg T_2^{-1}$, the expression simplifies to give a saturation-associated refractive index change:

$$n_2(\text{sat}) = -\hbar^{-3}\left(\frac{ep}{\hbar\omega}\right)^4 \frac{T_1}{T_2} \frac{2\pi}{15n_0^2 c} \left(\frac{2m_r}{\hbar}\right)^{3/2} (\omega_G - \omega)^{-3/2}. \tag{5.21}$$

The non-linear refractive index so obtained is negative as required and resonates as $(\omega_G - \omega)^{-3/2}$. In evaluating the magnitude of n_2, the effective masses and the matrix element p are well known empirically using, for example, linear absorption (Houghton and Smith 1966). The position of the energy gap and the ratio T_1/T_2 are questionable parameters. Figure 25 shows a comparison of theory and experiment (Miller and Smith 1979). Comparing the three approaches

$$n_2(\text{BM}) = \frac{5}{3} n_2(\text{sat}) = \frac{\omega_G}{4(\omega_G - \omega)} n_2(\text{plasma}). \tag{5.22}$$

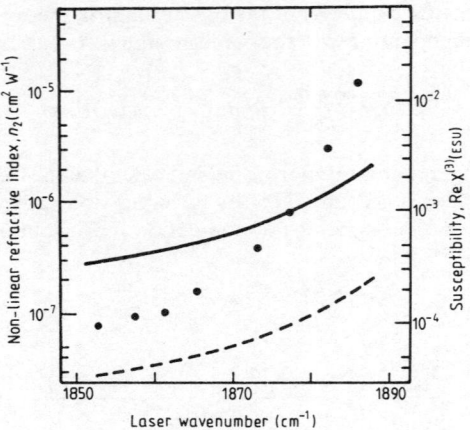

Figure 25. Comparison of theory for InSb (taking $E_G = 1899$ cm^{-1}) with experimental results at 5 K (from Miller et al 1980b). ●, experiment; ——, theory: $T_1/T_2 = 100$; ---, theory: $T_1/T_2 = 10$.

Descriptions (i) and (iii) are in some sense alternatives applying to the slowest and fastest situation, respectively. The plasma contribution should be present for either.

The largest possible value of T_1 is τ_R and a value 500 ns (and $T_2 \simeq 8 \times 10^{-11}$ s) in (5.21) predicts the large experimental value (Weaire et al 1979) of n_2 (-6×10^{-2} cm^2 kW^{-1}). In this limit (i) and (iii) are essentially the same. For frequencies well away from the band edge, model (iii) satisfies all the present requirements for n_2; namely (a) a resonant negative n_2, (b) a saturation of this non-linearity (figure 26) at high intensity, and (c) a mechanism for absorption, which is essential even if the Burstein–Moss effect is to be invoked.

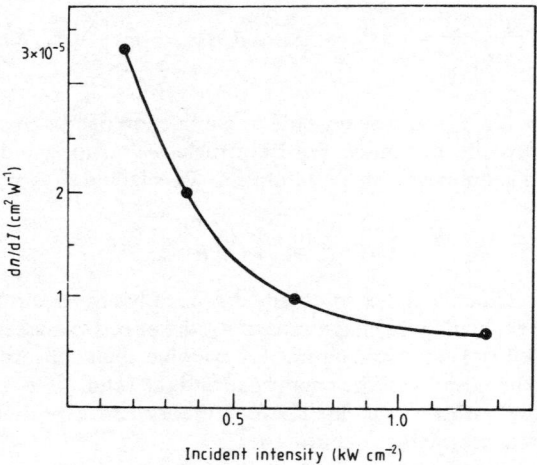

Figure 26. Saturation of non-linear refraction (dn/dI) with increasing intensity derived from non-linear Fabry–Perot results (the line joining points is for clarity only) (from Miller et al 1980b).

(iv) Screening or saturation of excitons: all absorption near ω_G is in principle modified by Coulomb effects. This is the primary mechanism in GaAs (Gibbs et al 1979c) but in small gaps (<0.5 eV) the effects are small. In InSb the Bohr radius of the exciton is 700 Å and 1×10^{14} carriers totally screen the exciton, corresponding to the initial, unexcited density.

The generality of the concept of interband saturation-induced non-linear refraction introduced by Miller et al implies that similar effects should be observable in other semiconductor materials and at higher temperatures (cf Miller et al 1978). A more detailed study of the origin of these non-linearities was done by Wherrett and Higgins (1981) who analyse the Burstein–Moss and direct saturation models. More recently, Miller et al (1981c) presented experimental measurements of the resonance of the refractive index in InSb near the band gap at 77 K. A semi-empirical theory was developed which gives a good fit to resonance using Boltzmann statistics to account for the thermalisation of the excited carriers. This treatment thus takes a step towards overcoming the deficiencies of model (i) (which assumes complete thermalisation at 5 K) and (iii) (which assumes no thermalisation) providing a theory nearer to the practical case. No explanation is, however, given for the 'band-tail excitation mechanism', which is included empirically. To date only the 'T_2 broadening' of (iii) provides an initial theory. Phonon, carrier collision and/or impurity effects may be involved.

5.2. Non-linearity in GaAs

This semiconducting material is known to have an intrinsic exciton with a sharp atomic-like resonance (Sell et al 1973) with low saturation intensity (Shah et al 1977). Prior to the observation of OB in GaAs by Gibbs et al (1979b), these authors have studied the non-linear spectroscopy of this material (Gibbs et al 1979a). To this end, the transmission of 500 ns pulses through GaAlAs–GaAs—GaAlAs heterostructures has been investigated as a function of light intensity and wavelength (figure 27). The light intensity ranged from 0.015 to 50 kW cm^{-2} and the wavelength from 810 to 840 nm.

Gibbs et al found that the intrinsic excitonic absorption at $\lambda_{ex}=818$ nm, $T=10$ K, can be modelled as a sum of a small unsaturable background and a dominant term which saturates as a Bloch resonance with a saturation parameter I_s of ~ 150 kW cm^{-2}:

$$\alpha L_{ex} = A[1+B(1+I/I_s)^{-1}]. \quad (5.23)$$

These authors argue that the form of the saturation curve does not imply that the exciton behaves as a saturable two-level system although it seems to saturate as a homogeneously broadened line.

From this determination of the absorption coefficient it is possible to represent the contribution of the exciton resonance to the refractive index as

$$n_{ex}(\lambda) = \frac{(\alpha_{ex}\lambda/4\pi)\Delta\lambda/\delta\lambda}{1+(\Delta\lambda/\delta\lambda)^2+I_1/I_s} \quad (5.24)$$

where $\Delta\lambda = \lambda - \lambda_{ex}$ and $\delta\lambda$ is the width (HWHM) of the exciton resonance with peak absorption $\alpha_{ex}L$. The non-linear refractive index n_2 can be determined by

$$n_2 = \frac{dn_{ex}}{dI_I}\bigg|_{I_1=0} = -\frac{\lambda\alpha_{ex}(\Delta\lambda/\delta\lambda)}{4\pi I_s[1+(\Delta\lambda/\delta\lambda)^2]^2}. \quad (5.25)$$

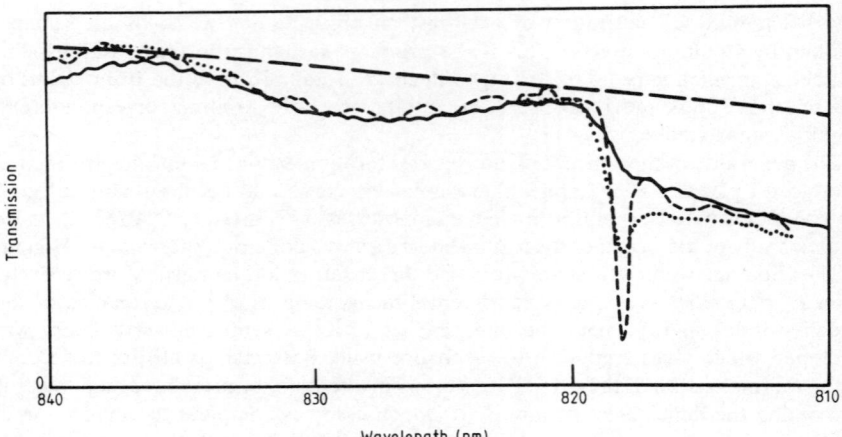

Figure 27. Non-linear transmission of 500 ns pulses through a GaAs sample for $I_{max} = 43$ kW cm^{-2} in the centre of the focal spot on the sample (from Gibbs *et al* 1979a). ——, I_{max}; — —, I_{max} (×0.7), no sample; – – –, $3.1 \times 10^{-4} I_{max}$, · · · , $0.062 I_{max}$; temperature; 10 K.

When $\Delta\lambda = \pm \bar{\delta}\lambda$, extreme values for n_2 are obtained:

$$|n_2|_{max} = \lambda \alpha_{ex}/16\pi I_s. \quad (5.26)$$

With $\alpha_{ex} = 4\mu$ m^{-1}, $I_s = 150$ kW cm^{-2} substituted in (5.26), $|n_2|_{max} = 4 \times 10^{-4}$ cm^2 kW^{-1}.

These excitonic non-linearities that ultimately led to the observation of OB in GaAs by Gibbs *et al* (1979b) has become the subject of much theoretical investigation. Koch *et al* (1981) calculated refractive-index intensity-dependent changes based on a many-body theory of the gain and absorption spectra of highly excited direct-gap semiconductors they previously developed. Koch and Haug (1981) showed that the fusion of two excited polaritons into an excitonic molecule gives rise to an intensity-dependent dielectric function. They suggest that this non-linearity may be used for OB with platelets of direct band-gap semiconductors. Hanamura (1981) studied the transient behaviour of an OB system consisting of a medium like CuCl with three states: the ground, exciton and biexciton. He suggests a response time for switching-off in the region of 1 ps but high holding powers ($\sim$ GW) are needed. Goll and Haken (1980) have showed that an excitonic bistability, absorptive and dispersive, is possible. It remains to reconcile these theoretical approaches with the experimental values for the various non-linearities.

6. Experiments on intrinsic optically bistable systems

As mentioned in § 1.2 these systems are characterised by having an optical feedback. Experiments have been performed on free atoms and molecules (Na, Rb, NH$_3$), Kerr media (MBBA, nitrobenzene, CS$_2$) and solids (ruby, InSb, GaAs and Te). In separate subsections we shall refer to the contrasting properties which vary over a large range of speeds (μs to ps), intensities (W cm^{-2} to MW cm^{-2}) and cavity and geometrical characteristics.

6.1. Free atoms and molecules

The first successful observation of OB by Gibbs et al (1976) has contributed enormously to the subject. These authors demonstrated the physical principles of a bistable device and succeeded in observing single-beam differential gain, bistability, clipping and limiting actions. Their system consisted of a 2 cm cell containing Na vapour at 10^{-4}–10^{-5} Torr pressure in an 11 cm separation Fabry–Perot interferometer with 90% reflecting mirrors. The dye laser was 50 mW single-mode and tunable to any frequency within either of the D lines; this was used to excite the $^2S_{1/2}$ ($F=2$) to $^2P_{3/2}$ ($F=2$) transition. Their results are shown in figure 28.

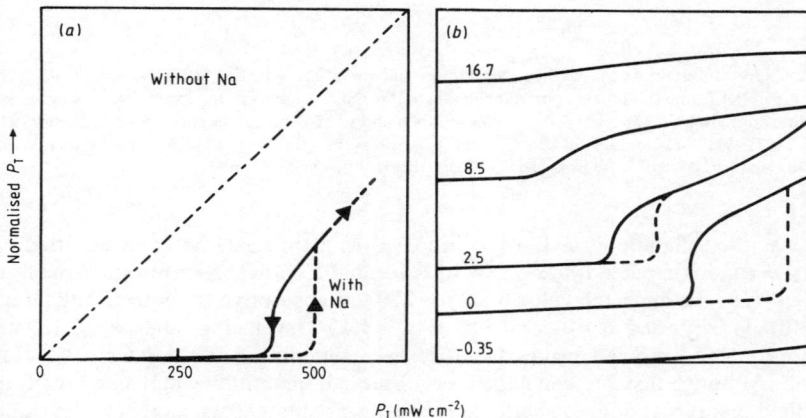

Figure 28. (a) Transmitted power P_T against incident power P_I in the Na experiment. (b) As (a) but exhibiting asymmetries with plate separation (in MHz) (from Gibbs et al 1976).

This experiment showed asymmetries with plate separation. In terms of the discussion of §§ 2 and 3 this amounts to changes in the sign of θ. Equation (2.3) shows in that simple model that, assuming $n_2 > 0$, the sign of θ causes Φ to be a multiple of 2π at different values of I_c. In addition, the mean field state equation (3.35), which also includes non-linear absorption, shows asymmetry with θ. This behaviour can only show up if there is dispersion, i.e. the atomic detuning $\Delta \neq 0$. Gibbs et al set $\Delta = 150$ MHz and the contribution to the non-linear refractive index, via hyperfine optical pumping, to bistability was noticed. This discovery was of prime importance for future progress in OB.

In an attempt to test mean field theories Sandle et al (1981) have investigated OB in Na vapour. They chose the $3^2S_{1/2}$–$3^2P_{1/2}$ D_1 line of atomic sodium—note the difference with Gibbs et al—as their two-level system; the sample was placed at the waist region of a focusing cavity. Up to 80 Torr of Ar was added in order to homogeneously broaden the line. The intensities ranged above 1 kW cm^{-2} at 80 μm waist. The cavity had a finesse of 55 and 30 cm radius mirrors; the free spectral range was 250 MHz.

As opposed to the experiment of Gibbs et al, here the input power was fixed and the laser frequency (i.e. Δ and θ) changed. This, of course, should also exhibit a hysteretic behaviour shown in figure 29 for the dispersive and absorptive cases. As comparison with a plane-wave theory is difficult because experimentally a Gaussian

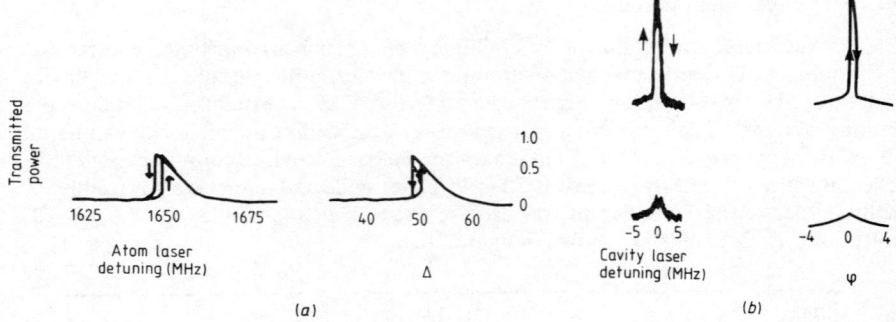

Figure 29. Optical bistability for fixed incident power and varying frequency. (a) Dispersive. (b) Absorptive: in the upper frame the power is marginally above threshold whereas in the lower frame is marginally below threshold. In both cases left-hand curves are data and right-hand curves correspond to theory (from Sandle et al 1981). (a) Na density = 9×10^{11} cm^{-3}, Ar pressure = 0.5 Torr, incident power = 60 mW. (b) Na density = 4.5×10^{11} cm^{-3}, Ar pressure = 20 Torr, incident power = 135 mW.

beam was used, Sandle et al left $|y|^2$ in (3.35) as a free parameter to be fitted. In dispersive OB, for expected values $C = 4.5 \times 10^3$ and $\Delta/\gamma_2 = 165$ corresponds the fitted value $|y| = 65$; the expected value was $|y| \sim 270$. In absorptive OB both C and $|y|$ had to be fitted; they used $\Delta = 0$, $C = 5$ with $|y| = 5.134$ (switching) and $|y| = 6.134$ (no switching). The expected values from the experimental conditions were $C \sim 110$, $|y| \sim 90$. Although bistable behaviour was observed, quantitative agreement with the mean field theory was poor. A better agreement was obtained with a theory developed by these authors (cf § 5).

Weyer et al (1981) reported a systematic study of absorptive OB in Na. The experiment was performed by means of a system of 50 highly collimated atomic beams entering a Fabry–Perot cavity perpendicular to its axis. High collimation of the atomic beams guaranteed a resonance condition for all atoms: the absorber's linewidth was reduced to 14 MHz close to the natural linewidth of 9.6 MHz. This multiple-beam system was necessary in order to meet the conditions for OB, namely $C = \alpha L/2T > 4$ (cf § 3); each beam was 1 mm in diameter. The laser was tuned to the $F = 2 \to F' = 3$ hyperfine transition of the sodium D_2 line ($3^2S_{1/2} \to 3^2P_{3/2}$), thus achieving a two-level pumping with the light circularly polarised. In this experiment $\alpha L = 0.8$, $R = 99\%$, and the free spectral range 300 MHz. The radiation was supplied by a CW dye laser of up to 6 mW and 2 mm (collimated) diameter. The laser intensity was swept from maximum to zero and back by means of an acousto-optic modulator with a modulation pulse length of 6 μs.

A comparison with mean field theory was made. The time-dependent mean field equations (3.28), appropriately scaled, were integrated numerically using a Gaussian input:

$$y^2(\tau) = y_0^2\{1 - \exp[-(\tau - \tau_0)^2/\bar{\tau}^2]\}$$

with $\bar{\tau} = 15 t_c$, $\tau_0 = 40 t_c$ and $t_c = 120$ ns. A good qualitative agreement was obtained (figure 30). Weyer et al expect that with this experimental set-up they will be able to resolve the controversy about the influence of cooperation on the fluorescence spectrum (Lugiato 1980b).

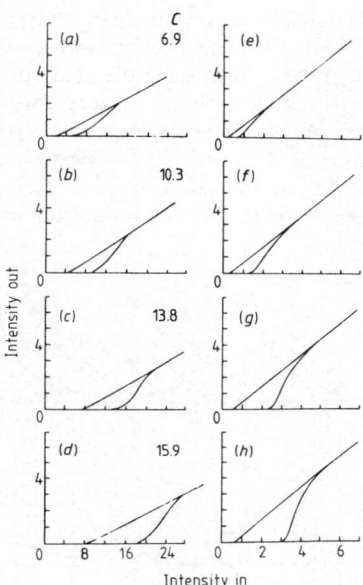

Figure 30. Dynamic absorptive bistability for different values of C and $t_c = 120$ ns. $(a)-(d)$ experimental results; $(e)-(h)$ theory for a Gaussian input pulse (see text) with $\bar{\tau} = 15 t_c$ and $\tau_0 = 40 t_c$. Intensities are in arbitrary units (from Weyer et al 1981).

Grischkowsky (1978) reported optical switching in Rb vapour contained in a Fabry–Perot cavity. The pumping light was offset from resonance by $\Delta\omega/2\pi c = 0.2$ cm^{-1}. The response time of the (electronic) non-linearity of Rb was faster than 0.2 ns; the cavity lifetime was less than 0.1 ns. The incident radiation came from a ruby-laser-pumped dye laser with a pulse width of 5 ns, peak intensity of 4.5 kW cm^{-2} and circularly polarised. The finesse was ~10 and the cavity was sealed inside a Pyrex glass Rb cell. The non-linearity was essentially dispersive with $n = n_0 + n_2 I$ in the low intensity limit; $n_2 = -3 \times 10^{-6}$ ESU. The output intensity was monitored for different increasing temperatures (or equivalently increasing Rb number density) of the cell. At 110 °C the output and input pulse shapes were the same, but this no longer occurred at 130 °C. At even higher temperatures a sharp switching action was observed on the leading edge of the pulse but not on the trailing one. A difficulty in this experiment was the 1 ns risetime of the detector compared with the ~2 ns risetime of the input pulse. Limiting action was also observed as the top of the output pulse became flat.

In the above experiments a one-photon non-linearity was used. A different possibility, already suggested theoretically (Arecchi and Politi 1978) making use of *two-photon* absorption or dispersion, was carried out experimentally by Giacobino et al (1980). This is particularly useful for a Fabry–Perot geometry where the existence of two-way propagating fields can cancel out the Doppler effect. As a result, rapid changes of the bistable behaviour can be expected in a range of a few MHz. Giacobino et al chose the $5S_{1/2}-5D_{5/2}$ two-photon transition ($\lambda = 7779$ Å) of ^{85}Rb as this has large oscillator strengths—f_{5S-5P} and f_{5P-5D}—and the energy detuning from the one-photon transition is small (~35 cm^{-1}).

The basic idea is to obtain optical hysteresis by fixing the input and atomic detuning, but to change the cavity mistuning θ. This is equivalent to keeping the slope of the straight line of figure 3 constant while displacing the peaks of the Airy function. This is possible for dispersive OB as shown in figure 5 where, for example, for $\theta > \theta_{crit}$ the Airy function appears shifted. As we change θ for a fixed I_I, a point A (figure 3) will be reached and then a jump to B occurred; as θ is swept back, the jump will occur from C to D and hysteretic behaviour observed. A plot of the transmission of the cavity in terms of θ will display sudden changes at different points (figure 31).

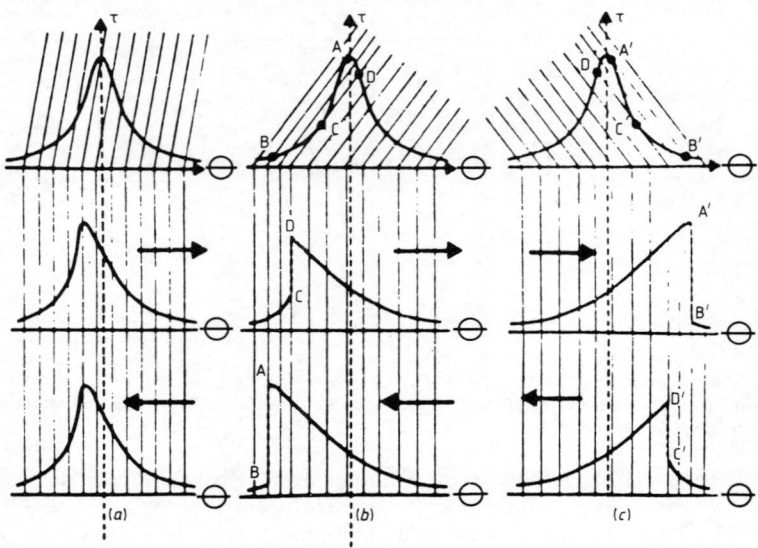

Figure 31. Upper frame: as in figure 3. Parallel straight lines are obtained by changing θ through L. In (b) a smaller slope corresponds to a larger I_1 than in (a); (c) is as (b) but the sign of n_2 is different. Middle frame: τ against θ as L is increased (right-pointing arrow); only in (b) and (c) is I_1 high enough for multiple intersection (cf figure 5) and a discontinuity in τ occurs from C to D and A' to B' respectively. Lower frame: as in middle frame but L is decreased; a discontinuity appears from A to B in (b) and from C' to D' in (c). Note both in (b) and (c) hysteresis effects as the discontinuities occur at different values of θ (from Giacobino et al 1980).

A single-mode oxazine dye laser pumped by a Kr^+ laser was used in this experiment. The power at 7779 Å was 150 mW. The Fabry–Perot cavity was 15 cm long with 96% reflectivity mirrors, one of which was mounted on a piezoceramic transducer (PZT) that continuously scanned the length of the cavity when a saw-tooth waveform was applied to it. Two peaks were observed on the oscilloscope that correspond to the increase and decrease of the length of the cavity; there is no overlap owing to the hysteresis. In figure 32 we show the experimental results at three different temperatures (165, 180 and 200 °C) for different atomic detunings from the two-photon resonances. Note that the transmission curves become more asymmetric at higher temperatures and lower detunings. The sharp edge is always on the same side of the trace which is indicative of the constancy of the sign of the non-linearity—positive in this case.

Arimondo et al (1981) observed microwave bistability in NH_3 placed in a confocal Fabry–Perot cavity. The radiation was on resonance with the (3,3) inversion line

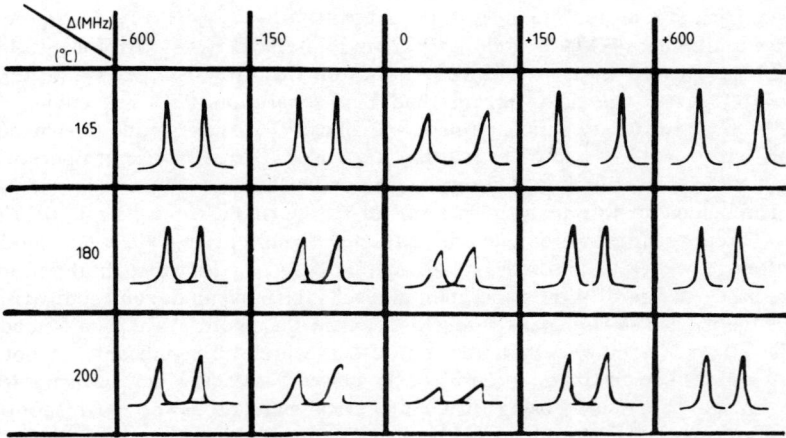

Figure 32. Oscilloscope traces of the transmission when the length of cavity is swept with a saw-tooth waveform (cf figure 31). The detuning Δ is measured with respect to the two-photon resonance. In each frame the first trace from the left corresponds to an increasing length of the cavity, the second trace to a decreasing length (from Giacobino *et al* 1980).

($\lambda = 1.256$ cm). Their cooperation parameter C (~40) was chosen by varying the pressure on which α depends: $\alpha = 7 \times 10^{-4}(1+1.5/p^2)^{-1/2}$ cm^{-1} for low pressures (measured in mTorr). The bistability is studied with constant C and input power while Δ and θ ($\theta \propto \Delta$ for $\omega_c = \omega_0$) are changed by a continuous sweeping of ω. In figure 33 we show a comparison between the theory they developed and data. The spacing between mirrors was 72.8 cm and ω was changed with a klystron of sufficient modulation depth to sweep Δ and θ simultaneously. The incident power ranged from 20–50 mW (31.75 mW in figure 33) and $p = 1.4$ mTorr; the agreement with theory is qualitatively good.

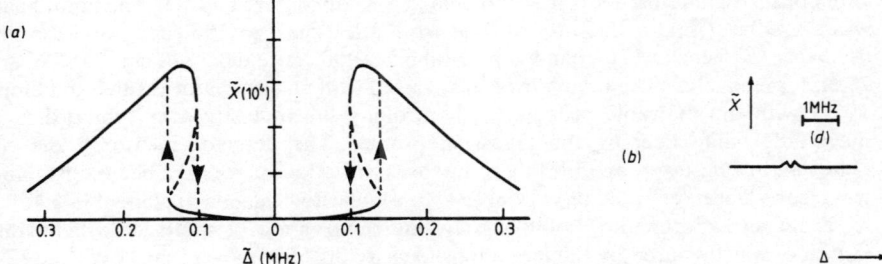

Figure 33. Dispersive bistability with ammonia molecules by changing the input frequency at fixed input power. Plot of normalised transmitted power $\tilde{X}$ against molecular detuning Δ (MHz). (*a*) Theory, (*b*) experiment (from Arimondo *et al* 1981).

6.2. Kerr media

The first dynamic experiment on optical bistability, namely with a pulsed input field of duration τ_p, was done by Bischofberger and Shen (1978, 1979a, b). Here the non-linearity is purely dispersive and linear absorption is also present. For the liquid crystalline material MBBA in the isotopic phase at 45 °C, $n_2 = 1.5 \times 10^{-9}$ ESU, the Debye relaxation time $\tau_D = 174$ ns and $\alpha = 0.22$ cm^{-1}, at 65 °C $\tau_D = 15$ ns; for nitrobenzene at 20 °C, $\tau_D = 45$ ps, $n_2 = 1.1 \times 10^{-11}$ ESU; for CS$_2$ at 20 °C, $\tau_D = 2$ ps and $n_2 = 2.2 \times 10^{-11}$ ESU. The cavity mirrors had 1 cm separation, 98% reflectivity, and the finesse ~ 13; the cavity build-up time $t_c = 0.55$ ns. Two single-mode Q-switched ruby lasers with $\tau_p = 14$ ns and 62 ns (FWHM) were used. Three modes of operation were investigated: power limiter, differential gain and bistable.

The behaviour was analysed in terms of τ_p, the round-trip time $t_R(\sim 0.11$ ns) and τ_D. Three regimes were studied: extremely transient ($\tau_D \gg \tau_p \gg t_R$), moderately transient ($\tau_p \gg \tau_D \gg t_R$) and quasi-steady-state ($\tau_p \gg t_R \gg t_D$) for which the different media were used. In figure 34 the experimental results (bistable mode) and comparison with their theory are shown; peak intensities are on the figures. The quasi-steady state with $\tau_p/t_R = 140$ showed that transient effects were still significant but not so for $\tau_p/t_R = 620$. The maximum intensity of the pulses was 25 MW cm^{-2} which is less than the threshold for self-focusing. Their numerical simulations show overshooting and ringing, also found by Abraham and Hassan (1980) and Goldstone and Garmire (1981).

6.3. Solids: semiconductors and ruby

In search for a solid-state device, Venkatesan and McCall (1977) did an experiment on ruby which showed bistable action at room temperature. This was a rather surprising result. They expected the refractive index to arise from the R_1 and R_2 excitonic transitions whose contributions at that temperature—with the CW laser emitting 6934 Å radiation—cancelled each other out. It was then found that the refractive index arose from dispersive contributions from differential population effects. They observed differential gain, discriminator and limiter actions. This experiment encouraged research into other solid-state materials.

The first (simultaneous but independent) observations of OB in semiconductor etalons were done by Miller et al (1979) on InSb and Gibbs et al (1979b) on GaAs. In the experiments of Miller et al the crystal was 580 μm thick (200 μm diameter) with polished plane-parallel faces and held at 5 K (and later at 77 K). The input beam was a CO laser that had 60 to 70 lines from 1930 cm^{-1} to 1660 cm^{-1} and the line used was 1895 cm^{-1}. (The band gap of InSb is 1900 cm^{-1} and 1840 cm^{-1} at 5 K and 77 K, respectively.) The input power was varied with an attenuator (Miller and Smith 1978) with the desirable property of controlling the intensity over four orders of magnitude while keeping the Gaussian profile. The detector measured the full Gaussian beam power, and in spite of intensity variations across the beam, bistability was clearly observed. The physics of this non-linearity has been explained in § 4.

From an experimental point of view the observation of multistable behaviour (figure 4) was favoured by the large non-linearity of InSb: $n_2 = -1$ cm^2 kW^{-1} at 77 K and photons of 1840 cm^{-1} (cf § 4). The increased intensity spacing between bistable regions in figure 4 is qualitatively consistent with a self-defocusing effect that causes the medium to become a divergent lens. As a result, electromagnetic energy could be lost and a greater input intensity would be necessary for the desired effects. However, analysis shows that such divergence of beam is small in such a thin crystal and the effect is primarily due to a $\chi^{(5)}$ term in which n_2 is itself not a constant with intensity but reduces as the band-gap resonant saturation responsible becomes stronger (Miller et al 1981a,b,c). In other words this is to say that in (5.18) the expansion in powers of Ω^2 must be truncated at powers higher than the first: Ω^4, Ω^6, etc.

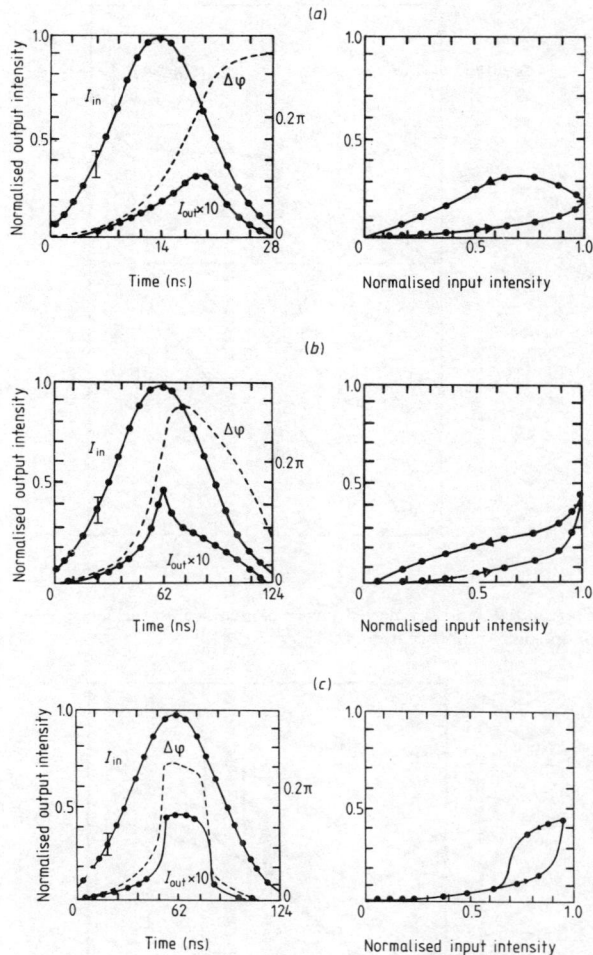

Figure 34. Dynamic response of a non-linear Fabry–Perot interferometer in the bistable mode. The dots correspond to measurements and the full curves to theoretical predictions. (a) Extremely transient: $t_R = 0.11$ ns, $\tau_p = 14$ ns, $\tau_D = 145$ ns; (b) moderately transient: $t_R = 0.11$ ns, $\tau_D = 15$ ns, $\tau_p = 62$ ns; (c) quasi-steady state: $\tau_D = 2$ ps, $t_R = 0.11$ ns and $\tau_p = 62$ ns. I_0 is the peak intensity of the input pulse; $\Delta\varphi$ is the phase increment of the output field as caused by the non-linearity of the medium. $\Delta\varphi$ is measured in submultiples of π on the right-hand vertical axes of the left-hand frames (from Bischofberger and Shen 1979a, b); $\theta = -0.2\pi$ and (a) $I_0 = 4.42$ MW cm^{-2}, (b) $I_0 = 1.74$ MW cm^{-2}, (c) $I_0 = 9.66$ MW cm^{-2}.

In figure 35 we show multistability in both the transmitted and reflected powers. In figure 36 we show the effect of a large non-linearity at 77 K ($\chi^{(3)} \sim 10^{-1}$ ESU) that combined with 70% reflecting coating gives clear bistability. This measurement also suggests that the mechanism is not excitonic as the latter would have meant a reduction in the non-linearity with increasing temperature (cf § 4). The sample used was 130 μm thick and made of impure polycrystalline InSb.

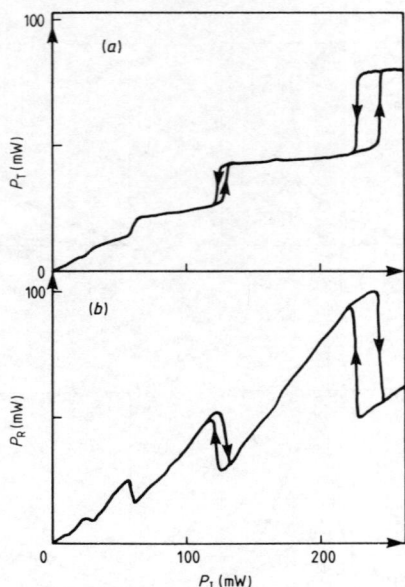

Figure 35. (a) Transmitted and (b) reflected power plotted against incident power. Data as in figure 4 but θ adjusted for optimum resonance (from Miller et al 1981b).

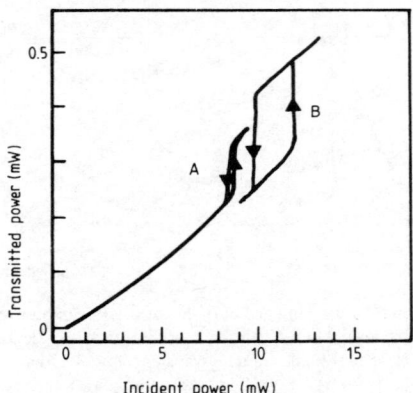

Figure 36. Hysteresis cycles for a laser beam ($\lambda^{-1} = 1827$ cm^{-1}) passing through a polished polycrystalline InSb slice (5 mm × 5 mm × 130 μm thick) coated to $R \sim 0.7$ on both faces and held at 77 K. Traces A and B differ in the cavity detuning (from Miller et al 1981b).

We mentioned in the introduction (§ 1.2) that a three-port device was possible. Miller and Smith (1979) observed optical transistor action, namely that a weak intensity laser controlled the output of a strong input beam. Their scheme is illustrated in figure 37(a) where the basic idea is to take advantage of the large refractive index of InSb ($n_0 = 4$) which allows two beams incident at slightly different angles to traverse the same crystal volume. The probe beam—whose electronic counterpart would be

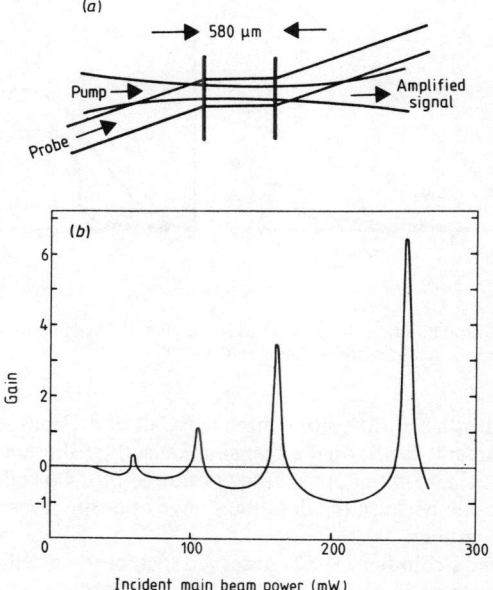

Figure 37. Differential amplification between CO laser beams at 1895 cm^{-1} in an InSb Fabry–Perot interferometer. (a) A weak probe beam incident at a slight angle controls the main beam ('transphasor' action). (b) Allowing for different mistuning the strongly amplified regions correspond to near vertical steps in figure 4 (from Miller and Smith 1979).

the transistor's base—is obtained either by splitting off a part of the main incident beam or by using a second laser. Figure 37(b) shows a plot of the gain (defined as the ratio between the increments of output and probe powers) against input power; the peaks correspond to the steep slope input values of figure 4. Miller and Smith called this system a 'transphasor' since modulation of the weak beam 'transfers' a small optical path change, causing a much larger change in the intensity of the main beam. The physics of the transphasor is affected by degenerate four-wave mixing (Miller et al 1980a) by which power is transferred between the various beams and new beams generated.

In the experiment of Gibbs et al (1979b) the bistable etalon consisted of a GaAlAs–GaAs–GaAlAs molecular-beam epitaxially-grown sandwich with 90% reflecting coatings; the temperature was varied between 5 and 120 K. The input laser beam was a tunable (770–870 nm) dye laser which was converted to produce 1 μs triangular pulses by an acousto-optic modulator; the holding intensity was 1 mW μm^{-2}. In figure 38 we show the results. In this system the contribution to the non-linear refractive index stems (cf § 4.2) from the free-exciton resonance, and although non-linear absorption is also present, Gibbs et al were unable to observe purely absorptive OB. The device had a sub-nanosecond switch-on time (figure 38) with a switching energy of 0.24 pJ μm^{-2} and a switch-off time of ~40 ns.

Bistability, being due to an excitonic effect, was observed in the range 15–107 K and the effect of using different wavelengths studied. At 150 K bistable behaviour was no longer due to excitonic effects but to thermal ones (Gibbs et al 1979c) which

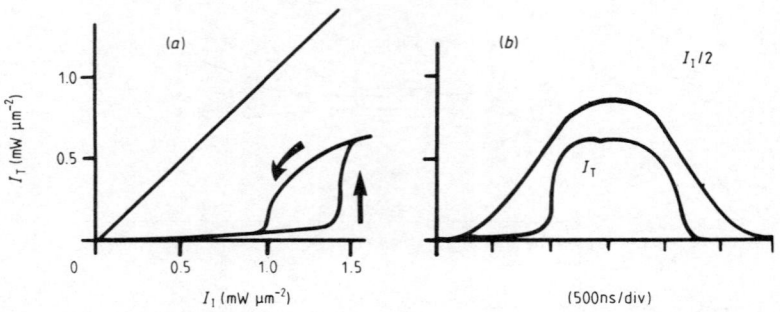

Figure 38. Excitonic optical bistability in GaAs at 15 K. (a) Transmitted intensity I_T against input intensity I_I. (b) Time dependence of I_I and I_T (from Gibbs et al 1979b).

was also observed in a 57 μm colour filter (McCall and Gibbs 1978a,b). The latter effect occurs as a result of thermal expansion caused by the absorption of light: the optical path changes, and with it, the tuning of the cavity. This effect is distinguishable from the excitonic one because the detunings have opposite signs: $dn/dI > 0$ (thermal) and $dn/dI < 0$ (excitonic).

Staupendhal and Schindler (1982) observed bistability in tellurium. Their experimental scheme was intended for optical–optical modulation, i.e. modulation of a laser beam by optical tuning of a Fabry–Perot interferometer. Here, the variation of refractive index depends on the change of density of electron–hole pairs, which in turn are created by multiphoton absorption—this is the origin of the non-linearity. The plane-parallel slab of tellurium is irradiated with a CO_2 laser ($\lambda = 10.6\,\mu$m, $\tau_p \sim 300$ ns) of peak intensity 6.6 MW cm^{-2}. The experimental set-up we are interested in is what these authors call self-controlled: the laser beam that changes the carrier concentration (the control beam) and the beam to be modulated (the probe beam) are exactly the same (this is essentially figure 3). The slab was temperature-stabilised in the region of 300 K. The observation is shown in figure 39. In table 1 of § 8 we show a comparison between the three semiconductors.

6.4. Non-linear interface and other geometries

We explained the physical principles in § 4. The first observation of OB at a non-linear interface was achieved by Smith et al (1979). The input beam was provided by a mode-locked ruby laser generating 1 ns pulses whose reflected fraction was detected; the Kerr medium was CS_2 ($n_2 = 3 \times 10^{-8}$ cm^2 mW^{-1}) and kept at room temperature. The measured threshold intensity was 7.5×10^9 W cm^{-2}. In figure 40 we show the experimental time dependence of intensities and optical hysteresis. Given the high response time of detection (~ 300 ps) these authors were unable to check the fast response expected from this non-linearity (i.e. ~ 2 ps).

Another cavity-less intrinsic device has been proposed and experimentally demonstrated by Bjorkholm et al (1981). This is based on self-focusing of light which occurs, for example, when a laser beam with a Gaussian profile propagates through a non-linear medium with $n_2 > 0$: as the larger refractive index appears at the centre of the beam, the medium behaves like a convergent lens. The beam is focused onto the input face of the medium. If the input power $P_I > P_{cr}$, where P_{cr} is called the critical power, then

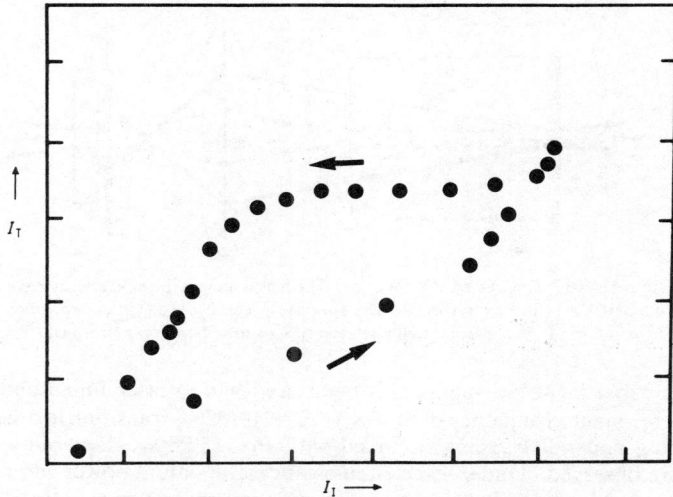

Figure 39. Optical bistability in tellurium at ~300 K (from Staunpendhal and Schindler 1982).

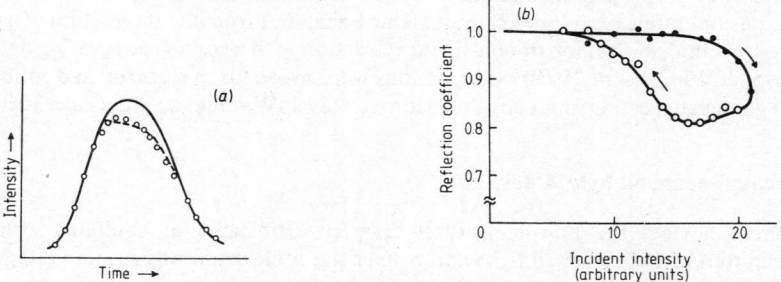

Figure 40. Non-linear interface experiment (cf figure 22). (a) Time dependence of incident and reflected pulses and (b) optical hysteresis in the reflection coefficient obtained with data from (a) (from Smith et al 1979). ——, incident pulse; ○, reflected pulse (experiment); – – –, reflected pulse (theory).

self-focusing occurs. For $P_I = P_{cr}$ the beam propagates with no change in spot size (self-trapping). The input focal spot size can be chosen so that for $P_I \ll P_{cr}$ there is appreciable divergence as shown by the full lines of figure 41. After the radiation is imaged by a lens L it finds an aperture in front of a partially transmitting mirror M. The aperture has to be small enough so as to provide very little feedback in the absence of self-focusing, i.e. low transmission, and large enough to reflect all the light under conditions of self-trapping (high transmission). For $P_I \geqslant P_{cr}$ strong feedback occurs which in turn increases the self-focusing effect. In order to regain divergence—and therefore low transmission—an input power $P_I < P_{cr}$ is necessary since both the incident and the reflected powers contribute to the non-linearity: optical hysteresis is obtained.

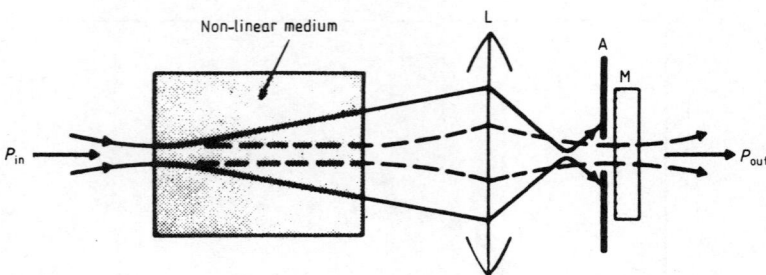

Figure 41. Optical bistability based on self-focusing. The lens L images the field after it leaves the non-linear medium onto a partially reflecting mirror M. The aperture A causes feedback to depend on the laser beam spot size at the exit face of the medium. Full line corresponds to self-trapping (from Bjorkholm et al 1981).

In this experiment Na vapour has been used with a laser tuned about 1.2 GHz above the resonance frequency of the $3S_{1/2}(F=2) \to 3P_{1/2}$ transition at 5896 Å. Strong self-focusing and self-trapping occurred with $P_I \sim 150$ mW. Bistability and limiter action were observed. Under some circumstances the output power was noisy by way of a rapid switching back and forth between different transmission states. The authors believe that the latter effect is related to chaos (cf § 3.3).

Other devices which do not require a cavity have been proposed. Kaplan (1981b) predicts bistability based on self-focusing (or self-defocusing) and self-bending caused by two counter-propagating beams in a non-linear medium. G Stegeman and D Sarid (private communication) predict multistable behaviour from the interaction of a guided wave travelling in a region of non-linear refraction with a corrugated grating deposited on top (cf Winful et al 1979). Calculations were made for two corrugated InSb-based waveguides suggesting input powers as low as a few mW would be practical possibilities.

7. Experiments on hybrid devices

In these devices the non-linearity is created 'artificially' by sampling either the transmitted or the reflected light and converting it electronically into a voltage. The latter is applied to the medium which responds by changing its refractive index. The advantage of these devices seems to lie in their arbitrarily large non-linearity and broad range of wavelengths: they can also function at low powers. The switching time is, however, limited by the detector and amplifier response times. Experimental hybrid devices have been pioneered by Smith and Turner (1977) but many workers in the field have constructed variations that work on the same principle. The macroscopic theory of the 'non-linearity' is formally that of § 2.

The first electro-optic device (Smith and Turner 1977) is shown in figure 42. The beamsplitter takes part of the light which is 'transformed' into a static electric field applied to the crystal whose refractive index changes accordingly. The transmission of the cavity is given by (2.1) where $\Phi/2 = 2\pi n_0 L/\lambda + \varphi(P_T)$ (intensities are replaced by powers). In this case (Smith et al 1978c), $\varphi(P_T) = AP_T$ where A is a constant that depends on the detector and electro-optic characteristics. The experiment of Smith and Turner was done with a He–Ne laser at 6328 Å. The cavity was a 10 cm confocal resonator $R \sim 80\%$ containing a KDP electro-optic modulator with a half-wave voltage of 1200 V; the finesse was 7. The familiar modes of operation were observed (figure 43). In the bistable mode the switching occurs with less than 1 μW on the detector.

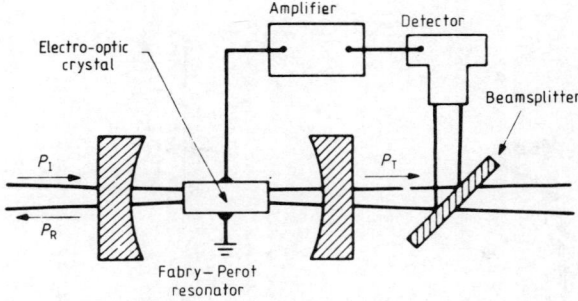

Figure 42. Electro-optic bistable device; data in figures 42–47 are obtained with this scheme (from Smith and Turner 1977).

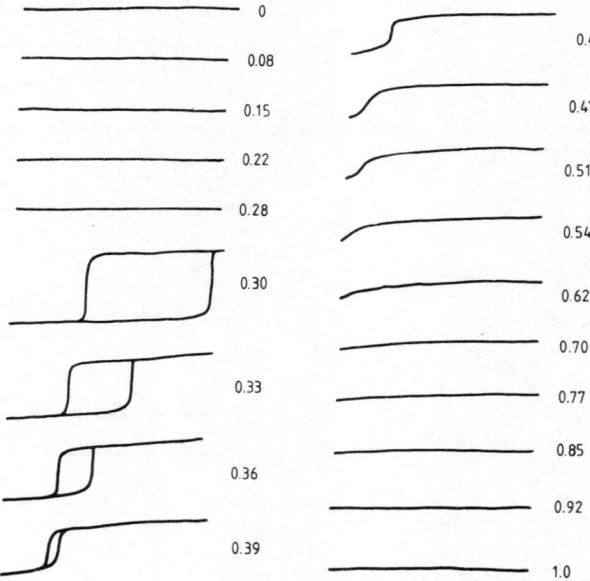

Figure 43. Characteristic curves for different cavity detunings in units of resonator intermode spacing (from Smith and Turner 1977).

Further work by Smith *et al* (1978c) using basically the set-up of figure 42 showed pulse-shaping, optical limiter and optical logic operations were proposed. In figure 44 we show differential gain (cf figure 14) and in figure 45 the 'optical triode' operation. The latter is obtained by sending through the beamsplitter a weak control beam that reaches the detector and modulates the main beam. The control beam can be of a different frequency and even incoherent. The resonator contained a $LiTaO_3$ electro-optic modulator with a half-wave voltage of 180 V; the gain was 7. Smith *et al* have also seen a high non-linearity behaviour by either increasing the incident powers or the gain in the feedback path which exhibited multistability.

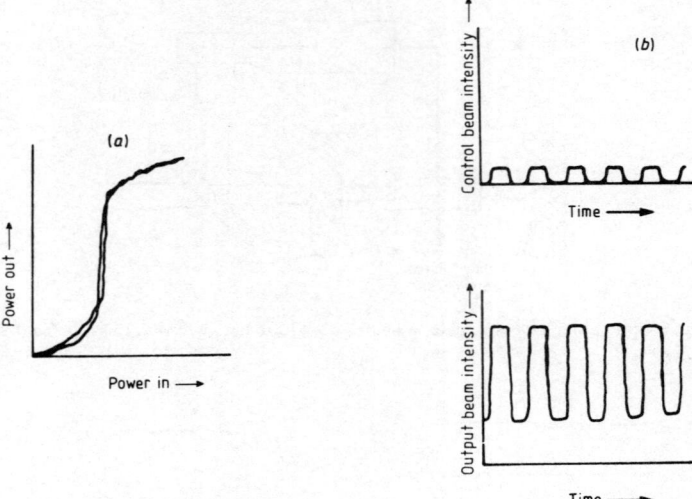

Figure 44. Triode operation. (a) Characteristic curves; (b) input and output powers against time showing differential gain (from Smith et al 1978c).

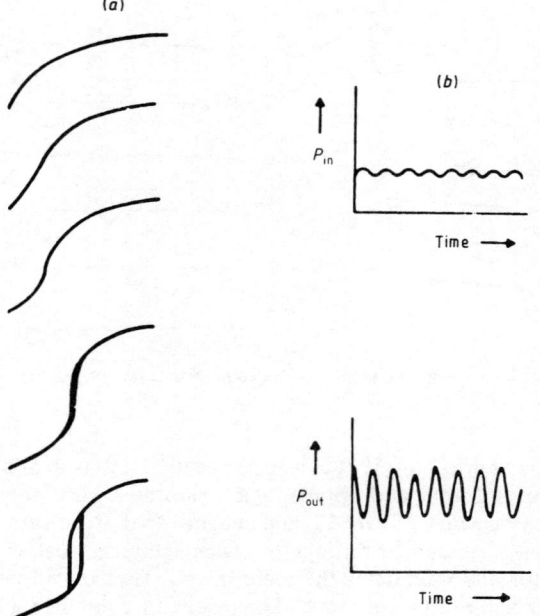

Figure 45. (a) Output characteristics as a function of cavity detuning; (b) input and output powers against time showing differential gain (from Smith et al 1978c).

If instead of sampling the transmitted light the reflected light is sampled, a different bistable behaviour results (Smith *et al* 1978d); the set-up is again given in figure 42. The resonator contained a LiTaO$_3$ electro-optic phase modulator with a half-wave voltage of ~250 V; the rest is unchanged. The fraction ρ (Smith *et al* 1978d) of reflected power is

$$\rho = \frac{P_R}{P_I} = \frac{4R \sin^2 \Phi/2}{(1-R)^2 + 4R \sin^2 \Phi/2}$$

with $\Phi/2 = 4\pi n_0 L/\lambda + \varphi(P_R)$ where $\varphi(P_R) = BP_R$; B depends on the same as A above. A graphical solution (cf § 2) leads to a hysteresis cycle (figure 46); the experimental results are in figure 47.

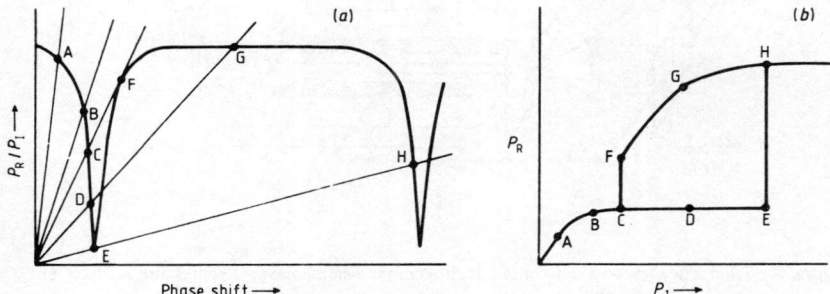

Figure 46. Graphical solution similar to figure 3 exhibiting hysteresis in the reflected light with the latter as feedback. (*a*) Fraction of reflected light against phase shift for several values of incident power; (*b*) corresponding hysteresis cycle. P_R is the power reflected and P_I is the incident power (from Smith *et al* 1978d).

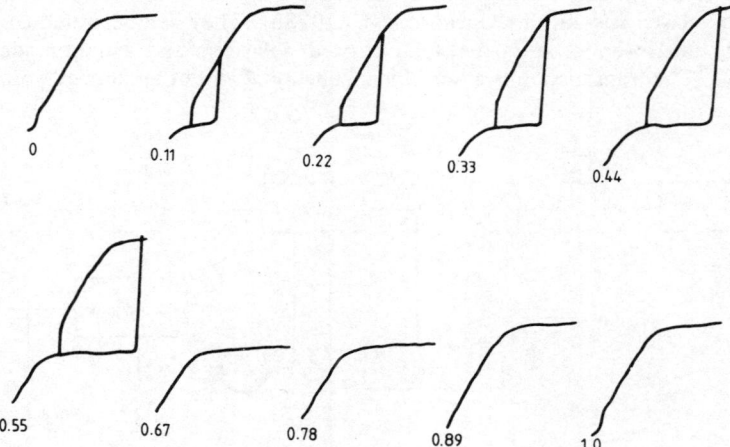

Figure 47. Experimental plots corresponding to figure 46; numbers indicate cavity detunings in units of the cavity mode spacing (from Smith *et al* 1978d).

In pursuit of integrability Smith et al (1978a,b) have built a self-contained bistable device (figure 48) with Ti in-diffused waveguides in $LiNbO_3$ proposed by Kaminow et al (1975). The ends of this crystal were cleaved and dielectric mirrors attached to form a Fabry–Perot resonator. Feedback from a photovoltaic detector that takes part of the transmitted light and the voltage so produced is applied to the modulator electrodes. The single-mode waveguide had a width of 4 μm, a length of 1.5 cm and electrode separation of 9 μm. They observed optical hysteresis for optical powers of 50 μW at the detector; the switching energy was 0.5 μJ and the switching time 10 ms.

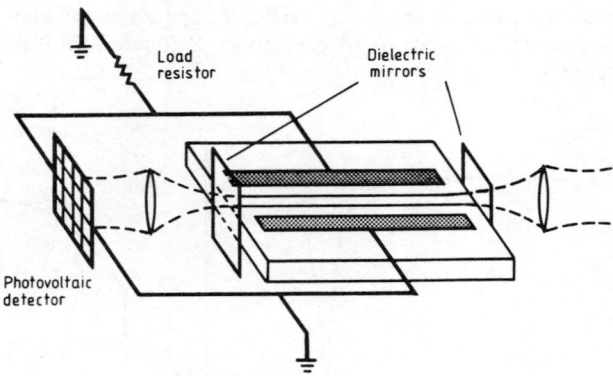

Figure 48. Schematic view of a self-contained, integrated bistable device (from Smith et al 1978b).

Cross et al (1978) built an integrated device based on a waveguide directional coupler. It has four all-optical parts such that all optical data processing or remote optical switching can be achieved. Switching times of ~300 μs and 3 pJ energies were observed.

A device of extreme simplicity with no resonator and which did not require a single-mode laser was built by Garmire et al (1978b). They demonstrated OB using a $LiNbO_3$ electro-optic crystal between crossed polarisers and a multimode laser (figure 49). The transmission is a non-linear function $T(V)$ of an applied voltage so

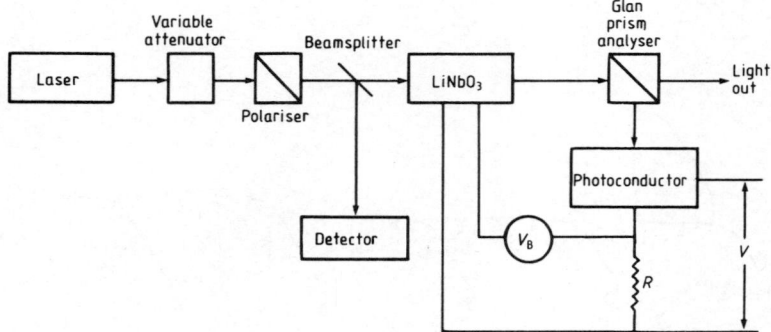

Figure 49. Experimental set-up for an incoherent mirrorless bistable device (see text) using an electro-optic polarisation modulator (from Garmire et al 1978b).

that $P_T = P_I T(V)$. The output is converted into this voltage $V = \alpha P_T$ which is fed back into the LiNbO$_3$, then $P_T/P_I = T(V_B + \alpha P_T)$ where V_B is the bias voltage (independent of P_T). By knowing $T(V)$ a graphical solution is possible (cf § 2). In this experiment the 'modulator' was a c-cut crystal 1×1 mm^2 and 2 cm long with a half-wave voltage of 260 V; the laser was a multimode He–Ne of 0.5 mW. The data are in figure 50. The operating powers and switching times are of the same order as those of Smith et al. The elimination of a resonator is a step forward towards integrability. Garmire et al (1978a) have also constructed an OB switch using a LiNbO$_3$ phase modulator—active length 2 mm—in a non-cavity configuration. Using a multimode He–Ne laser they observed bistability and discriminator action.

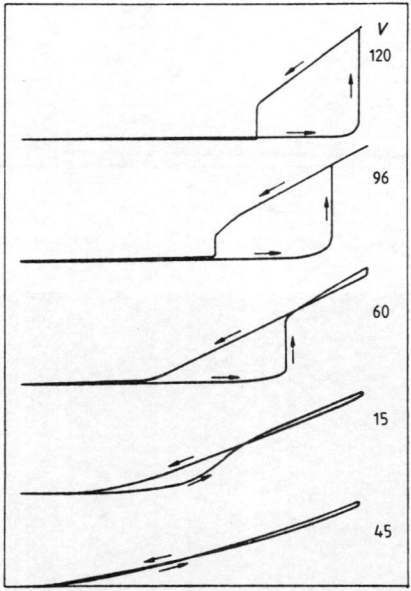

Figure 50. Hysteresis cycles with the set-up of figure 49 for a number of bias voltages (V_B) (from Garmire et al 1978b).

The first observation of critical slowing down in OB was done by Garmire et al (1979). They studied the response of a hybrid device to stepwise excitation and found a dependence of the switching time on the applied light increment (figure 51). Critical slowing down both limited response times and also increased switching energies at low input intensities. Another dynamical experimental study by Goldstone et al (1980) shows a remarkable behaviour of a hybrid device that consisted of a LiNbO$_3$ polarisation modulator driven by the output of a photoconductor monitoring the transmission of the device. Two time constants are in play: the risetime of the photoconductor and the time constant of a variable capacitance placed across the resistor of the modulator. It was observed that it is possible for the switching of an OB device to overshoot at input intensities well below the critical steady-state value (figure 52). It was also observed that an incident train of pulses on an OB device that displays overshoot switching may cause a regime in which the device switches alternately with

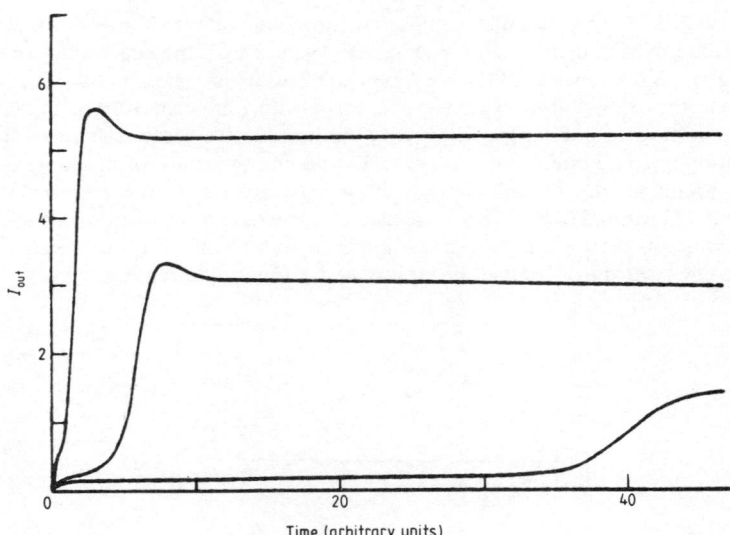

Figure 51. Experimental observation of 'critical slowing down' (cf § 3.4) (from Garmire et al 1979).

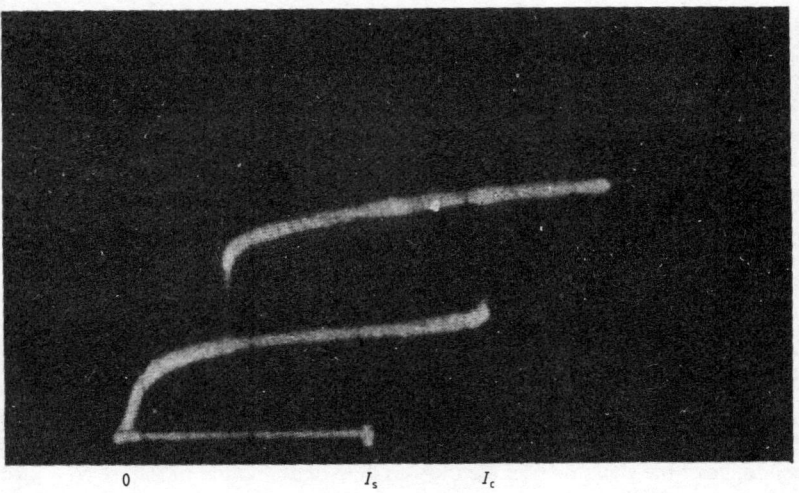

Figure 52. Experimental characteristic curve showing overshoot switching represented by the difference between the steady-state switching intensity I_c and the minimum overshoot switching intensity I_s (from Goldstone et al 1980).

an output which is a subharmonic of the input pulse train. They also showed that the difference between the critical and the actual switching intensities increases as the ratio of time constants approaches unity.

Okada and Takizawa (1979) observed multistability in a mirrorless electro-optic device. This multistable operation was controllably achieved by averaging the polariser and/or analyser at appropriate angles with respect to the crystallographic axis of a $LiNbO_3$ electro-optic crystal. Okada (1980a) and Okada and Takizawa (1980) have also studied optical regenerative oscillations and monostable pulse generation in hybrids. Okada (1980b) also investigated means of reducing the switching energy in electro-optic OB devices.

Kompanets et al (1981a,b) developed an optical multistable spatial device which uses a spatial light modulator. This contains a photosensitive electro-optical liquid crystal layer where an internal feedback loop is established. Feldman (1978, 1979) has built optical bistable devices based on a Pockels cell. Schnaper et al (1979) observed bistability, differential gain (up to 50) and clipping with an integrable two-arm interferometer. Schnaper et al (1981) also reported results on the remote control of a bistable directional coupler. They show that the output from the coupler can be switched from one channel to another by means of information in the incident light itself. Switching times of 100 µs were limited by the electronic feedback loop of the system. Vincent and Otis (1981) developed a device that can operate at 10.6 µm. A CW CO_2 laser was used and an electro-optic CdTe amplitude modulator. Ito et al (1981) developed an integrated optical multivibrator using an electro-optically controlled OB device. Tarucha et al (1981) obtained bistable switching and triode operation using an optical directional coupler switch. Sohler (1980) has reported an OB device that with the adequate input power can be used as a Schmitt trigger (electrical bistability). With electrical feedback it becomes an electro-optical multivibrator.

A hybrid device was also used by McCall (1978) to obtain regenerative pulsations that proved the instability of the upper branch as predicted by this very author: a CW laser was converted into a train of pulses. Finally, the first observation of chaos in OB and laser systems was done by Gibbs et al (1981) with a hybrid device. The theory of this system had the same mathematical structure as that of the Maxwell–Debye equations of Ikeda et al (1980); good qualitative agreement was obtained. In figure 53 we show I_T plotted against I_I for two cases: $t_R \ll \tau$ and $t_R \gg \tau$ (condition for chaos). Three domains are noticeable: stable, periodic and chaotic, as predicted (cf figure 16(b)).

8. Switching times and energies

Optically bistable devices are expected to be used in the future as essential parts of optical integrated circuits and optical computers (see, for example, Smith and Miller 1980a,b), in which the light beam plays the role of the electron current in electronic integrated circuits in electronic computers. In order to get more effective logic and memory devices than very large scale integration or Josephson junctions (see, for example, McDonald 1981), the system would be required to switch on and off in a few ps and using only a few pJ for each operation. One problem already faced by Miller and Smith and Gibbs et al is that the 'switch-off' is comparatively long, e.g. tens or hundreds of nanoseconds in GaAs and InSb. Other requirements are micron dimensions, which ensures cavity build-up times of ~1ps (see below) and small holding power (mW). Fast response (~1 ps) of the non-linearity is required ideally.

Semiconductors already offer the possibility of an arbitrarily fast 'switch on' of refractive non-linearity since only the intensity and the effective absorption coefficient (cf § 5) for carrier generation controls the rate of excitation. To 'switch off' the non-linearity natural relaxation processes are needed. Considerable control is possible

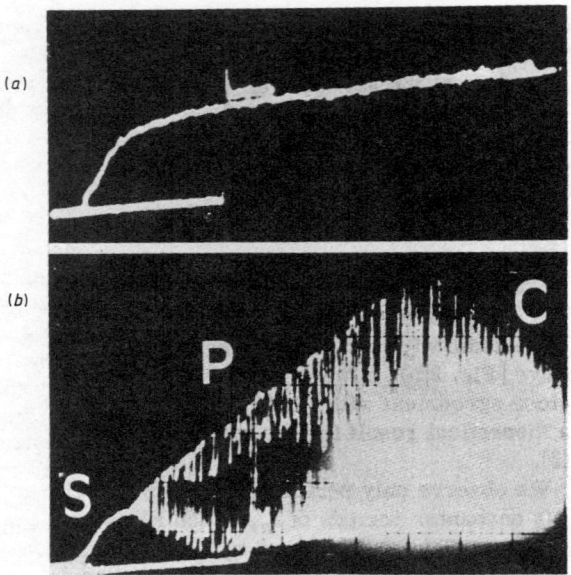

Figure 53. Observation of Ikeda-type instabilities (cf § 3.3) with a hybrid device. (a) Characteristic curve as the input intensity is cycled slowly from zero to some maximum and back over a time of 30 s. Here $t_R = 160\ \mu s \ll T_1 = 1$ ms. (b) As (a) but $t_R = 40$ ms $\gg T_1 = 1$ ms where S, P and C stand for stable, periodic and chaotic regions (from Gibbs et al 1981).

of these de-excitation processes, e.g. the following:

(a) rapid sweep-out with electric field;

(b) trapping, with introduced disorder or use of amorphous material, e.g. α-Si;

(c) at high enough intensities upper-state lifetime limiting by stimulated emission to a lower state;

(d) carrier diffusion out of a small μm-sized element, possibly aided by recombination.

There is already evidence (e.g. dynamic reflectivity recovery in Si) that relaxation can be on a time scale of ~ 10 ps.

The length of the cavity (or sample) poses one of the limitations. In an empty cavity when the field is suddenly switched on, the field inside the Fabry–Perot cavity rises exponentially:

$$E_c(t) = \frac{E_I}{(1-R)^{1/2}}[1 - \exp(-t/t_c)]$$

where the risetime (cf § 3) $t_c = 2L/c(1-R)$. An optimum value for R is ~ 0.7 so $t_c = 2L/0.3c$; thus $t_c = 10$ ps for $L = 0.5$ mm or $t_c = 1$ ps for $L = 50\ \mu$m (these lengths are to be divided by the linear refractive index). On the other hand, the condition for observation of significant non-linear interferometric action is the requirement that the change in optical thickness contribution should be of the order of the cavity linewidth, i.e.

$$n_2 IL \geq (\lambda/2) F^{-1}$$

where F is the finesse, and the non-linearity should respond within t_c. These last two requirements are not so far trivially met although there are examples of sufficient speed of non-linearity, using Kerr liquids (cf § 6) where, however, the cavities are too long. Staupendahl and Schindler (1982), on the other hand, reported that a Te resonator ($L = 1.3$ mm) switched up in less than 1 ns with a $t_c \sim 100$ ps. We can summarise some of these characteristic times and non-linearities from present experiments in table 1.

Table 1.

	GaAs	InSb	Te
L (μm)	4	130	1300
	$R = 0.9$	$R = 0.7$	$R = 0.4$
t_c (calc)	~ 1 ps	~ 10 ps	~ 100 ps
n_2 (cm^2 kW^{-1})	10^{-4}	1.0	10^{-6}
α (cm^{-1})	$>10^3$	~ 1	~ 1
$\beta = n_2/\lambda\alpha$	10^{-6}	0.2	10^{-6}

β is the figure of merit (cf § 2). None of the above systems have been optimised; in both InSb and Te the resonator length L can be shortened.

Using the direct-saturation mechanism of § 5 we can understand the limiting physical process for a fast response of the non-linearity. Power broadening theory gives significant effects when

$$\Omega^2 T_1 T_2 > 1 + \Delta^2 T_2^2.$$

The significant quantity here is the Rabi frequency Ω which in our context represents (Ω^{-1}, that is) the minimum time to completely change the population from the lower to the upper level of the transition. The square of Ω must therefore be comparable with $T_1 T_2$. With a high laser intensity the system can be 'driven' as fast as required and therefore have a fast non-linear response. On reducing the field, however, relaxation must take place and its speed will be controlled by T_1. Approximate values $T_1 \sim 100$ ps, $T_2 \sim 1$ ps have been given from saturation of cyclotron resonance by Gornik et al (1978). This states the problem for the 'state filling' case; the conclusion is essentially the same for 'band filling' although the thermalising time must also be considered in relation to the switching time in this case.

Using the ideas and theory of § 5 and this one we can deduce from experiments what the device speculations are. Using n_2 and α from the above table and assuming it is possible to utilise an element 10 μm × 10 μm in dimension, we can estimate the following.

(a) Holding power, i.e. the steady-state power required to hold the resonator near to the switching intensity ~ 250 nW for InSb.

(b) Switching energy. One photon per cm^3 gives $\Delta n = 10^{-17}$ in InSb; this yields a switching energy of 0.01 pJ.

(c) Switching time. Switch up with an energy 0.1 pJ implies that 100 mW will switch the device in 1 ps. Switch down will depend on relaxation times.

In the next section we give an example of a scaling possibility.

9. Summary and conclusions

Optical systems which are said to be bistable if two transmission states correspond to the same input power have been realised in practice in several different forms. Conceptually the simplest bistable system consists of a passive non-linear medium within a Fabry–Perot cavity that provides optical feedback. Different modes of operation are possible, such as differential gain, power limiter, clipper and so on, which can find useful applications in optical circuits. The first suggestions for optical bistable systems seemed only achievable with saturable absorbers; the first successful experiment on sodium vapour demonstrated the role of non-linear refraction, thereby releasing the restriction on absorptive methods. This important discovery led to an ever-increasing interest in the subject as it has shown that bistability can be observed with powers much lower than those needed for saturation of the absorption and subsequently to a whole series of different systems and processes.

The simplest theoretical approach to optical bistability consists of introducing a non-linear refractive index in the equation for the transmission of a Fabry–Perot cavity; this refractive index takes the form $n = n_0 + n_2 I$. Early experiments were performed on sodium, Kerr media and ruby. However, the most promising materials for device purposes are semiconductors: to date, observations have been achieved in indium antimonide, gallium arsenide and tellurium.

If devices derived from the family of bistable devices, but particularly non-linear Fabry–Perot interferometers, are to be practically useful in optical processing and computing, switching and response times need to approach picoseconds. A scaling example is as follows: the experiments on InSb gave a change in refractive index $\Delta n = 10^{-17}$ for 1 photon cm^{-3} and this is the most favourable case to consider at present. The area d^2 of the sample is limited by diffraction so d must satisfy $d^2 > (\lambda/n_0)^2$; then for $\lambda = 5$ μm, $n_0 = 4$, we have $d \sim 2.0$ μm. For a thickness of 6 μm, $R = 95\%$, the cavity build-up time (cf § 8) $t_c \sim 2$ ps. With the former choice of parameters the resulting switching pulse energy is 1 pJ. Sufficient carriers can be excited with an approximate mW input power for less than 1 ns 'switch-up' time; the latter can be reduced provided the power is increased in order to keep the same 1 pJ pulse energy. Recovery is determined by the recombination time τ_R and for the mechanisms discussed in § 5 it is shown that n_2 is proportional to τ_R. For example, if $n_2 = 1.0$ cm^2 kW^{-1} ($\chi^{(3)} \sim 1$ ESU) when $\tau_R \sim 100$ ns, then for $n_2 \simeq 10^{-5}$ cm^2 kW^{-1} ($\chi^{(3)} \sim 10^{-5}$ ESU) for $\tau_R \sim 1$ ps; this last value of n_2 still represents a very large non-linearity. The size of τ_R itself can be reduced for faster recovery by making use of the 'micro-engineering' methods discussed in § 8. Consequently there is hope that optically bistable devices may be realisable on picosecond time scales. The question as to whether they can compete with present devices to process information can then be considered.

Additional reading

For completeness we add a list of references relevant to the following sections.

Section 1.

Abraham E, Seaton C T and Smith S D 1982 *Sci. Am.* to appear
Abraham E and Smith S D 1982 *J. Phys. E: Sci. Instrum.* **15** 33–9

Collins S A Jr and Wasmundt K C 1980 *Opt. Engng* **19** 478–87
Gibbs H M, McCall S L and Venkatesan T N C 1979 *Opt. News* **5** 6–10
Greene W P, Gibbs H M, Passner A, McCall S L and Venkatesan T N C 1980 *Opt. News* **6** 16–9
Smith S D and Abraham E 1982 *Proc. Int. School on Semiconductor Opto-Electronics, CETNIEWO '81, Wladiswowo, Poland* ed M A Hermann (New York: Wiley)
Smith S D and Miller D A B 1980 *New Scientist* **21** 554–6

Section 3.

Agarwal G S, Narducci L M, Feng D H and Gilmore R 1980 *Phys. Rev.* A **21** 1029–38
Agarwal G S and Tewari S P 1980 *Phys. Rev.* A **21** 1638–47
Arecchi F T and Politi A 1979 *Opt. Commun.* **29** 361–3
Benza V, Lugiato L A and Meystre P 1980 *Opt. Commun.* **33** 113–8
Bowden C M and Sung C G 1979 *Phys. Rev.* A **19** 2392–400
Casagrande F and Lugiato L A 1980 *Nuovo Cim.* B **55** 173–90
Chrostowski J and Zardecki A 1979 *Opt. Commun.* **29** 230–4
Drummond P D 1981 *Opt. Commun.* **40** 224–8
Farina J D, Narducci L M, Yuan J M and Lugiato L A 1980 *Opt. Engng* **19** 469–77
Gronchi M, Benza V, Lugiato L A, Meystre P and Sargent M III 1981 *Phys. Rev.* A **24** 1419
Gronchi M and Lugiato L A 1978 *Lett. Nuovo Cim.* **23** 593–8
Hasegawa H, Nakagomi I, Mabuchi M and Kondo K 1980 *J. Stat. Phys.* **23** 281–313
Hermann J A and Elgin J N 1981 *Phys. Lett.* **86A** 461–3
Lugiato L A 1975 *Physica* **81A** 565
—— 1976 *Physica* **82A** 1
—— 1980 *Lett. Nuovo Cim.* **29** 375–80
—— 1981 *Z. Phys.* B **41** 85–94
Lugiato L A and Milani M 1976 *Physica* A **85a** 1–17
Narducci L M, Gilmore R and Feng D H 1979 *Phys. Rev.* A **20** 545–9
Schwendiman P 1979 *J. Phys. A: Math. Gen.* **12** L39–42
Tewari S P 1980 *Opt. Commun.* **34** 273–7
Tewari S P and Tewari S P 1979 *Optica Acta* **26** 145–8
—— 1980 *Optica Acta* **27** 129–33

Section 4.

Averbukh I Sh, Kovarskii V A and Perel'man N F 1980 *Zh. Eksp. Teor. Fiz. Pis Red.* **32** 277–81
Baker H C and Armstrong L Jr 1981 *Opt. Lett.* **6** 357–9
Drummond P D, McNeil K J and Walls D F 1979 *Opt. Commun.* **28** 255–8
Drummond P D and Walls D F 1980 *J. Phys. A: Math. Gen.* **13** 725–41
Hassan S S, Tewari S P and Abraham E 1982 *Opt. Commun.* **40** 461–5
Kaplan A E and Meystre P 1982 *Opt. Commun.* **40** 229–32
Li Chun-Fi 1980 *Wuli (China)* **9** 99–101
Lugovoi V N 1980 *Sov J. Quantum Electron.* **9** 1207–12
Marcuse D 1980 *Appl. Opt.* **19** 3130–9
Nishiyama Y 1980 *Phys. Rev.* A **21** 1618–23
—— 1980 *Phys. Rev.* A **22** 2723–5
Rozanov N N 1981 *Zh. Eksp. Teor. Fiz.* **80** 96–108
Rozanov N N and Semenov V E 1980 *Opt. Spectrosc.* **48** 59–63
Sarid D 1981 *Opt. Lett.* **6** 552–3
Selloni A and Schwendiman P 1979 *Optica Acta* **26** 1541–7
Selloni A, Quattropani A, Schwendiman P and Baltes H P 1981 *Optica Acta* **28** 125–30
Walls D F, Zoller P, Drummond P D and Kunasz C V 1981 *Chaos and Order in Nature* ed H Haken (Berlin: Springer-Verlag) pp 102–10

Section 5.

Koch S W, Schmitt-Rink S and Haug H 1981 *Phys. Stat. Solidi* b **106** 135–40

Section 6.

Eichler H J, Heritage J P and Beisser F A 1981 *IEEE J. Quantum Electron.* **QE-17** 2351–5
Gibbs H M, McCall S L and Venkatesan T N C 1980 *Opt. Engng* **19** 463–8
Pepper D M and Klein M B 1979 *IEEE J. Quantum Electron.* **QE-15** 1362–9
Sengupta U K, Gerlach U H and Collins S A 1978 *Opt. Lett.* **3** 199–201

Section 7.

Carenco A and Menigaux L 1980 *Appl. Phys. Lett.* **37** 880–2
Ito H, Ogawa Y, Makita K and Inaba H 1979 *Electron Lett.* **15** 791–3
Laulicht I 1981 *Opt. Quantum Electron.* **13** 295–300
Rozanov N N 1980 *Sov. Tech. Phys. Lett.* **6** 77–8
Smith P W 1980 *Opt. Engng* **19** 456–62
Smith P W, Tomlinson W J, Maloney P J and Hermann J-P 1981 *IEEE J. Quantum Electron.* **QE-17** 340–8
Sohler W 1980 *Appl. Phys. Lett.* **36** 351–3

Section 8.

Hopf F A and Meystre P 1979 *Opt. Commun.* **31** 245–50
—— 1979 *Opt. Commun.* **29** 235–8
—— 1980 *Opt. Commun.* **33** 225–30
Hopf F A, Meystre P, Drummond P D and Walls D F 1979 *Opt. Commun.* **31** 245
Lugiato L A, Milani M and Meystre P 1982 *Opt. Commun.* **40** 307–11

For a collection of recent papers—some of them already cited in the main text—see 1981 *Optical Bistability* ed C M Bowen, M Cliftan and H R Robl (New York: Plenum).

References

Abraham E 1979 *PhD Thesis* University of Manchester
Abraham E and Bullough R K 1980 *Proc. 4th National Quantum Electronics Conf., Edinburgh 1979* ed B S Wherrett *Laser Advances and Applications* (New York: Wiley) pp 245–50
Abraham E, Bullough R K and Hassan S S 1979 *Opt. Commun.* **29** 109–14
Abraham E, Firth W J and Wright E M 1982 *Proc. 5th National Quantum Electronics Conf., Hull* ed P L Knight (New York: Wiley)
Abraham E and Hassan S S 1980 *Opt. Commun.* **35** 291–7
Abraham E, Hassan S S and Bullough R K 1980 *Opt. Commun.* **33** 93–8
Agarwal G S 1980 *Opt. Commun.* **35** 149–52
Agarwal G S, Narducci L M, Feng D H and Gilmore R 1978a *Coherence and Quantum Optics IV* ed P Mandel and E Wolf (New York: Plenum) pp 281–92
—— 1978b *Phys. Rev.* A **18** 620–34
Agrawal G P and Carmichael H J 1979 *Phys. Rev.* A **19** 2074–86
—— 1980 *Optica Acta* **27** 651–60
Agrawal G P and Flytzanis C 1980 *Phys. Rev. Lett.* **44** 1058–61
—— 1981 *IEEE J. Quantum Electron.* **QE-17** 371–80
Agrawal G P, Flytzanis C, Frey R and Pradere F 1981 *Appl. Phys. Lett.* **38** 492–4
Agrawal G P and Lax M 1979 *J. Opt. Soc. Am.* **69** 1717–9
Allen L and Eberly J H 1975 *Optical Resonance and Two-Level Atoms* (New York: Wiley)
Arecchi F T and Politi A 1978 *Lett. Nuovo Cim.* **23** 65–9
Arimondo E, Gozzini A, Lovitch L and Pistelli E 1981 *Proc. Int. Conf. on Optical Bistability, Ahseville 1980. Optical Bistability* ed C M Bowden, M Cliftan and H R Robl (New York: Plenum) pp 151–71
Austin J W and De Shazer L 1971 *J. Opt. Soc. Am.* **61** 650
Ballagh R J, Cooper J, Hamilton M W, Sandle W J and Warrington D M 1981 *Opt. Commun.* **37** 143–8
Benza V and Lugiato L A 1979a *Lett. Nuovo Cim.* **26** 405–9
—— 1979b *Z. Phys.* B**35** 383–9

Bischofberger T and Shen Y R 1978 *Appl. Phys. Lett.* **32** 156–8
—— 1979a *Opt. Lett.* **4** 40–1
—— 1979b *Phys. Rev.* A **19** 1169–76
Bjorkholm J E, Smith P W, Tomlinson W J and Kaplan A E 1981 *Opt. Lett.* **6** 345–7
Bonifacio R and Lugiato L A 1975a *Phys. Rev.* A **11** 1507
—— 1975b *Phys. Rev.* A **12** 587
—— 1976 *Opt. Commun.* **19** 172–6
—— 1978a *Phys. Rev.* A **18** 1129–44
—— 1978b *Lett. Nuovo Cim.* **21** 505–9
—— 1978c *Lett. Nuovo Cim.* **21** 517–21
—— 1978d *Lett. Nuovo Cim.* **21** 510–16
—— 1978e *Phys. Rev. Lett.* **40** 1023–6
Bonifacio R, Lugiato L A, Farina J D and Narducci L M 1981 *IEE J Quantum Electron.* **QE-17** 357–65
Bonifacio R, Gronchi M and Lugiato L A 1978 *Phys. Rev.* A **18** 2266–79
—— 1979a *Nuovo Cim* B **53** 311
—— 1979b *Opt. Commun.* **30** 129–33
—— 1979c *Proc. 4th Conf. on Laser Spectroscopy* ed H Walther and K Rothe (Berlin: Springer-Verlag) p 426
Bonifacio R and Meystre P 1978 *Opt. Commun.* 147
—— 1979 *Opt. Commun.* **29** 131
Born M and Wolf E 1965 *Principles of Optics* (Oxford: Pergamon)
Brand H and Schenzle A 1978 *Phys. Lett.* **68A** 427–9
Bulsara A R, Schieve W C and Gragg R F 1978 *Phys. Lett.* **68A** 294–6
Carmichael H J 1980 *Optica Acta* **27** 147–58
Carmichael H J and Agrawal G P 1980 *Opt. Commun.* **34** 293–9
Carmichael H J and Hermann J A 1980 *Z. Phys.* B **38** 365
Casagrande F, Lugiato L A and Asquini M L 1980 *Opt. Commun.* **32** 492–6
Cross P S, Schmidt R V, Thornton R L and Smith P W 1978 *IEEE J. Quantum Electron* **QE-14** 577–80
De Giorgio V and Scully M O 1970 *Phys. Rev.* A **2** 1170–6
Drummond P D 1981 *IEEE J Quantum Electron.* **QE-17** 301–6
Felber F S and Marburger J H 1976 *Appl. Phys. Lett.* **28** 731–3
Feldman A 1978 *Appl. Phys. Lett.* **33** 143–5
—— 1979 *Opt. Lett.* **4** 115–7
Firth W J 1981 *Opt. Commun.* **39** 343–6
Firth W J and Wright E M 1982 *Opt. Commun.* **40** 233–8
Flytzanis C 1975 *Quantum Electronics* vol 1A, ed H Rabin and C L Tang (New York: Academic) p 9
Franken P A and Ward J F 1963 *Rev. Mod. Phys.* **35** 23
Garmire E, Allen S D, Marburger J H and Verber C M 1978a *Opt. Lett.* **3** 69–71
Garmire E, Marburger J H and Allen S D 1978b *Appl. Phys. Lett.* **32** 320–1
Garmire E, Marburger J H, Allen S D and Winful H G 1979 *Appl. Phys. Lett.* **34** 374–5
Giacobino E, Devaud M, Biraben F and Grynberg G 1980 *Phys. Rev. Lett.* **45** 434–7
Gibbs H M, Gossard A C, McCall S L, Passner A, Wiegman W and Venkatesan T N C 1979a *Solid St. Commun.* **30** 271–5
Gibbs H M, Hopf F A, Kaplan D L and Shoemaker R L 1981 *Phys. Rev. Lett.* **46** 474–7
Gibbs H M, McCall S L and Venkatesan T N C 1976 *Phys. Rev. Lett.* **36** 1135–8
Gibbs H M, McCall S L, Venkatesan T N C, Gossard A C, Passner A and Wiegman W 1979b *Appl. Phys. Lett.* **35** 451–3
—— 1979c *Appl. Phys. Lett.* **35** 451–3
—— 1979d *Laser Spectroscopy IV* ed H Walther and K Rothe (Berlin: Springer-Verlag) pp 441–50
Goldstone J A and Garmire E M 1981 *IEEE J. Quantum Electron.* **QE-17** 366–74
Goldstone J A, Ho P T and Garmire E 1980 *Appl. Phys. Lett* **37** 126–8
Goll J and Haken H 1980 *Phys. Stat. Solidi* B **101** 489–501
Gornik E, Chang T Y, Bridges T J, Nguyen V T and McGee J B 1978 *Phys. Rev. Lett.* **40** 1151
Gragg R F, Schieve W C and Bulsara A R 1979 *Phys. Rev.* A **19** 2052–5
Graham R and Haken H 1968 *Z. Phys.* **213** 420
—— 1970 *Z. Phys.* **273** 31
Graham R and Schenzle A 1981 *Phys. Rev.* A **23** 1302–21
Grischkowsky D 1978 *J. Opt. Soc. Am.* **68** 641–2
Gronchi M and Lugiato L A 1980 *Opt. Lett.* **5** 108–10

Haken H 1970 *Handbuch der Physik* vol XXV (Berlin: Springer-Verlag)
—— 1978 *Synergetics, An Introduction* (Berlin: Springer-Verlag)
Hanamura E 1981 *Solid St. Commun.* **38** 939–42
Hanggi P, Bulsara A R and Janda R 1980 *Phys. Rev.* A **22** 671–83
Hassan S S, Drummond P and Walls D F 1978 *Opt. Commun.* **27** 480–3
Hermann J A 1981 *Opt. Commun.* **37** 431–6
Hermann J A and Thompson B V 1980 *Phys. Lett.* **79A** 153–5
Hilsum C and Rose-Innes A C 1961 *Semiconducting III-V Compounds* (Oxford: Pergamon)
Houghton J T and Smith S D 1966 *Infrared Physics* (Oxford: Clarendon)
Ikeda K 1979 *Opt. Commun.* **30** 257–61
Ikeda K, Daido H and Akimoto O 1980 *Phys. Rev. Lett.* **45** 709–12
Ito H, Ogawa Y and Inaba H 1981 *IEEE J. Quantum Electron.* **QE-17** 325–31
Jain R K and Klein M B 1979 *Appl. Phys. Lett.* **35** 454
Kaminow I P, Stulz L W and Turner E H 1975 *Appl. Phys. Lett.* **27** 555
Kane E O 1957 *J. Phys. Chem. Solids* **1** 249
Kaplan A E 1976 *JETP Lett.* **24** 114–9
—— 1977 *Sov. Phys.–JETP* **72** 896–905
—— 1981a *IEEE J. Quantum Electron.* **QE-17** 336–40
—— 1981b *Appl. Phys. Lett.* **38** 67–9
—— 1981c *Opt. Lett.* **6** 360–2
Kastal'skii A A 1973 *Sov. Phys.–Semicond.* **7** 645
Kitano M, Yabuzaki T and Ogawa T 1981 *Phys. Rev. Lett.* **46** 926–9
Koch S W and Haug H 1981 *Phys. Rev. Lett.* **46** 450–2
Koch S W, Schmitt-Rink S and Haug H 1981 *Solid St. Commun.* **38** 1023–6
Kogelnik H and Li T 1966 *Appl. Opt.* **5** 1550
Kompanets I N, Parfenov A V and Popov Y M 1981a *Opt. Commun.* **36** 415–6
—— 1981b *Opt. Commun.* **36** 417–8
Kondo K, Mabuchi M and Hasegawa H 1980 *Opt. Commun.* **32** 136–40
Lamb W E Jr 1964 *Phys. Rev.* **134** A 1429
Louisell W H 1973 *Quantum Statistical Properties of Radiation* (New York: Wiley)
Lugiato L A 1979 *Nuovo Cim.* B **50** 89–133
—— 1980a *Opt. Commun.* **33** 108–12
—— 1980b *Lett. Nuovo Cim.* **29** 375–80
Lugiato L A, Farina J D and Narducci L M 1980 *Phys. Rev.* A **22** 253–60
Lugovoi V N 1977 *Optica Acta* **24** 743–56
—— 1979 *Phys. Stat. Solidi* b **94** 79–86
—— 1981 *IEEE J. Quantum Electron.* **QE-17** 384–6
Marburger J H and Felber F S 1978 *Phys. Rev.* A **17** 335–42
Mattar F P 1981 *Proc. Int. Conf. on Optical Bistability. Optical Bistability* ed C M Bowden, M Cliftan and H R Robl (New York: Plenum) pp 503–55
May R M 1976 *Nature* **261** 459–67
McCall S L 1974 *Phys. Rev.* A **9** 1515–23
—— 1978 *Appl. Phys. Lett.* **32** 284–6
McCall S L and Gibbs H M 1978a *Appl. Phys. Lett.* **32** 284–6
—— 1978b *J. Opt. Soc. Am.* **68** 1378
—— 1980 *Opt. Commun.* **33** 335–9
McDonald D G 1981 *Physics Today* **34** 37–47
Meystre P 1978 *Opt. Commun.* **26** 277
Miller D A B 1981 *IEEE J. Quantum Electron.* **QE-17** 306–11
Miller D A B, Harrison R G, Johnston A M and Seaton C T 1980a *Opt. Commun.* **32** 478
Miller D A B, Mozolowski M H, Miller A and Smith S D 1978 *Opt. Commun.* **27** 133
Miller D A B, Seaton C T, Prise M E and Smith S D 1981a *Phys. Rev. Lett.* **47** 197–200
Miller D A B and Smith S D 1978 *Appl. Opt.* **17** 3804
—— 1979 *Opt. Commun.* **31** 101
Miller D A B, Smith S D and Johnston A 1979 *Appl. Phys. Lett.* **35** 658
Miller D A B, Smith S D and Seaton C T 1981b *IEEE J. Quantum Electron.* **QE–17** 312
—— 1981c *Proc. Int. Conf. on Optical Bistability, Asheville 1980. Optical Bistability* ed C M Bowden, M Ciftan and H R Robl (New York: Plenum) pp 115–26
Miller D A B, Smith S D and Wherrett B S 1980b *Opt. Commun.* **35** 221

Narducci L M, Gilmore R, Feng D H and Agarwal G S 1978a *Opt. Lett.* **2** 88–90
—— 1978b *Phys. Rev.* A **18** 1571–6
Okada M 1980a *Opt. Commun.* **34** 153–8
——1980b *Opt. Commun.* **35** 31–6
Okada M and Takizawa K 1979 *IEEE J. Quantum Electron.* **QE-15** 82–5
—— 1980 *IEEE J. Quantum Electron.* **QE-16** 770–6
Okuda M and Onaka K 1977a *Jap. J. Appl. Phys.* **16** 303–9
—— 1977b *Jap. J. Appl. Phys.* **16** 769–73
—— 1978 *Jap. J. Appl. Phys.* **17** 1105–10
Okuda M, Onaka K and Sakai K 1978 *Jap. J. Appl. Phys.* **17** 2123–8
Okuda M, Toyota M and Onaka K 1976 *Opt. Commun.* **19** 138–42
Rabinovich M I 1978 *Sov. Phys.–Usp.* **21** 443
Roy R and Zubairy M S 1980a *Opt. Commun.* **32** 163–8
—— 1980b *Phys. Rev.* A **21** 274–80
Sandle W J, Ballagh R J and Gallagher A 1981 *Proc. Int. Conf. on Optical Bistability, Asheville 1980. Optical Bistability* ed C M Bowden, M Ciftan and H R Robl (New York: Plenum) pp 93–108
Sargent M III, Scully M O and Lamb W E Jr 1974 *Laser Physics* (New York: Addison-Wesley) p 164
Schenzle A and Brand H 1978 *Opt. Commun.* **27** 485–8
—— 1979 *Opt. Commun.* **31** 401–7
Schnaper A, Papuchon M and Puech C 1979 *Opt. Commun.* **29** 364–8
—— 1981 *IEEE J. Quantum Electron.* **QE-17** 332–5
Scully M O and Lamb W E Jr 1967 *Phys. Rev.* **159** 208
Sell D D, Stokowski S E, Dingle R and DiLorenzo J V 1973 *Phys. Rev.* B **7** 4568
Shah J, Leheny R F and Wiegmann W 1977 *Phys. Rev.* B **16** 1577
Smith P W, Hermann J P, Tomlinson W J and Maloney P J 1979 *Appl. Phys. Lett.* **35** 846–8
Smith P W, Kaminow J P, Maloney P J and Stulz L W 1978a *Appl. Phys. Lett.* **33** 24–6
—— 1978b *Appl. Phys. Lett.* **34** 62–4
Smith P W and Turner E H 1977 *Appl. Phys. Lett* **30** 280–1
Smith P W, Turner E H and Moloney P J 1978c *IEEE J. Quantum Electron.* **QE-14** 207–12
Smith P W, Turner E H and Mumford B B 1978d *Opt. Lett.* **2** 55–7
Smith S D and Miller D A B 1980a *J. Phys. Soc. Japan* **49** Suppl. A 597
—— 1980b *New Scientist* 554–6
Smith S D, Miller D A B and Wherrett B S 1980 *Proc. 2nd Int. Symp., Reinhardsbrunn, GDR. Ultrafast Phenomena in Spectroscopy* vol 2, ed J Schwarz, W Triebel and G Wiederhold, pp 425–35
Sohler W 1980 *Appl. Phys. Lett.* **36** 351–3
Spiller E 1971 *J. Opt. Soc. Am.* **61** 669
Staupendahl G and Schindler K A 1982 *Opt. Quantum Electron.* **14** 157–67
Szöke A, Daneu V, Goldhar J and Kurnit N A 1969 *Appl. Phys. Lett.* **15** 376–9
Tarucha S, Minakata M and Noda J 1981 *IEEE J. Quantum Electron.* **QE-17** 321–4
Venkatesan T N C and McCall S L 1977 *Appl. Phys. Lett.* **30** 282–4
Vincent D and Otis G 1981 *IEEE J. Quantum Electron.* **QE-17** 318–20
Walls D F, Drummond P D, Hassan S S and Carmichael H J 1978 *Suppl. Prog. Theor. Phys.* **64** 307–20
Walls D F and Zoller P 1980 *Opt. Commun.* **34** 260–4
Walls D F, Zoller P and Steyn-Ross M L 1981 *IEEE J. Quantum Electron.* **QE-17** 380–3
Weaire D, Wherrett B S, Miller D A B and Smith S D 1979 *Opt. Lett.* **4** 331–3
Weidlich W and Haake F 1965 *Z. Phys.* **185** 30
Weyer K G, Wiedenman H, Rateike M, MacGillivray W R, Meystre P and Walther H 1981 *Opt. Commun.* **37** 426–30
Wherrett B S and Higgins N A 1981 *Proc. R. Soc.*
Willis C R 1977 *Opt. Commun.* **23** 151–4
—— 1978 *Opt. Commun.* **26** 62–5
Willis C R and Day J 1979 *Opt. Commun.* **28** 137–41
Winful H G and Marburger J H 1980 *Appl. Phys. Lett.* **36** 613–4
Winful H G, Marburger J H and Garmire E 1979 *Appl. Phys. Lett.* **35** 379–81
Wysin G M, Simon H J and Deck R T 1981 *Opt. Lett.* **6** 30–2
Yariv A 1975 *Quantum Electronics* (New York: Wiley) p 153
Zardecki A 1980 *Phys. Rev.* A **22** 1664–71
—— 1981 *Phys. Rev.* A **23** 1281–9

Further references:

Bistability experiments with semiconductors and molecular gases

Spatial hysteresis in optical bistability
W.J. Firth, C.T. Seaton, E.M. Wright and S.D. Smith
Appl. Phys. B28, 131-133 (1982).

Optical bistability in InSb at room temperature with two photon excitation
A.K. Kar, J.G.H. Mathew, S.D. Smith, B. Davis and W. Prettl
Appl. Phys. Lett., 42, 334 (1983).

The realisation of an InSb bistable device as an optical AND gate and its use to measure carrier recombination times
C.T. Seaton, S.D. Smith, F.A.P. Tooley, M.E. Prise and M.R. Taghizadeh
Appl. Phys. Lett., 42(2), 131-133 (1983).

Optical bistability in semiconductors
S.D. Smith and B.S. Wherrett
Lecture Notes in Physics 182, pp. 1-13, Springer-Verlag Berlin, Heidelberg, New York, Tokyo (1983).

Optical bistability and the optical transistor using semiconductors
S.D. Smith and D.A.B. Miller
NATO ASI Series B: Physics, 91, 239-246 (1983), eds. S. Martellucci and A.N. Chester, "Integrated Optics: Physics and Applications", Plenum.

High gain signal amplification in an InSb transphasor at 77 K
F.A.P. Tooley, S.D. Smith and C.T. Seaton
Appl. Phys. Lett., 43(9), Nov., 1983, pp. 807-809.

Room temperature bistability, logic gate operation, incoherent switching and high signal gain with InSb devices
S.D. Smith and F.A.P. Tooley
in Optical Bistability 2, Edited by Charles M. Bowden, Hyatt M. Gibbs and Samuel L. McCall (Plenum Publishing Corporation, 1984), pp. 215-222.

Room temperature, visible wavelength optical bistability in ZnSe interference filters
S.D. Smith, J.G.H. Mathew, M.R. Taghizadeh, A.C. Walker, B.S. Wherrett and A. Hendry
Opt. Comm., 51, 357 (1984).

Observation of optical hysteresis in an all-optical passive ring cavity containing molecular gas
R.G. Harrison, W.J. Firth, C.A. Emshary and I.A. Al-Saidi
Appl. Phys Lett., 44, 716-718 (1984).

Optical modulators and bistable devices using molecular gases
R.G. Harrison, W.J. Firth and I.A. Al-Saidi
Proc. AGARD Meeting, "Digital Optical Circuit Technology", Sept., 1984 AGARD-NATO.

Observation of optical bistability in millimetre gas cells
R.G. Harrison, I.A. Al-Saidi, E.J.D. Cummins and W.J. Firth
Appl. Phys. Lett. (to appear).

Observation of instabilities in passive resonators

Observation of period-doubling in an all-optical resonator containing NH$_3$ gas
R.G. Harrison, W.J. Firth, C.A. Emshary and I.A. Al-Saidi
Phys. Rev. Lett., 51, 562-565 (1983).

Observation of period doubling to chaos in all-optical Fabry-Perot resonators
R.G. Harrison, W.J. Firth and I.A. Al-Saidi
JOSA B1, 488-9 (1984).

Observation of bifurcation to chaos in a passive all-optical Fabry-Perot resonator
R.G. Harrison, W.J. Firth and I.A. Al-Saidi
Phys. Rev. Lett., 53, 258-261 (1984).

Observation of Ikeda instabilities and optical bistability in an all-optical resonator containing NH$_3$ gas
R.G. Harrison, C.A. Emshary, I.A. Al-Saidi and W.J. Firth
Optical Bistability 2: C.M. Bowden, H.M. Gibbs and S.L. McCall, Plenum (1984), pp. 9-16.

Theory

Instabilities and transverse effects in nonlinear Fabry-Perot resonators
E. Abraham, W.J. Firth and E.M. Wright
Quantum Electronics and Electro-Optics, P.L. Knight (ed.), Wiley (1983), pp. 213-216.

Oscillation and chaos in a Fabry-Perot bistable cavity with Gaussian input beam
W.J. Firth and E.M. Wright
Phys. Lett., 92A, 211-216 (1982).

Multi-parameter universal route to chaos in a Fabry-Perot resonator
E. Abraham and W.J. Firth
Optical Bistability 2: C.M. Bowden, H.M. Gibbs and S.L. McCall (eds.), Plenum Press, pp. 119-126.

Connection between Ikeda instability and phase conjugation
W.J. Firth, E.M. Wright and E.J.D. Cummins
(ibid), pp. 111-118.

Periodic oscillations and chaos in a Fabry-Perot cavity containing nonlinearity of finite response time
E. Abraham and W.J. Firth
Opt. Acta, 30, 1541-1560 (1983).

Physical interpretation of the route to chaos in nonlinear resonators
W.J. Firth, E.M. Wright and E. Abraham
JOSA B1, 489 (1984).

Beam propagation in a Kerr-type medium with diffusive nonlinearity
E.M. Wright, W.J. Firth and I. Galbraith
JOSA B (to appear).

Technological applications

Towards the optical computer
S.D. Smith
Nature, 307:5949, 315-6, Jan., 1984.

Optical bistability and its application to computing
S.D. Smith and A.C. Walker
ICO Conference, Sapporo, Japan, August 1984, Conference Digest, pp. 448-551.

Optical bistable devices using InSb
S.D. Smith, F.A.P. Tooley, A.C. Walker, A.K. Kar, J.G.H. Mathew and B.S. Wherrett
Proceedings of the 48th Specialist Meeting on Digital Optical Circuit Technology (Schliersee), Paper 1, AGARD-NATO (1984), p. 1-8.

All-optical logic gates with external switching by laser and incoherent radiation
S.D. Smith, F.A.P. Tooley, A.C. Walker, J.G.H. Mathew, M.R. Taghizadeh and B.S. Wherrett, ibid, Paper 9, 9-1 to 9-7.

Optical bistability strides toward realisation as a new technology: Field is fruitful for laser and materials research
S.D. Smith and A.C. Walker
Laser Focus, August 1984, p. 18-28.

Reviews, books and proceedings

Theory of Optical Bistability
L.A. Lugiato
in **Progress in Optics XXI**, ed. E. Wolf (Elsevier, 1984), pp. 71-216.

Controlling Light with Light
H.M. Gibbs and S.L. McCall
(Academic Press, 1985).

Optical Bistability 2
eds. C.M. Bowden, H.M. Gibbs and S.L. McCall (Plenum, 1984).
(Note in particular the papers on other solid state systems).

Proceedings of the Royal Society Meeting on Optical Bistability, Dynamic Nonlinearity and Photonic Logic to appear in Phil. Trans. Royal Soc. (1985).

Coherence and Quantum Optics V
eds. L. Mandel and E. Wolf (Plenum, 1984).

Non-classical effects in the statistical properties of light

RODNEY LOUDON

Physics Department, Essex University, Colchester CO4 3SQ, UK

Abstract

Much experimental and theoretical work on the statistical properties of light is concerned with the correlation of two readings of the intensity or photon number. The normalised correlation function is the degree of second-order coherence of the light. The properties of the classical degree of second-order coherence, defined in terms of classical intensity measurements, are here contrasted with the properties of the quantum degree of second-order coherence, defined in terms of photon counts. It is shown that the classical quantity has to satisfy three inequalities which do not apply to its quantum analogue. Light whose degrees of second-order coherence fall outside the ranges allowed by classical theory is said to be non-classical for the purposes of the present review.

Measurements of non-classical effects fall into two categories depending on whether the two intensities or photon counts to be correlated refer to a single light beam or to two different beams. The main experimental work in the single-beam case is on resonance fluorescence; the relevant theory is here reviewed in some detail, with particular emphasis on the simpler aspects of the calculations. Alternative proposals for generating non-classical single beams by non-linear optical experiments are reviewed more briefly. A similar coverage of double-beam experiments is given, with a detailed account of the theory of non-classical correlations between the light beams generated in two-photon cascade emission, and a briefer survey of corresponding non-linear optical experiments.

It is concluded that, although the classical theory often reproduces the predictions of quantum theory, there are various measurements of degrees of second-order coherence that lie outside the domain of classical theory. There are direct experimental violations of two of the three classical inequalities and indirect violations of the third. All these experimental results are in good agreement with the predictions of quantum theory.

1. Introduction

The classical and quantum theories provide quite different descriptions of light and its properties. For the present review, a classical theory is defined to be one in which the light is specified by classical vector fields that obey Maxwell's equations. More particularly, the intensity of a beam of light at a point in space at time t is described by a positive single-valued function $I(t)$. It is assumed in classical theory that the intensity can be measured to arbitrary precision and the value of $I(t)$ is not affected by the measurement process. Thus repeated measurements of the intensity do not interfere in any way, but are all determined by the same function $I(t)$. By contrast, the quantum theory represents the vector fields and intensity of the light by operators. There is the usual quantum-theory separation between the specification of the state of the light, represented by a ket vector, and the act of measurement, represented by an application of the appropriate operator. The measurement process in general changes the state of the light, and successive measurements may register different results, in typical quantum fashion. Intensity measurements usually involve destruction of photons, so that progressively fewer photons remain to give progressively smaller intensities in a series of measurements.

An intriguing feature of the theory of light is the large measure of agreement between the predictions of classical and quantum theory, despite the fundamental differences between the two approaches. This occurs to some extent because even classical light takes on some apparent quantum properties when viewed via the intermediary of an atomic detection system controlled by the laws of quantum mechanics. There is a continuing lively debate on the extent to which quantum theory needs to be applied to the light itself, independently of the quantisation of the atoms with which it interacts. This has stimulated careful experiments seeking to distinguish between the predictions of different theories. Many of the varied aspects of this work are very clearly reviewed by Milonni (1976) and Mandel (1976) (see also the shorter articles by Mandel (1974) and Wódkiewicz (1978)); the related, but somewhat separate, question of the possibility of hidden-variable theories of light is included in a review by Clauser and Shimony (1978).

The present review is concerned with the contributions from the study of the statistical and fluctuation properties of light towards a fuller understanding of the relations between classical and quantum theories. The measurements involve the correlation of the intensity of the light at two space–time points, the normalised form of the intensity correlation function being called the degree of second-order coherence. It was foreseen in early work by Glauber (1963a, b) that measurements of the degree of second-order coherence could distinguish classical and quantum predictions, and the first experimental detection of non-classical light was made nearly ten years ago (Burnham and Weinberg 1970). Activity in this field has, however, intensified in the last four or five years. There have been several experiments, proposed and/or performed, that seek non-classical properties of light beams. Although different classical and quantum predictions occur for intensity correlation functions of higher order than the second, it is possible to give a unified discussion of the measurements made to date in terms of the degree of second-order coherence.

An appreciation of the various experiments and their interpretations requires a fair knowledge of the classical and quantum theories of light. The following two sections give introductions to these theories with particular emphasis on their predictions for the degree of second-order coherence. The cases are considered in which the correlation function is formed from intensity readings on the same light beam or on two different light beams. It is shown that there are no quantum analogues of three inequalities that the classical degree of second-order coherence must satisfy.

Subsequent sections review a range of single-beam and double-beam experiments in which the classical inequalities are violated. Most attention is given to the single-beam resonance fluorescence measurements and to the observations of double-beam intensity correlations in two-photon cascade emission. These are linear experiments that can be analysed by similar theoretical techniques. Non-linear experiments, many of which should in principle produce light with non-classical statistical properties, are more difficult to analyse and have not been so successful in practice; they are given a more superficial review. The main aim of the review is the illustration of the theoretical ideas behind the various experiments in terms of simple models of the interacting atoms and light beams. The more sophisticated models that are needed to make detailed agreement with the experimental data, particularly in the resonance fluorescence measurements, are carefully treated in the literature references given in the appropriate sections of the text.

The statistical experiments that have so far been performed give results in very good agreement with quantum theory. It should, however, be remarked that non-classical light is the exception, rather than the rule, and that the classical theory gives essentially identical results to the quantum theory for a wide range of statistical measurements that are beyond the scope of the present review.

2. Classical degree of second-order coherence

Consider a beam of light described by a time-dependent cycle-averaged classical intensity $I_1(t)$. The intensity may show random fluctuations, as in the chaotic light from a thermal source, and it may have discontinuities, as in sharply pulsed light. It is assumed throughout the review that all light beams have stationary and ergodic statistical properties, so that ensemble averages over the probability distribution are equivalent to long-time averages over the beams in a single experiment.

The degree of second-order coherence of the light is defined to be

$$G_{11}^{(2)}(t) = \langle I_1(t)I_1(0)\rangle/\bar{I}_1^2 \tag{2.1}$$

where the angle brackets denote an average over a long series of pairs of intensity readings separated by a fixed positive time delay t. The average intensity is

$$\bar{I}_1 = \langle I_1 \rangle \tag{2.2}$$

where there is no need to attach a time to the intensity for a beam of stationary statistical properties. The detector ideally measures the intensity instantaneously but much of the theory remains valid if the readings are degraded by a finite resolving time. In a more general experiment that correlates the intensities $I_1(t)$ and $I_2(0)$ of two different light beams measured by detectors at positions r_1 and r_2, the degree of second-order coherence becomes

$$G_{12}^{(2)}(t) = \langle I_1(t)I_2(0)\rangle/\bar{I}_1\bar{I}_2 \tag{2.3}$$

where the delay time is again positive, and the other conditions are also the same as for (2.1).

The classical degree of second-order coherence satisfies various inequalities. Since the intensity is always positive

$$G_{12}^{(2)}(t) \geq 0. \tag{2.4}$$

Further properties follow from some standard inequalities (see chapter 2 of Hardy et al (1951)). For the single-beam case, the average of products of simultaneous intensity measurements satisfies

$$\langle I_1^2 \rangle \geq I_1^2 \tag{2.5}$$

and it therefore follows from (2.1) that

$$G_{11}^{(2)}(0) \geq 1. \tag{2.6}$$

In the more general case of a series of intensity measurements on two beams, Cauchy's inequality gives

$$\langle I_1^2 \rangle \langle I_2^2 \rangle \geq \langle I_1 I_2 \rangle^2. \tag{2.7}$$

The inequality holds for the correlation of pairs of measurements made at any random times, but when there is a fixed time delay t between the readings on the two beams, the correlation function on the right of (2.7) is the same as that in (2.3), and it follows that

$$G_{11}^{(2)}(0) G_{22}^{(2)}(0) \geq [G_{12}^{(2)}(t)]^2. \tag{2.8}$$

Returning to the single-beam case, the square root of (2.8) yields

$$G_{11}^{(2)}(0) \geq G_{11}^{(2)}(t). \tag{2.9}$$

Further more complicated inequalities satisfied by degrees of coherence of various orders have been derived by Titulaer and Glauber (1965) but the simpler results given above are adequate for the present review.

The degrees of second-order coherence (2.1) and (2.3) simplify when the time delay t is very much longer than any correlation times within or between the light beams. The measured intensities are then essentially uncorrelated and it follows that

$$G_{11}^{(2)}(\infty) = G_{12}^{(2)}(\infty) = 1. \tag{2.10}$$

Unlike the inequalities (2.6), (2.8) and (2.9), this result is not universally valid, even in classical theory, since some models of light beams include correlation times of indefinitely long duration.

We now consider some examples of single light beams that illustrate the general properties of the classical degree of second-order coherence. One of the simplest cases is the train of rectangular pulses shown in figure 1(a). Suppose that the intensity I_0 within each pulse is constant and that the duration and separation of the pulses give a mean intensity that can be written

$$I_1 = n I_0 \quad 0 < n \leq 1. \tag{2.11}$$

Here n is the number of pulses per unit time multiplied by the pulse duration τ_0. It is a trivial exercise to show from (2.1) that

$$G_{11}^{(2)}(0) = 1/n \tag{2.12}$$

and only a brief calculation is needed to verify the time dependence of the degree of

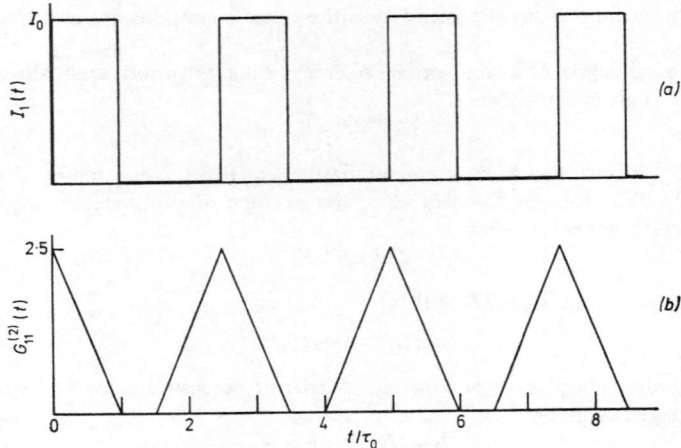

Figure 1. (a) Intensity variation for a square wave of 'on' period τ_0 and 'off' period $1 \cdot 5 \tau_0$ giving $n = 0 \cdot 4$. (b) Degree of second-order coherence of the square wave.

second-order coherence shown in figure 1(b). This degree of coherence correctly satisfies the inequalities (2.6) and (2.9).

The special case of a beam of constant intensity I_0 is obtained for $n = 1$, when it is easily shown that

$$G_{11}^{(2)}(t) = 1 \qquad (2.13)$$

independent of the time delay t. A beam of constant cycle-averaged intensity is the classical representation of coherent light.

Except in the limit of coherent light, the degree of second-order coherence shown in figure 1 does not have the property (2.10). The discrepancy is caused by the infinitely long time correlations implied by the rigorous periodicity of the train of pulses; any randomness in the pulse train tends to modify its degree of second-order coherence to one that does satisfy (2.10). Consider a generalisation of the model to a light beam consisting of the same rectangular pulses but now distributed with random times of initiation. We replace n in (2.11) by $\bar{n}$, the average number of pulses present simultaneously, which can take all positive values. It is not difficult to evaluate the average in the numerator of (2.1) for a Poisson distribution of identical rectangular pulses, and the degree of second-order coherence is

$$G_{11}^{(2)}(t) = 1 + \left(1 - \frac{t}{\tau_0}\right) \frac{1}{\bar{n}} \qquad \text{for } t < \tau_0 \qquad (2.14)$$

$$= 1 \qquad \text{for } t > \tau_0.$$

The unit contribution to the degree of coherence that occurs for all times t comes from correlations between *different* pulses; the additional contribution for $t < \tau_0$ comes from the correlations within *individual* pulses. The latter are dominant at short times for sparse pulses ($\bar{n} \ll 1$), but they become negligible for crowded, multiply-overlapping pulses ($\bar{n} \gg 1$). The two contributions in (2.14) are denoted, respectively, as 'background' and 'anomalous'. The degree of coherence (2.14) does now satisfy (2.10).

A further development of the pulse model provides a theory of the chaotic light emitted by a collision-broadened light source. In the simple theory of such a source (Loudon 1973), the elastic collisions break up the wave radiated by each atom into discrete sections. Each section has a constant phase that changes randomly when a collision occurs, and there is a distribution of section lengths whose average τ_0 is the mean time between collisions. The pulse model must be generalised to include the different phases and lengths of the sections of electromagnetic wave.

Suppose first that the light contains contributions of intensity I_0 from n radiating atoms, the phase of the wave from atom i being ϕ_i. The instantaneous average value of the square of the intensity is

$$\overline{I_1^2(0)} = I_0^2 \; \overline{|\sum_{i=1}^{n} \exp(i\phi_i)|^4} = I_0^2(2n^2 - n) \tag{2.15}$$

where the bar denotes an average over the random phase angles. The number n of radiating atoms in the source is assumed to fluctuate with a Poisson distribution of mean number $\bar{n}$. A further average of (2.15) gives

$$\langle \overline{I_1^2(0)} \rangle = I_0^2(2\bar{n}^2 + \bar{n}) \tag{2.16}$$

and the degree of second-order coherence obtained from (2.1) with the help of (2.11) is

$$G_{11}^{(2)}(0) = 2 + (1/\bar{n}). \tag{2.17}$$

It is seen by comparison with the zero time delay value of (2.14) that the random phases produce an additional unit contribution to the degree of second-order coherence.

The time dependence of the degree of second-order coherence is governed by the distribution of discrete section lengths. If the fluctuations in n are arbitrarily given the same time scale τ_0 as the collisions, a suitable average of the intensity correlation gives

$$G_{11}^{(2)}(t) = \underset{\text{(background)}}{1} + \underset{\text{(HBT)}}{\exp(-2\gamma' t)} + \underset{\text{(anomalous)}}{(1/\bar{n}) \exp(-2\gamma' t)} \tag{2.18}$$

where

$$\gamma' = 1/\tau_0. \tag{2.19}$$

This change in notation is convenient since $2\gamma'$ is the linewidth of the Lorentzian frequency distribution of the collision-broadened chaotic light. The first two terms in (2.18) are the contributions from correlations between the light from *different* atoms, and that labelled (HBT) results from their random phases. The third term is the contribution from correlations within the light from *individual* atoms. The result (2.18) satisfies all the classical conditions (2.6), (2.9) and (2.10). Its form is shown in figure 2.

The standard theory of chaotic light (see, for example, Loudon 1973) considers a large number of radiating atoms ($\bar{n} \gg 1$). The summation of phase factors in (2.15) is treated by random-walk theory to give a probability distribution

$$P(I_1) = (1/\bar{I}_1) \exp(-I_1/\bar{I}_1) \tag{2.20}$$

for the instantaneous intensity. Integration over the continuously variable intensity gives the well-known result

$$G_{11}^{(2)}(0) = 2 \tag{2.21}$$

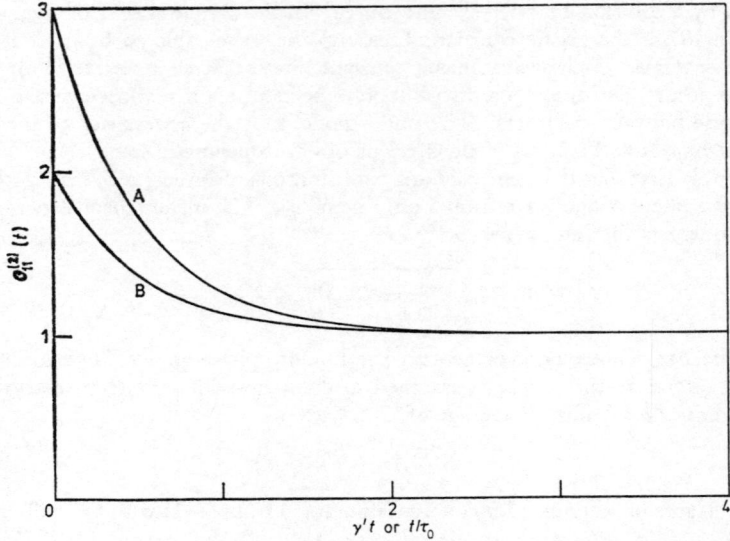

Figure 2. Classical degree of second-order coherence for collision-broadened light radiated by an average number $\bar{n}$ of atoms. A, $\bar{n}=1$; B, $\bar{n} \gg 1$.

in agreement with (2.17) for $\bar{n} \gg 1$. More generally, the time dependence of the coherence is shown to be

$$G_{11}^{(2)}(t) = 1 + \exp(-2\gamma' t) \tag{2.22}$$

in agreement with (2.18) for $\bar{n} \gg 1$. The function is shown in curve B of figure 2.

The increase in the intensity correlations or the degree of second-order coherence of chaotic light for delay times t shorter than the coherence time τ_0 was first observed by Hanbury Brown and Twiss (see Hanbury Brown (1974) for a review). Only chaotic light was available at the time of their experiments in 1956-8, but similar measurements have since been made on other kinds of light. Figure 3(a) shows the schematic arrangement of the experimental components. The beam of light to be

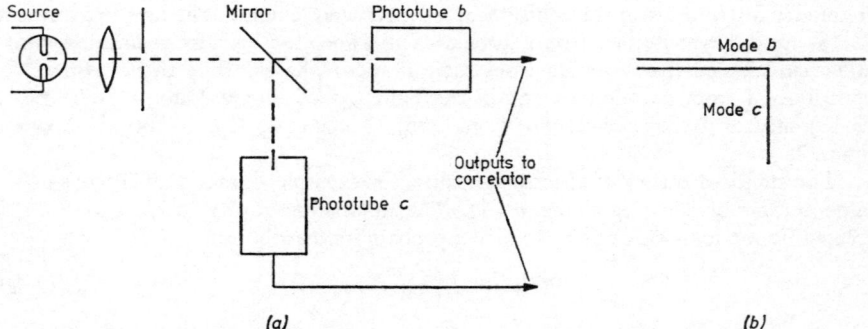

Figure 3. Experiment of Hanbury Brown and Twiss. (a) Arrangement of main components and (b) the light paths of the modes corresponding to transmission and reflection at the mirror.

studied is split into two parts by a semitransparent mirror. The intensities of the separate parts are measured by two phototubes, whose readings are correlated and averaged electronically with a fixed time delay t. According to classical theory, the half-silvered mirror splits the incident light into two identical beams which show exactly the same intensity fluctuations. The arrangement thus provides a convenient way of taking the two intensity readings to be correlated. Various factors of $\frac{1}{2}$ produced by the splitting of the incident intensity cancel when the degree of second-order coherence (2.1) is formed from the phototube readings.

It should be emphasised that the optical coherence considered here is of the temporal variety. However, the performance of experiments of the Hanbury Brown and Twiss type requires a careful consideration of spatial coherence effects. Thus the variation in optical path length from different parts of the light source to different sections of the detector area can remove the intensity correlations represented by the second term on the right of (2.18) or (2.22). To minimise these effects, the original experiment had a light source with an angular diameter of order 10^{-4} rad viewed from the detectors, and the detectors subtended a solid angle of order 5×10^{-6} sr at the light source. The main interest of Hanbury Brown and Twiss was indeed concerned with spatial coherence since its fall-off with lateral separation of the detectors at zero time delay provides a measure of the angular diameter of the source, leading to important applications in astronomy (Hanbury Brown 1974). This book also treats the effects of finite detector response times.

The degrees of second-order coherence derived above are compared in later sections with the quantum theory predictions and with experimental results. However, it should be emphasised here that some features of the classical theory do not agree with experiment. Thus, while the expression (2.22) for a large number of radiating atoms has been verified by Hanbury Brown and Twiss experiments and is also reproduced in the quantum theory, the additional anomalous contribution in (2.18) is not observed experimentally. The entire collision-interrupted wave train from a single radiating atom corresponds in quantum theory to a single emitted photon; the photon either passes through or is reflected from the Hanbury Brown and Twiss mirror and cannot as an individual produce any intensity correlations. These matters are discussed in the following section.

3. Quantum degree of second-order coherence

The electromagnetic fields associated with a light beam are represented by operators in the quantum theory. We consider only linearly polarised light where the electric field operator is a scalar of the form (see, for example, Loudon 1973)

$$\hat{E}(rt) = \hat{E}^-(rt) + \hat{E}^+(rt) \qquad (3.1)$$

with

$$\hat{E}^-(rt) = -\mathrm{i} \sum_k (\hbar\omega_k/2\epsilon_0 V)^{1/2} \hat{a}_k{}^\dagger \exp\,(\mathrm{i}\omega_k t - \mathrm{i} k.r) \qquad (3.2)$$

and the conjugate relation for $\hat{E}^+$. Here k is the wavevector of a plane-wave mode of the optical system, $\hat{a}_k{}^\dagger$ is the photon creation operator for the mode, and V is a quantisation volume.

The degree of second-order coherence is defined analogous to (2.3) in terms of the correlation of light intensities at two space–time points. It is obtained in the quantum theory by evaluating the transition rate for a joint absorption of photons at the two points with an appropriate delay time t (Glauber 1963a). The resulting quantum degree of second-order coherence is

$$G_{12}^{(2)}(t) = \langle \hat{E}_2^-(0)\hat{E}_1^-(t)\hat{E}_1^+(t)\hat{E}_2^+(0)\rangle / \langle \hat{E}_1^-\hat{E}_1^+\rangle\langle \hat{E}_2^-\hat{E}_2^+\rangle \qquad (3.3)$$

where the subscripts denote field operators (3.2) evaluated at the appropriate space–time points. The angle brackets now denote quantum expectation values evaluated according to

$$\langle \hat{O} \rangle = \mathrm{Tr}\{\hat{\rho}\hat{O}\} \qquad (3.4)$$

where $\hat{\rho}$ is the density operator of the light and $\hat{O}$ is any arbitrary operator. The density operator describes the probability distribution of the radiation field amongst its various quantum states (see, for example, Loudon 1973). Its diagonal matrix elements with respect to the states that have definite numbers of photons give the photon-number distribution of the light, examples of which are given in (3.15) and (3.17) below.

The inequalities that restrict the possible values of the quantum degree of second-order coherence are much less severe than for the classical case. Because of the occurrence of products of Hermitian conjugate operators in (3.3), it is possible to show that (Glauber 1963a)

$$G_{12}^{(2)}(t) \geqslant 0 \qquad (3.5)$$

similar to (2.4). There are, however, no inequalities analogous to (2.6), (2.8) and (2.9) in the quantum case. The present review is mainly concerned with the kinds of light beam whose existence is allowed by the quantum theory, but not by the classical theory, as shown by their violation of some or all of the classical inequalities.

We now consider some examples of light beams to illustrate the quantum theory, and it is convenient to begin with the single-beam case where (3.3) reduces to

$$G_{11}^{(2)}(t) = \langle \hat{E}_1^-(0)\hat{E}_1^-(t)\hat{E}_1^+(t)\hat{E}_1^+(0)\rangle / \langle \hat{E}_1^-\hat{E}_1^+\rangle^2 \qquad (3.6)$$

analogous to the classical definition (2.1). A considerable further simplification occurs in the very special case where the light beam excites only a single mode of the optical system. Then most factors cancel when (3.2) and its conjugate are inserted into (3.6), leaving

$$G_{11}^{(2)}(t) = \langle \hat{a}_1^\dagger \hat{a}_1^\dagger \hat{a}_1 \hat{a}_1 \rangle / \langle \hat{a}_1^\dagger \hat{a}_1 \rangle^2 = \langle n_1(n_1-1)\rangle / \bar{n}_1^2 \qquad (3.7)$$

where n_1 is the photon number in the mode with mean value $\bar{n}_1$. There is no time dependence in the single-mode case.

The quantity to be averaged on the right of (3.7) exemplifies the quantum principle that the act of measurement can interfere with the measured system; a measurement of the number n_1 of photons itself reduces the photon number by 1, so that the second measurement finds only $n_1 - 1$ photons. This effect lies at the root of the differences between the degrees of second-order coherence in classical and quantum theories.

The simplest example of the non-classical light is provided by a beam that contains exactly n_1 photons, corresponding to an eigenstate of the photon number operator. The degree of second-order coherence from (3.7) is

$$G_{11}^{(2)}(t) = (n_1 - 1)/n_1 \qquad (3.8)$$

which clearly violates (2.6), being smaller than unity at all times. The reason for the disagreement between classical and quantum theories can be seen more clearly from an examination in the quantum picture of the way in which the Hanbury Brown and

Twiss experiment determines the degree of second-order coherence. The workings of the experiment are quite different in quantum terms since each photon that strikes the mirror is *either* reflected *or* transmitted, and can only be registered in one of the phototubes. The two beams that arrive at the detectors are therefore not identical, contrary to the assumption made in the classical theory of the experiment, and it is not immediately obvious how their photon counts are related to the properties of the unsplit beam to the left of the mirror. It should be emphasised that the purpose of the experiment is the determination of the coherence properties of the single incident beam of light; its splitting into two beams in the apparatus facilitates the measurements of the intensity or photon-number correlations.

The quantum theory of the Hanbury Brown and Twiss experiment is readily analysed by a method used by Walls (1977) to treat Young's interference experiment. Figure 3(b) shows the modes b and c of the optical system that correspond respectively to transmission and reflection at the mirror. The incident photons have equal probabilities of transmission or reflection and their creation operator can be written

$$\hat{a}_1{}^\dagger = (\hat{b}^\dagger + \hat{c}^\dagger)/2^{1/2} \tag{3.9}$$

where $\hat{b}^\dagger$ and $\hat{c}^\dagger$ create photons in modes b and c of the optical system. Then with the usual properties for the photon creation operator, the state with n_1 incident photons can be written

$$|n_1\rangle = (n_1!)^{-1/2}(\hat{a}_1{}^\dagger)^{n_1}|0\rangle = (n_1!2^{n_1})^{-1/2}(\hat{b}^\dagger + \hat{c}^\dagger)^{n_1}|0\rangle \tag{3.10}$$

where $|0\rangle$ is the vacuum state for all kinds of photon. The probability that n_b photons are detected at phototube b and n_c at phototube c (with $n_b + n_c = n_1$) is

$$P_{n_b, n_c} = |\langle n_b, n_c | n_1\rangle|^2 = n_1!/n_b!n_c!2^{n_1}. \tag{3.11}$$

The probability distribution can be used to obtain average properties of the photon counts in the two arms of the apparatus. The mean numbers of counts are

$$\bar{n}_b = \bar{n}_c = n_1/2 \tag{3.12}$$

and the correlation is

$$\langle \hat{b}^\dagger \hat{c}^\dagger \hat{c} \hat{b} \rangle = \langle n_b n_c \rangle = n_1(n_1 - 1)/4. \tag{3.13}$$

Thus the degree of second-order coherence computed from

$$G_{11}{}^{(2)}(t) = \langle n_b n_c\rangle/\bar{n}_b\bar{n}_c \tag{3.14}$$

does indeed reproduce the quantum result (3.8) for a photon-number-state incident beam. Table 1 shows the photon count distributions and the degrees of second-order coherence for small numbers n_1 of incident photons. It is seen that non-classical light with degree of second-order coherence smaller than unity occurs in the Hanbury Brown and Twiss experiment because the photons must make a choice between reflection and transmission at the mirror. A classical-type analysis in which each photon is represented by a constant intensity pulse that is half-transmitted and half-reflected produces a degree of second-order coherence equal to unity for any given number of incident photons.

The quantum analysis of the Hanbury Brown and Twiss experiment is easily extended to incident beams that have a probability distribution P_{n_1} over the photon number states of a single mode. The degree of second-order coherence obtained from

Table 1. The columns show, respectively, the number of incident photons in a Hanbury Brown and Twiss experiment, the possible numbers detected by the two phototubes, their mean, their correlation, and the degree of second-order coherence.

n_1	n_b	n_c	$\bar{n}_b = \bar{n}_c$	$\langle n_b n_c \rangle$	$G_{11}^{(2)}(t)$
1	1 0	0 1	$\frac{1}{2}$	0	0
2	2 1 1 0	0 1 1 2	1	$\frac{1}{2}$	$\frac{1}{2}$
3	—	—	$\frac{3}{2}$	$\frac{3}{2}$	$\frac{2}{3}$
4	—	—	2	3	$\frac{3}{4}$

(3.14) then reproduces the general single-mode result (3.7). Two illustrations (see, for example, Loudon 1973) are the coherent distribution

$$P_{n_1} = \bar{n}_1^{n_1} \exp(-\bar{n}_1)/n_1! \qquad (3.15)$$

giving

$$G_{11}^{(2)}(t) = 1 \qquad (3.16)$$

and the chaotic distribution

$$P_{n_1} = \bar{n}_1^{n_1}/(1+\bar{n}_1)^{1+n_1} \qquad (3.17)$$

giving

$$G_{11}^{(2)}(t) = 2. \qquad (3.18)$$

In these particular cases, which correspond respectively to the light obtained from a single-mode laser well above threshold and from a single mode of a thermal source, the averaging over the photon-number distribution produces degrees of second-order coherence that no longer violate the classical inequality (2.6).

A full treatment of the chaotic light emitted by a thermal source requires a generalisation of the above discussion to include a multi-mode incident beam. It is not difficult to show that the degree of second-order coherence (3.6) is correctly obtained from the Hanbury Brown and Twiss correlation according to

$$G_{11}^{(2)}(t) = \langle n_b(t) n_c(0) \rangle / \bar{n}_b \bar{n}_c \qquad (3.19)$$

for a beam of stationary statistical properties, where the correlation between the two detectors is made with a fixed time delay t. A standard quantum treatment of the degree of second-order coherence of the chaotic light radiated by a large number of atoms gives an expression identical to the classical result (2.22) illustrated in curve B of figure 2. The enhanced photon-counting correlations for delay times less than $\tau_0 = 1/\gamma'$ are a manifestation of the *photon bunching* property of chaotic light, illustrated schematically in figure 4(a). Figure 4(b) shows the analogous random distribution of counts that corresponds to the constant degree of second-order coherence (3.16) for single-mode coherent light.

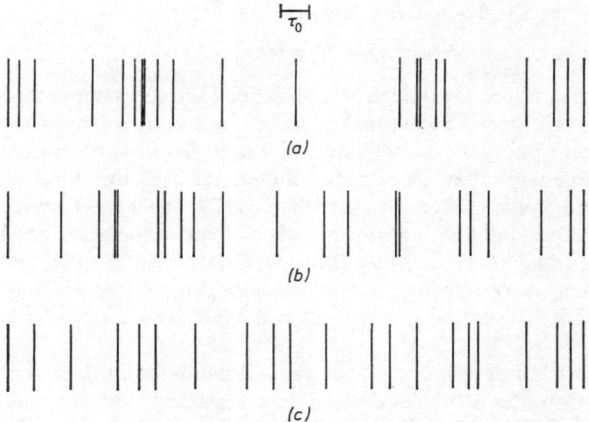

Figure 4. Schematic representations of photon counts as functions of the time for light beams that are (a) bunched with $G_{11}^{(2)}(0) > 1$, (b) random with $G_{11}^{(2)}(0) = 1$, and (c) antibunched with $G_{11}^{(2)}(0) < 1$.

In contrast to these two examples, the following section is mainly concerned with light whose degree of second-order coherence falls outside the domain of classical theory. One possibility is a violation of the classical inequality (2.6), when the light is said to be *antibunched* and the pattern of photon counts is illustrated in figure 4(c); the separation of counts in this case is reminiscent of the spacings that occur in beams of electrons (Every 1973, 1975a). The light beam with a definite number of photons treated above is a naive example of non-classical light. More sophisticated examples are considered by Baltes *et al* (1979) and Helstrom (1979), while the following section reviews the practical realisation of non-classical light.

The degree of second-order coherence for two-beam experiments is reviewed in §5. It is a simple matter to find examples for which the quantum degrees of second-order coherence fail to satisfy the classical inequality (2.8). For a pair of single-mode light beams (3.3) reduces in an obvious notation to

$$G_{12}^{(2)}(t) = \langle \hat{a}_2^\dagger \hat{a}_1^\dagger \hat{a}_1 \hat{a}_2 \rangle / \bar{n}_1 \bar{n}_2 \tag{3.20}$$

with the time dependence again cancelling. If the beams have definite photon numbers n_1 and n_2, then (3.20) gives

$$G_{12}^{(2)}(t) = 1 \tag{3.21}$$

but the individual beams both have degrees of coherence similar to (3.8), so that

$$G_{11}^{(2)}(0) G_{22}^{(2)}(0) = (n_1 - 1)(n_2 - 1)/n_1 n_2 \tag{3.22}$$

and (2.8) is violated. More practical examples of double-beam experiments are treated in §5. However, the examples have the common feature that non-classical effects tend to be most marked for beams with small well-defined numbers of photons.

4. Non-classical single-beam statistics

4.1. Radiation by a single driven atom: special case

One way of producing small fairly well-defined numbers of photons is to have them emitted by a single atom. Such photons leave the atom in a stream with time intervals between adjacent photons resulting from the time required to re-excite the atom and emit a subsequent photon. Resonance fluorescence of this kind was proposed by Carmichael and Walls (1976a, b) (see Walls (1979) for a brief review) as a practical means of generating light with non-classical coherence properties, and they derived the statistical properties to be expected in a particular case. In the present section we derive the analogous results for a somewhat more general atomic model. The methods used are similar to those of Carmichael and Walls and were originated by Mollow (1969).

We begin with a description of the general method and then apply it to a simple special case. Consider a two-level atom with a ground state $|1\rangle$ and an excited state $|2\rangle$ of excitation energy $\hbar\omega_0$. The atomic transitions brought about by the application of an incident beam of coherent light of frequency ω are governed by the optical Bloch equations, whose properties are reviewed in the appendix. The calculation of the degree of second-order coherence of the light re-radiated by the atom falls into two parts. In the first part, we derive the correlation functions of products of the operators $\hat{\pi}^\dagger$ and $\hat{\pi}$ that excite and de-excite the two-level atom. These operators are formally defined by (A1.2) in the appendix, and in terms of the atomic density matrix of (A1.4) their expectation values in the interaction representation are

$$\langle \hat{\pi}^\dagger(t) \rangle = \exp(i\omega t)\rho_{12}(t) \tag{4.1}$$

$$\langle \hat{\pi}(t) \rangle = \exp(-i\omega t)\rho_{21}(t) \tag{4.2}$$

$$\langle \hat{\pi}^\dagger(t)\hat{\pi}(t) \rangle = \rho_{22}(t) = 1 - \langle \hat{\pi}(t)\hat{\pi}^\dagger(t) \rangle. \tag{4.3}$$

In general, the values of the density matrix elements at time t depend upon the initial values of all the matrix elements. Thus, using the above connections between operator expectation values and the density matrix we can write

$$\langle \hat{\pi}^\dagger(t)\hat{\pi}(t) \rangle = \alpha_1(t) + \alpha_2(t)\langle \hat{\pi}(0) \rangle + \alpha_3(t)\langle \hat{\pi}^\dagger(0) \rangle + \alpha_4(t)\langle \hat{\pi}^\dagger(0)\hat{\pi}(0) \rangle. \tag{4.4}$$

The $\alpha_i(t)$ are obtained by solution of the Bloch equations; usually the state of the system after a very long time is independent of the initial conditions, and we have

$$\langle \hat{\pi}^\dagger(\infty)\hat{\pi}(\infty) \rangle = \alpha_1(\infty) \quad \alpha_2(\infty) = \alpha_3(\infty) = \alpha_4(\infty) = 0. \tag{4.5}$$

The single-time expectation value (4.4) can be converted to a two-time correlation function by means of the quantum regression theorem proved by Lax (1968) (see also Mollow 1969), which shows that if

$$\langle \hat{A}(t) \rangle = \sum_i \alpha_i(t) \langle \hat{A}_i(0) \rangle \tag{4.6}$$

then

$$\langle \hat{B}(0)\hat{A}(t)\hat{C}(0) \rangle = \sum_i \alpha_i(t) \langle \hat{B}(0)\hat{A}_i(0)\hat{C}(0) \rangle \tag{4.7}$$

where the capitals denote any operators. It follows from (4.4) that

$$\langle \hat{\pi}^\dagger(0)\hat{\pi}^\dagger(t)\hat{\pi}(t)\hat{\pi}(0) \rangle = \alpha_1(t)\langle \hat{\pi}^\dagger(0)\hat{\pi}(0) \rangle \tag{4.8}$$

since

$$\langle \hat{n}^\dagger(0)\hat{n}(0)\hat{n}(0)\rangle = \langle \hat{n}^\dagger(0)\hat{n}^\dagger(0)\hat{n}(0)\rangle = \langle \hat{n}^\dagger(0)\hat{n}^\dagger(0)\hat{n}(0)\hat{n}(0)\rangle = 0 \quad (4.9)$$

and bearing in mind that a pure number is a special case of an operator.

The second part of the calculation relates the correlation of atomic operators to the correlation functions of electric-field operators needed for the degree of second-order coherence (3.6). The two kinds of correlation function are simply related since the quantised field radiated by an electric dipole with transition moment proportional to $\hat{n} + \hat{n}^\dagger$ satisfies an expression similar to that relating the analogous classical field to the classical dipole that radiates it (Mollow 1969, Durrant 1977). For evaluating correlations of normally ordered operators, the equivalences between atomic and field operators are

$$\hat{E}^+(rt) = f(r)\hat{n}(t - r/c)$$
$$\hat{E}^-(rt) = f^*(r)\hat{n}^\dagger(t - r/c) \quad (4.10)$$

where $f(r)$ is the classical function that describes the spatial distribution of a dipole field. It is thus a simple matter to convert from atomic to radiative correlation functions, and the functions $f(r)$ cancel in forming the degree of second-order coherence (3.6).

The degree of second-order coherence is now obtained with the use of (4.8) and (4.10). In steady-state conditions it is appropriate to substitute the steady-state value (4.5) for the expectation value that appears on the right of (4.8) and in the denominator of the degree of second-order coherence. The result is

$$g_{11}^{(2)}(t) = \frac{\langle \hat{n}^\dagger(0)\hat{n}^\dagger(t)\hat{n}(t)\hat{n}(0)\rangle}{[\langle \hat{n}^\dagger(\infty)\hat{n}(\infty)\rangle]^2} = \frac{\alpha_1(t)}{\alpha_1(\infty)} \quad (4.11)$$

where we use the small g to denote the degree of coherence of the light radiated by a single atom. It is seen from (4.3) and (4.4) that $\alpha_1(t)$ is equal to $\rho_{22}(t)$ for initial conditions in which $\rho_{12}(0)$, $\rho_{21}(0)$ and $\rho_{22}(0)$ are all zero.

The general method outlined above is easily applied to a simple special case. It is shown in the appendix that the time dependence of the excited-state population ρ_{22} is sometimes described by a rate equation

$$d\rho_{22}/dt = R - 2\gamma\rho_{22} \quad (4.12)$$

where R is an optical pumping rate and 2γ is the radiative decay rate. The solution is

$$\rho_{22}(t) = (R/2\gamma)[1 - \exp(-2\gamma t)] + \rho_{22}(0)\exp(-2\gamma t) \quad (4.13)$$

and by comparison with (4.3) and (4.4),

$$\alpha_1(t) = (R/2\gamma)[1 - \exp(-2\gamma t)]. \quad (4.14)$$

Substitution in (4.11) gives a degree of second-order coherence

$$g_{11}^{(2)}(t) = 1 - \exp(-2\gamma t). \quad (4.15)$$

This function is illustrated by the broken curve in figure 5; it clearly violates the classical inequalities (2.6) and (2.9) and the light is antibunched with a photon distribution similar to figure 4(c). The theory thus predicts the generation of non-classical light for a very simple atomic model that is valid in the rate equation limit for excitation by light of arbitrary coherence properties. The model can also be applied to other methods of atomic excitation, for example electron bombardment.

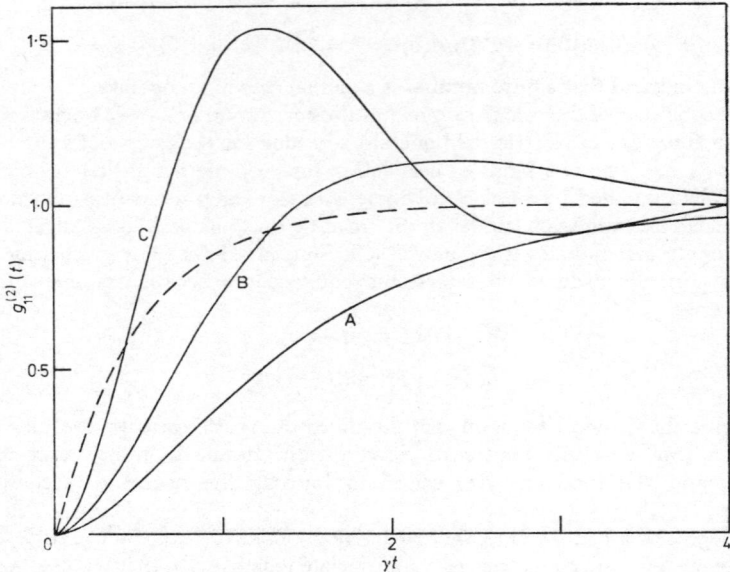

Figure 5. Time dependence of the degree of second-order coherence. Broken curve: the rate-equation result (4.15). Full curves: the weak-beam, zero collision broadening result (4.29) for A, $\Delta=0$; B, $\Delta=\gamma$ and C, $\Delta=2\gamma$.

We are not directly interested in the degree of first-order coherence, but its form is needed in §4.3 to obtain the degree of second-order coherence of an assembly of two or more atoms. The single-operator expectation value (4.2) has a general form similar to (4.4),

$$\langle \hat{\pi}(t) \rangle = \exp(-i\omega t)[\beta_1(t) + \beta_2(t)\langle \hat{\pi}(0)\rangle + \beta_3(t)\langle \hat{\pi}^\dagger(0)\rangle + \beta_4(t)\langle \hat{\pi}^\dagger(0)\hat{\pi}(0)\rangle] \quad (4.16)$$

where the $\beta_i(t)$ are obtained by solution of the Bloch equations for $\rho_{21}(t)$. The steady-state result analogous to (4.5) is

$$|\langle \hat{\pi}(\infty)\rangle| = |\beta_1(\infty)| \quad \beta_2(\infty) = \beta_3(\infty) = \beta_4(\infty) = 0. \quad (4.17)$$

It follows from the regression theorem (4.6) and (4.7) that

$$\langle \hat{\pi}^\dagger(0)\hat{\pi}(t)\rangle = \exp(-i\omega t)[\beta_1(t)\langle \hat{\pi}^\dagger(0)\rangle + \beta_2(t)\langle \hat{\pi}^\dagger(0)\hat{\pi}(0)\rangle] \quad (4.18)$$

and the remaining contributions vanish because of zero initial expectation values similar to (4.9). The same procedure used in second order now gives a complex degree of first-order coherence

$$g_{11}^{(1)}(t) = \langle \hat{\pi}^\dagger(0)\hat{\pi}(t)\rangle / \langle \hat{\pi}^\dagger(\infty)\hat{\pi}(\infty)\rangle$$
$$= \exp(-i\omega t)[\beta_1(t)\beta_1^*(\infty) + \beta_2(t)\alpha_1(\infty)]/\alpha_1(\infty) \quad (4.19)$$

where the steady-state expectation values are taken from (4.5) and (4.17). The first-order result is more complicated than the second-order result (4.11); it requires a calculation of $\rho_{21}(t)$ for initial conditions in which $\rho_{22}(0)$ can be set equal to zero but $\rho_{21}(0)$ (and hence $\rho_{12}(0)$) must be taken non-vanishing in order to obtain $\beta_2(t)$.

The normalised spectrum $S(\nu)$ of the light emitted in resonance fluorescence is given by the Fourier transform of the degree of first-order coherence according to

$$S(\nu) = (1/2\pi)2\text{Re} \int_0^\infty \exp(i\nu t) g_{11}^{(1)}(t) \, dt. \quad (4.20)$$

It is clear from the properties of the $\beta_i(t)$ in (4.16) that (4.19) gives

$$g_{11}^{(1)}(0) = 1 \quad (4.21$$

and it then follows from (4.20) that

$$\int_{-\infty}^{\infty} S(\nu) \, d\nu = 1. \quad (4.22)$$

The spectrum contains two distinct kinds of contribution corresponding to the parts of $\beta_1(t)$ and $\beta_2(t)$ in (4.19) that decay to zero as $t \to \infty$ and to the part $\beta_1(\infty)$ of $\beta_1(t)$ that has a constant amplitude. The Fourier transforms of the decaying parts of the degree of first-order coherence produce emitted light at frequencies that differ from the frequency ω of the exciting light; this is the *inelastic* (sometimes called incoherent) part of the fluorescent light. On the other hand, the Fourier transform of the constant-amplitude part of the degree of first-order coherence produces a frequency of emitted light equal to the frequency of the exciting light; this is the *elastic* (sometimes called coherent) part of the fluorescent light.

The spectrum of the elastic part of the fluorescence obtained from (4.20) is

$$S(\nu)_{\text{elastic}} = |g_{11}^{(1)}(\infty)| \delta(\nu - \omega) \quad (4.23)$$

where from (4.17) and (4.19)

$$|g_{11}^{(1)}(\infty)| = |\beta_1(\infty)|^2 / \alpha_1(\infty). \quad (4.24)$$

The form of the inelastic part of the fluorescent spectrum depends on the relative values of the experimental parameters and it is not possible to write down a general result. Much effort has been applied to calculations of the inelastic spectrum (Mollow 1969, 1970, 1975, Carmichael and Walls 1975, 1976a, b, Swain 1975, Kimble and Mandel 1976, and other papers). The details of the spectrum are not of direct interest for the present review, but some consideration of its division into elastic and inelastic parts is useful in interpreting the coherence properties of the fluorescent light derived in the following subsection.

4.2. *Radiation by a single driven atom: general case*

The methods of the previous subsection can be applied to more general models of resonance fluorescence described by the optical Bloch equations. The result (4.11) for the degree of second-order coherence holds in general, and it is only necessary to determine the form of $\rho_{22}(t)$ for the initial conditions given in (A1.11). The degree of first-order coherence is similarly given in general by (4.19). We first consider the steady-state and short-time behaviours, and then turn to the general-time behaviours in some special cases.

It is obvious from (4.11) that

$$g_{11}^{(2)}(\infty) = 1 \quad (4.25)$$

in agreement with the general property (2.10). The initial behaviour of the degree of second-order coherence for small times t is also given by a general expression that holds for all values of the parameters. Substitution of (A1.8) and (A1.12) into (4.11) gives

$$g_{11}^{(2)}(t) = [\gamma(\gamma'^2 + \Delta^2) + 2\gamma'|g\alpha|^2]t^2/\gamma' \qquad (4.26)$$

correct to order t^2. Thus the light radiated in resonance fluorescence by a single atom always has a degree of second-order coherence that initially increases from its starting value of 0. The light is always antibunched and the classical inequalities (2.6) and (2.9) are always violated. As discussed by Mollow (1975), Carmichael and Walls (1976a, b), Kimble and Mandel (1976) and Wódkiewicz (1978), the zero initial value of the degree of second-order coherence reflects the inability of a single two-level atom to emit a pair of photons simultaneously; successive photon emissions are separated by the time required for an atom arriving in its ground state after photon emission to be re-excited and emit a second photon.

The infinite time-delay value of the degree of first-order coherence is obtained from (4.24) with the help of (A1.8) and (A1.9),

$$|g_{11}^{(1)}(\infty)| = \frac{\gamma^2(\gamma'^2 + \Delta^2)/\gamma'}{\gamma(\gamma'^2 + \Delta^2) + 2\gamma'|g\alpha|^2}. \qquad (4.27)$$

As discussed in the previous subsection, this quantity is equal to the fraction of the fluorescent intensity in the elastic part of the spectrum, and its magnitude lies in the range 0–1. For small collision broadening ($\gamma' \approx \gamma$) and a weak incident beam ($|g\alpha| \ll \gamma$), almost all the fluorescent light is elastic; the light is approximately first-order coherent even for an infinite time delay, since $|g_{11}^{(1)}(\infty)| \approx 1$. In the opposite limits of large collision broadening ($\gamma' \gg \gamma$) or an intense incident beam ($|g\alpha| \gg \gamma$), almost all the fluorescent light is inelastic; the light is first-order incoherent for an infinite time delay since $|g_{11}^{(1)}(\infty)| \approx 0$. For intermediate values of the parameters, there are significant components of coherent elastic light and incoherent inelastic light, producing a partially coherent fluorescent emission. The zero time-delay value of the degree of first-order coherence has already been given in (4.21).

There is no simple analytical solution of the Bloch equations for arbitrary t in the general case, but such solutions do exist for a range of special cases that embrace many experiments of interest in the optical region. The equations have been thoroughly treated by Allen and Eberly (1975) and the discussion given in the appendix is restricted to a summary of the results in two particular cases of interest here.

Consider first the case where the driving beam of light is sufficiently weak for the Rabi frequency (A1.10) to be much smaller than the other characteristic frequencies in the Bloch equations. Then, using the method of the previous subsection together with the solution for $\rho_{22}(t)$ given in (A1.13), we find

$$g_{11}^{(2)}(t) = 1 + \frac{(2\gamma - \gamma')(\Delta^2 + \gamma'^2)}{\gamma'[\Delta^2 + (2\gamma - \gamma')^2]} \exp(-2\gamma t)$$

$$- \frac{2\gamma[\gamma'(2\gamma - \gamma') + \Delta^2]\cos \Delta t - 4\gamma\Delta(\gamma - \gamma')\sin \Delta t}{\gamma'[\Delta^2 + (2\gamma - \gamma')^2]} \exp(-\gamma' t) \qquad (4.28)$$

(weak beam).

This rather complicated result is best appreciated by considering a couple of further limiting cases. Suppose first that $\gamma' \gg \gamma$ and consider times such that $t \gg 1/\gamma'$. This limit is considered in (A1.16) and it is easily seen that (4.28) reduces in this case to the 'rate equation' result (4.15), illustrated by the broken curve in figure 5. The initial quadratic behaviour demanded by (4.26) extends over times of order $1/\gamma'$ and is not visible in the time scale of order $1/\gamma$ used in the figure, so that the initial time dependence becomes effectively linear.

Another limiting case occurs for zero collision broadening where (4.28) reduces to

$$g_{11}^{(2)}(t) = 1 + \exp(-2\gamma t) - 2\cos\Delta t \exp(-\gamma t) \qquad (\gamma' = \gamma). \qquad (4.29)$$

The zero-detuning result

$$g_{11}^{(2)}(t) = [1 - \exp(-\gamma t)]^2 \qquad (\gamma' = \gamma, \Delta = 0) \qquad (4.30)$$

was first given by Carmichael and Walls (1976a, b). The predicted variations of the degree of second-order coherence for different values of the detuning are shown by the full curves in figure 5.

The degree of first-order coherence in the weak-beam limit is obtained similarly from (4.19) with the help of (A1.8) and (A1.14), giving

$$g_{11}^{(1)}(t) = \exp(-i\omega t)\{(\gamma' - \gamma)\exp[-(\gamma' + i\Delta)t] + \gamma\}/\gamma' \qquad \text{(weak beam)}. \qquad (4.31)$$

The corresponding frequency spectrum of the fluorescent light obtained from (4.20) agrees with results of Huber (1969) and Omont et al (1972),

$$S(\nu) = \frac{\gamma' - \gamma}{\gamma'} \frac{\gamma'/\pi}{(\nu - \omega_0)^2 + \gamma'^2} + \frac{\gamma}{\gamma'}\delta(\nu - \omega) \qquad \text{(weak beam)} \qquad (4.32)$$

where the prefactors of the Lorentzian and the delta function give the integrated strengths of the inelastic and elastic contributions, the latter agreeing with the weak-beam limit of (4.27). It is again instructive to consider a couple of further limiting cases. Suppose first that the collision broadening is strong, when (4.31) reduces to

$$g_{11}^{(1)}(t) = \exp[-(\gamma' + i\omega_0)t] \qquad (\gamma' \gg \gamma) \qquad (4.33)$$

and most of the spectral intensity lies in the inelastic term of (4.32). The degree of first-order coherence (4.33) has the standard form for Lorentzian chaotic light (Loudon 1973). Note that the time dependence is determined by γ' in contrast to the role of γ in determining the time dependence of the degree of second-order coherence (4.15) in the same limit. Now suppose that there is no collision broadening, when (4.31) reduces to

$$g_{11}^{(1)}(t) = \exp(-i\omega t) \qquad (\gamma' = \gamma) \qquad (4.34)$$

and all the spectral intensity lies in the elastic term of (4.32). This is the standard result for the degree of first-order coherence of coherent light of frequency ω.

The second main case in which the Bloch equations can be solved is that of zero detuning. With $\rho_{22}(t)$ taken from (A1.20), the method of §4.1 gives

$$g_{11}^{(2)}(t) = 1 - \frac{2\gamma + \gamma' + \lambda}{2\lambda}\exp[-(2\gamma + \gamma' - \lambda)t/2] + \frac{2\gamma + \gamma' - \lambda}{2\lambda}\exp[-(2\gamma + \gamma' + \lambda)t/2]$$

$$(\Delta = 0) \quad (4.35)$$

where

$$\lambda = [(2\gamma - \gamma')^2 - 16|g\alpha|^2]^{1/2}. \qquad (4.36)$$

The zero collision broadening version of this result was derived by Carmichael and

Walls (1976a, b); it reproduces (4.30) in the limit of a weak driving beam. In the limit of a strong driving beam (4.35) reduces to

$$g_{11}^{(2)}(t) = 1 - \cos(2|g\alpha|t)\exp[-(2\gamma+\gamma')t/2] \quad (\Delta=0, \text{ strong beam}). \quad (4.37)$$

More general cases, where neither of the above limits is appropriate, can be treated by numerical solution of the Bloch equations. Kimble and Mandel (1976) give what are essentially graphs of $g_{11}^{(2)}(t)$ for some examples of such cases. Calculations of the resonance fluorescence process have been extended to excitation by finite-bandwidth incident light by Kimble and Mandel (1977), Knight et al (1978) and Agarwal (1978), and to multi-level atoms by Cohen-Tannoudji and Reynaud (1977).

4.3. Radiation by a distribution of atoms

Observations of the resonance fluorescence by a restricted number of atoms are most conveniently made by sending a weak atomic beam through the driving light beam, and this is the arrangement used in a series of experiments by Mandel and co-workers. It is not possible to make continuous observations of the radiation by a single atom. We consider the generalisations of the theory of the previous subsection to take account successively of radiation by several atoms, fluctuations in the number of radiating atoms, and the effect of finite atomic transit times through the observation region.

Consider first the effect on the degree of second-order coherence of combining the radiation by a fixed number N of atoms. The problem has been treated by Agarwal et al (1977), Jakeman et al (1977), Carmichael et al (1978) and Kimble et al (1978). For N identical atoms radiating independently, the correlation function required for the degree of second-order coherence becomes

$$\sum_{i,j,k,l} \langle \hat{\pi}_i^\dagger(0)\hat{\pi}_j^\dagger(t)\hat{\pi}_k(t)\hat{\pi}_l(0)\rangle = N\langle\hat{\pi}^\dagger(0)\hat{\pi}^\dagger(t)\hat{\pi}(t)\hat{\pi}(0)\rangle + N(N-1)\{\langle\hat{\pi}^\dagger(0)\hat{\pi}(t)\rangle$$
$$\times \langle\hat{\pi}^\dagger(t)\hat{\pi}(0)\rangle + \langle\hat{\pi}^\dagger(0)\hat{\pi}(0)\rangle\langle\hat{\pi}^\dagger(t)\hat{\pi}(t)\rangle\}. \quad (4.38)$$

Here i, j, k and l run over the different atoms, but the contributions of correlation functions that involve two or more atoms are negligibly small on account of the rapidly oscillating phase factors resulting from the spatial separations of the atoms. Thus the operators in each correlation function on the right of (4.38) refer to the same atom. The second and third terms are new contributions that arise in the presence of two or more atoms even though the radiation by different atoms is assumed uncorrelated (correlation effects are considered by Agarwal et al (1977)). We consider here only the temporal coherence of the light, but as in the Hanbury Brown and Twiss experiment, the spatial coherence determined by the arrangement of source and detector controls the extent to which the predicted correlations can be observed. Thus, in (4.38), observation of the contribution of the second term on the right-hand side requires a small solid angle of detection, and the references given above discuss the modification of this term required by spatial coherence effects.

The N-atom form of the two-operator correlation function is similarly given by

$$\sum_{i,j} \langle\hat{\pi}_i^\dagger(0)\hat{\pi}_j(t)\rangle = N\langle\hat{\pi}^\dagger(0)\hat{\pi}(t)\rangle. \quad (4.39)$$

Thus, replacing the correlation functions in the numerator and denominator of (4.11) by their N-atom forms obtained from (4.38) and (4.39), the degree of second-order coherence is

$$G_{11}^{(2)}(t) = [g_{11}^{(2)}(t) + (N-1)(|g_{11}^{(1)}(t)|^2 + 1)]/N. \qquad (4.40)$$

The N-atom degree of first-order coherence obtained from (4.19) is

$$G_{11}^{(1)}(t) = g_{11}^{(1)}(t). \qquad (4.41)$$

Some limiting values of the N-atom degree of second-order coherence are easily found. Thus (4.21) and (4.26) give

$$G_{11}^{(2)}(0) = 2(N-1)/N \qquad (4.42)$$

and the inequality (2.6) is no longer violated when the fluorescence of two or more atoms is observed. In the limit of a long time delay (4.25) and (4.27) give

$$G_{11}^{(2)}(\infty) = 1 + \frac{N-1}{N} \frac{\gamma^4(\gamma'^2 + \Delta^2)^2/\gamma'^2}{[\gamma(\gamma'^2 + \Delta^2) + 2\gamma'|g\alpha|^2]^2}. \qquad (4.43)$$

For a sufficiently small number of atoms, small collision broadening, and a weak incident beam, it is possible for this quantity to be larger than the zero time-delay value (4.42), in violation of (2.9). The requirement (2.10) is also violated, a consequence of the first-order coherence of the elastic component of the emitted light at infinite time delay, analogous to the violation of (2.10) shown by the classical example of figure 1.

In the limit of a large number of atoms ($N \gg 1$), (4.40) reduces with the help of (4.41) to

$$G_{11}^{(2)}(t) = |G_{11}^{(1)}(t)|^2 + 1 \quad (N \gg 1). \qquad (4.44)$$

This is similar to a well-known relation between the degrees of first- and second-order coherence of chaotic light (see, for example, (5.101) of Loudon (1973)) but (4.44) applies in the present calculation whatever the nature of the fluorescent light. It gives

$$G_{11}^{(2)}(0) = 2 \quad (N \gg 1) \qquad (4.45)$$

irrespective of the values of the atomic parameters. The kinds of time dependence predicted by (4.44) can be illustrated by the weak-beam case, where substitution of (4.31) gives

$$G_{11}^{(2)}(t) = [(\gamma' - \gamma)^2 \exp(-2\gamma' t) + 2\gamma(\gamma' - \gamma) \exp(-\gamma' t) \cos \Delta t + \gamma'^2 + \gamma^2]/\gamma'^2$$
$$(N \gg 1, \text{ weak beam}). \qquad (4.46)$$

The spectrum of the fluorescent light for this case is given by (4.32) and contains the Lorentzian inelastic contribution plus the delta-function elastic part. The inelastic part is dominant for large collision broadening where (4.46) gives

$$G_{11}^{(2)}(t) = 1 + \exp(-2\gamma' t) \quad (\gamma' \gg \gamma) \qquad (4.47)$$

identical to the standard expression (2.22) for chaotic light illustrated in curve B of figure 2. The elastic part is dominant in the opposite extreme of zero collision broadening where

$$G_{11}^{(2)}(t) = 2 \quad (\gamma' = \gamma). \qquad (4.48)$$

The light in this case is formed by superposition with random phases of the coherent single-mode contributions of the individual atoms and (4.48) is identical to the result (3.18) for a single-mode chaotic distribution. The classical inequalities (2.6) and (2.9) are satisfied by the fluorescent light from a large number of atoms showing again the need to seek non-classical effects in the radiation by small numbers of atoms.

For an atomic beam experiment where the number N of observed atoms fluctuates with a Poisson probability distribution of mean number $\bar{N}$, it is a simple matter to take appropriate averages of (4.38) and (4.39) (Jakeman et al 1977, Carmichael et al 1978, Kimble et al 1978). The degree of second-order coherence (4.40) is then replaced by the averaged quantity

$$\bar{G}_{11}^{(2)}(t) = (g_{11}^{(2)}(t)/\bar{N}) + |g_{11}^{(1)}(t)|^2 + 1. \quad (4.49)$$

It is obvious by comparison with (4.40) that the averaging makes no difference at all in the limit of a large number of atoms ($\bar{N} \gg 1$) and the remarks of the preceding paragraph continue to apply. The averaging does make a difference in general, however, and (4.42) is replaced by

$$\bar{G}_{11}^{(2)}(0) = 2 \quad (4.50)$$

independent of $\bar{N}$, while (4.43) is replaced by

$$\bar{G}_{11}^{(2)}(\infty) = 1 + \frac{1}{\bar{N}} + \frac{\gamma^4(\gamma'^2 + \Delta^2)^2/\gamma'^2}{[\gamma(\gamma'^2 + \Delta^2) + 2\gamma'|g\alpha|^2]^2}. \quad (4.51)$$

It is clearly not possible to violate the inequality (2.6) and antibunched light cannot be obtained. It is, however, simple in principle to violate (2.9) since, tor a sufficiently weak atomic beam with $\bar{N}$ smaller than unity, (4.51) is larger than (4.50) whatever the values of the atomic parameters.

The complete time dependence of the degree of second-order coherence (4.49) is obtained by substituting the appropriate single-atom degrees of coherence, for example (4.28) and (4.31) in the weak-beam limit. The resulting expression is rather complicated and we do not reproduce it here. However, it is not difficult to see that the quantum result contains no term with the same form as the anomalous final term in the comparable classical expression (2.18). Such a contribution would correspond in the quantum theory to the impossible process of counting, and hence destroying, the same photon twice.

A completely realistic theory of the time dependence of the degree of second-order coherence of the light radiated by an atomic beam requires a further ingredient, since the atoms are only observed for the length of time taken to cross the observation region. Suppose that all atoms are observed for the same period t_0. The time limit has different effects on the three terms on the right of (4.38). In the first term, if an average number $\bar{N}$ of atoms are present at time zero, the average number that remain at time t is $\bar{N}f(t)$, where

$$f(t) = 1 - (t/t_0) \quad \text{for } t < t_0$$
$$= 0 \quad \text{for } t > t_0. \quad (4.52)$$

Similarly, the second term on the right requires the survival of a pair of atoms in the observation region over a length of time t, leading to an additional factor $f(t)^2$. The third term involves only correlation functions evaluated at single times, and there is no additional factor. The required generalisation of (4.49) is accordingly

$$\bar{G}_{11}^{(2)}(t) = (g_{11}^{(2)}(t)/\bar{N})f(t) + |g_{11}^{(1)}(t)|^2 f(t)^2 + 1. \quad (4.53)$$

The transit-time factors make no difference to the behaviour of the degree of second-order coherence at times t much shorter than t_0, and the initial value (4.50) is unchanged. The time dependence is, however, radically changed for t close to or greater than t_0, and in the latter case (4.53) gives

$$G_{11}^{(2)}(t) = 1 \quad \text{for } t > t_0. \tag{4.54}$$

The property (2.10) therefore holds irrespective of the values of the other parameters. Curve A in figure 6 shows the time dependence of the degree of second-order coherence (4.53) for a weak beam with no collision broadening and for the parameter values given in the caption. The single-atom degrees of coherence are given by (4.30) and (4.34), and the time dependence shown in figure 6 is to be compared with that for a single atom shown in curve A of figure 5.

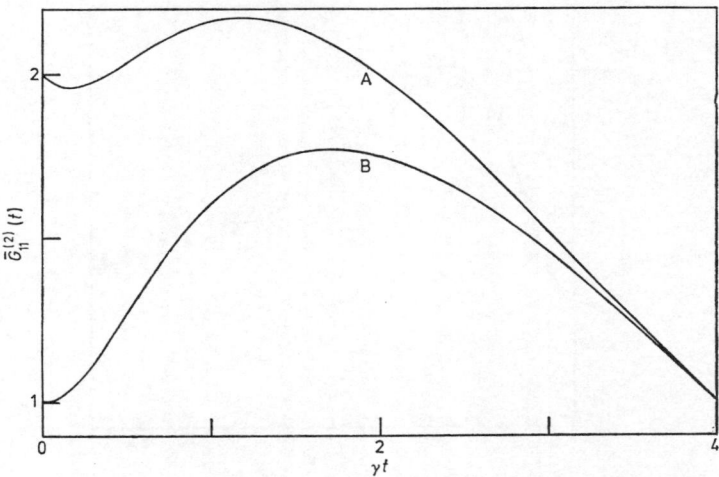

Figure 6. Degree of second-order coherence (4.53) for the case $\gamma' = \gamma$, $|g\alpha| \ll \gamma$, $\Delta = 0$, $\bar{N} = 0.5$ and $\gamma t_0 = 4$. A, the complete expression, and B, with the second term removed.

As was mentioned earlier in the subsection, observation of the contribution of the second term in (4.38) requires a small solid angle of detection. The contribution is lost for large detection angles, corresponding to the experimental conditions needed to collect a significant fraction of the fluorescent light. In this case, the theory should be modified by removal from (4.40), (4.49) and (4.53) of the terms that involve the degree of first-order coherence. Curve B of figure 6 shows the resulting modification of the time dependence of the degree of second-order coherence (4.53). It is a simple matter to construct more general forms of the modified degree of second-order coherence using the single-atom expressions (4.28) for a weak beam or (4.35) for zero detuning. However, the more general examples are similar to that shown in figure 6 in that the light is not antibunched, since (2.6) is satisfied but the degree of second-order coherence initially increases from its zero time-delay value in clear violation of the classical inequality (2.9).

Kimble et al (1978) give a more accurate treatment of the effect of finite observation times that allows for a distribution of atomic velocities. This paper and the others mentioned at the beginning of the subsection also discuss in more detail the spatial coherence requirements for observation of the contribution from the second term in (4.38), and they treat other effects not considered here such as the influence of direct detection of background incident light. Degrees of coherence are also computed for more general values of the atomic parameters than the weak-beam and zero-detuning limits used to illustrate the present review. With all these aspects taken into account, Kimble et al (1977, 1978) and Dagenais and Mandel (1978) find good agreement between their experiments and the theory. Thus figure 7 shows a comparison between measured values and the theoretical expression for the correlation function that occurs in the numerator of the degree of second-order coherence. The figure resembles curve B of figure 6 and represents the direct observation of a violation of inequality (2.9). Figure 8 shows two examples of single-atom degrees of second-order coherence

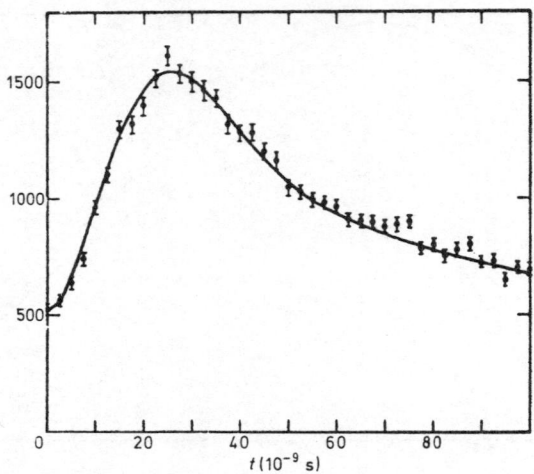

Figure 7. Unnormalised photon correlation as a function of the time delay t for resonance fluorescence of a beam of sodium atoms with $2|g\alpha| = 3 \cdot 3\gamma$ and $\Delta = 0$. The points are measured values and the curve shows the theoretical dependence (from Dagenais and Mandel 1978).

obtained by processing the measured data; the coherence has all the expected properties reviewed in §4.2, including the photon antibunching with $g_{11}{}^{(2)}(0) = 0$.

4.4. Non-linear optical processes

Non-classical single-beam statistics can in principle be generated by various processes in non-linear optics. There has been a great deal of theoretical work in the area but little outcome in terms of clear experimental observations, and we give here only a brief account of the main predictions.

The main process considered is that of two-photon absorption from a single beam of light (Agarwal 1970, Chandra and Prakash 1970, McNeil and Walls 1974, 1975, Tornau and Bach 1974, Simaan and Loudon 1975a, 1978, Bandilla and Ritze 1975, Every 1975b, Paul et al 1976). It is assumed that the width of the excited state of the two-photon absorbing atoms is much larger than the bandwidth of the incident light,

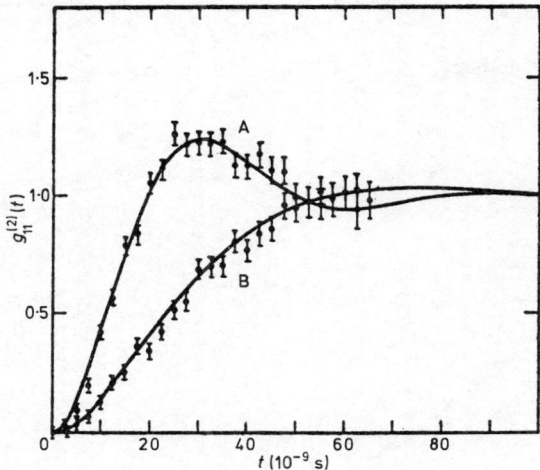

Figure 8. The measured and theoretical time dependences of the degree of second-order coherence of a single sodium atom for $\Delta=0$ and different strengths of driving. Curve A, $2|g\alpha|=3\cdot 3\gamma$; curve B, $2|g\alpha|=1\cdot 4\gamma$ (from Dagenais and Mandel 1978).

and rate equations are used to describe the exchange of energy between the light and the atoms. The calculations assume single-mode incident light, and Mohr and Paul (1979) show that the two-photon absorption of multimode light can sometimes be reduced to an effective single-mode problem.

The above references treat the situation where incident light of some well-defined statistical nature is subjected to two-photon absorption for some length of time T, and is then given a further statistical analysis. Figure 9 shows the dependence of the zero time-delay degree of second-order coherence on the length of time T for which the absorption took place. On the horizontal axis N is the number of absorbing atoms and J is the absorption rate defined so that in the presence of n photons the transition probability per unit time for a single atom is $Jn(n-1)$. The two curves show the time dependences for light that is initially single-mode chaotic or coherent and it is clear that the two-photon absorption has qualitatively the same effect in both cases. The reduction in the zero time-delay degree of second-order coherence is readily understood by reference to figure 4, since the absorption occurs most rapidly for pairs of coincident or nearly coincident photons. The removal of such photons from the bunched or random distributions in parts (a) and (b) of the figure produces distributions similar to that shown for antibunched light in part (c).

Bandilla and Ritze (1976) and Chaturvedi et al (1977) consider somewhat different experimental situations in which steady states are established, with the loss of photons from a cavity by the two-photon absorption balanced by a supply of new photons. Thus the latter reference shows that in the case where the supply is in the form of a beam of coherent light, the degree of second-order coherence is

$$G_{11}^{(2)}(t)=1-(1/3\bar{n})\exp(-3\bar{n}Jt) \qquad (4.55)$$

where $\bar{n}$ is the steady-state mean number of photons in the cavity. This result shows

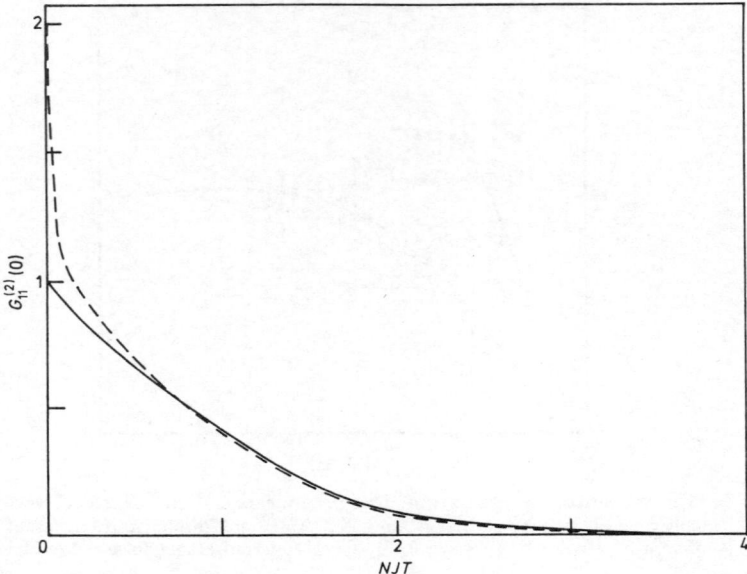

Figure 9. Zero time-delay degrees of second-order coherence of light beams after a period T of two-photon absorption. The beams are initially chaotic (-----) and coherent (———) (after Simaan and Loudon 1975a).

that the light is antibunched, the effect being greater the smaller the mean number of photons.

Two-photon absorption is a rather weak effect and it is difficult to observe the changes in beam statistics to which it gives rise. Also, any simultaneous single-photon absorption that occurs tends to remove or mask the sought antibunching effects (Bandilla 1977). There have been observations of a decreasing degree of second-order coherence brought about by two photon absorption (Krasinski and Dinev 1976, Krasinski 1978) but the results are so far confined to the classical region where the inequality (2.6) is obeyed. In addition to the production of antibunching, Hildred and Hall (1978) show that two-photon absorption can produce light whose degree of second-order coherence remains below the value unity even at infinite time delay, in disagreement with (2.10) and similar to the property of the photon number state shown in (3.8).

The above calculations assume in the main that the energy removed from the light beam is transferred to the atoms in some non-returnable way so that re-emission of photons can be ignored. Rather similar antibunching effects in the fundamental beam can be caused by other non-linear optical processes that depend on the conversion of pairs of photons into some other form of excitation. Thus Simaan (1978) shows that the statistics of the incident beam in the hyper-Raman effect can be made non-classical; in this process, the energy of two incident photons is shared by a single scattered photon and an excitation of the scattering medium.

Both two-photon absorption and the hyper-Raman effect involve an active role of the non-linear medium in absorbing energy from the radiation field. Non-classical beam statistics can also be generated in non-linear processes where the medium plays a more passive role, with no exchange of energy between medium and field. There has been a lot of theoretical work in this connection on the statistical effects of second-harmonic generation (Stoler 1974, Dewael 1975, Kozierowski and Tanaś 1977, Mostowski and Rzazewski 1978, Drummond et al 1979). Various kinds of statistical change are in principle possible depending upon the relative initial intensities and phases of the fundamental and second-harmonic beams, but as would be expected, the fundamental beam tends towards an antibunched nature in conditions where it loses energy to the second harmonic by removal of photons two at a time. In this, as in the other non-linear processes, observation of antibunching seems most favourable when the initial light beam is coherent, with its degree of second-order coherence already on the boundary of the non-classical region.

As an interesting complement to the calculations of the effects of two-photon absorption and second-harmonic generation on photon statistics, Wagner et al (1979) have carried out experiments that simulate photon antibunching. Their work uses an analogue arrangement in which a non-linear filter changes the intensity fluctuations in light beams in a similar way to the calculated changes in photon-number fluctuations for the non-linear optical processes.

Non-classical effects are predicted more generally for higher-order non-linear processes that require the simultaneous removal from a light beam of more than two photons, such as three-photon absorption or third-harmonic generation. However, the physical significance of the antibunching remains the same, and the practical difficulties of carrying out suitable experiments, already great for the second-order non-linear processes, presumably increase.

5. Non-classical double-beam coincidence statistics

5.1. Two-photon cascade emission

Various methods for observing violations of the classical inequality (2.8) rely on processes in which two photons of different frequency are emitted more or less simultaneously. If the characteristic period of a single such two-photon emission process is short compared to the time interval between successive two-photon emissions, the conditions are ripe for a coincidence correlation between the two beams that exceeds the correlation within each beam individually.

One arrangement of this kind uses the photons obtained in cascade emission associated with spontaneous transitions between three atomic energy levels. With the notation shown in figure 10, atoms excited to level 2 decay to level 1 emitting a photon of frequency ω_2, and subsequently decay to level 0 emitting a photon of frequency ω_1. Measurements of intensity correlations within and between the two beams of frequencies ω_1 and ω_2 provide values of the three degrees of second-order coherence required for an experimental test of the classical inequality (2.8).

The theory of second-order coherence in two-photon cascade emission outlined below is a straightforward extension of the theory of two-level resonance fluorescence reviewed in §4. We particularly have in mind an experiment performed by Clauser (1974) using the three Hg energy levels indicated in figure 10. The atoms were excited by electron bombardment and the light of frequencies ω_1 and ω_2 was collected from large solid angles. Each light beam was analysed by a half-silvered mirror and double phototube arrangement similar to that of Hanbury Brown and Twiss shown in figure 3. It was possible to record correlations between the readings of any pair of the

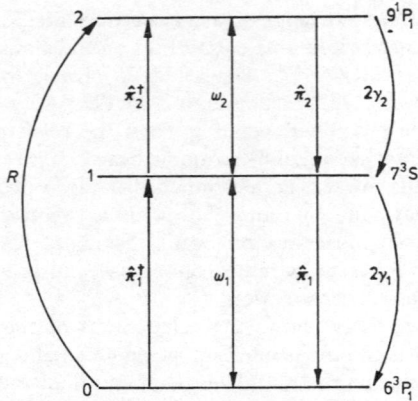

Figure 10. Atomic energy levels used in cascade emission showing transition frequencies, operators and rates used in the analysis.

four phototubes in the complete experiment. The three degrees of second-order coherence appearing in (2.8) could be measured in a symmetrical fashion that eliminated systematic differences between the responses of individual phototubes.

The transition operators $\hat{\pi}_1^\dagger$, $\hat{\pi}_1$ and $\hat{\pi}_2^\dagger$, $\hat{\pi}_2$, and the radiative decay rates $2\gamma_1$ and $2\gamma_2$ shown in figure 10 are defined by analogy with the corresponding quantities for the two-level atom treated in the appendix, and we need only a simple generalisation of the two-level theory. For excitation by electron bombardment, it is permissible to describe the time development of the atomic density matrix ρ_{ij} by rate equations similar to (A1.17). If the excitation takes atoms only from the ground state to the upper excited state at a rate R, then an obvious generalisation of the rate equation to the three-level case gives

$$d\rho_{22}/dt = R - 2\gamma_2 \rho_{22}$$
$$d\rho_{11}/dt = 2\gamma_2 \rho_{22} - 2\gamma_1 \rho_{11} \quad (5.1)$$
$$d\rho_{00}/dt = -R + 2\gamma_1 \rho_{11}.$$

The off-diagonal elements of the density matrix decay at much faster rates determined by the collision and Doppler broadening.

The solutions of (5.1) are

$$\rho_{22}(t) = [\rho_{22}(0) - (R/2\gamma_2)] \exp(-2\gamma_2 t) + (R/2\gamma_2) \quad (5.2)$$

$$\rho_{11}(t) = [\rho_{11}(0) - (R/2\gamma_1)] \exp(-2\gamma_1 t) + (R/2\gamma_1) - [2\gamma_2 \rho_{22}(0) - R][\exp(-2\gamma_1 t) - \exp(-2\gamma_2 t)]/2(\gamma_1 - \gamma_2). \quad (5.3)$$

These elements of the density matrix determine the atomic expectation values

$$\langle \hat{\pi}_2^\dagger(t)\hat{\pi}_2(t) \rangle = \rho_{22}(t) \quad \langle \hat{\pi}_1^\dagger(t)\hat{\pi}_1(t) \rangle = \rho_{11}(t) \quad (5.4)$$

similar to (4.3), and the steady-state values are

$$\langle \hat{\pi}_2^\dagger(\infty)\hat{\pi}_2(\infty) \rangle = R/2\gamma_2 \quad \langle \hat{\pi}_1^\dagger(\infty)\hat{\pi}_1(\infty) \rangle = R/2\gamma_1. \quad (5.5)$$

The various degrees of second-order coherence of the emitted light are obtained by an application of the method of §4.1. Thus the two-time atomic correlation functions are obtained from (5.2) and (5.3) with the use of the quantum regression theorem (4.6) and (4.7), whereas in (4.9) any products of equal-time operators that include $\hat{n}_i\hat{n}_i$ or $\hat{n}_i{}^\dagger\hat{n}_i{}^\dagger$ have zero value. The field and atomic operators are related by obvious generalisations of (4.10), and the degrees of second-order coherence are determined by the correlation functions of the radiating atoms, as for example in

$$g_{12}{}^{(2)}(t) = \frac{\langle \hat{n}_2{}^\dagger(0)\hat{n}_1{}^\dagger(t)\hat{n}_1(t)\hat{n}_2(0)\rangle}{\langle \hat{n}_1{}^\dagger(\infty)\hat{n}_1(\infty)\rangle \langle \hat{n}_2{}^\dagger(\infty)\hat{n}_2(\infty)\rangle}. \tag{5.6}$$

Similar to the two-level calculation, the initial values of correlation functions that occur in the evaluation of the numerator of (5.6) are set equal to their steady-state magnitudes.

The resulting single-atom degrees of second-order coherence given by (5.6) and the corresponding expressions for other choices of subscript are

$$g_{22}{}^{(2)}(t) = g_{21}{}^{(2)}(t) = 1 - \exp(-2\gamma_2 t) \tag{5.7}$$

$$g_{11}{}^{(2)}(t) = \{\gamma_1[1 - \exp(-2\gamma_2 t)] - \gamma_2[1 - \exp(-2\gamma_1 t)]\}/(\gamma_1 - \gamma_2) \tag{5.8}$$

$$g_{12}{}^{(2)}(t) = g_{11}{}^{(2)}(t) + (2\gamma_1/R)\exp(-2\gamma_1 t). \tag{5.9}$$

These degrees of second-order coherence are similar to the expression (4.15) for single-photon emission in that the first three vanish at $t=0$ and increase towards unity for long times t. The classical inequalities (2.6) and (2.9) are violated for each of the beams taken individually.

The fourth degree of second-order coherence (5.9), however, has an additional contribution that does not vanish at $t=0$. With the convention of (2.3) and (3.3) that the measurement of the intensity of the beam indicated by the first subscript on the degree of coherence *follows* the measurement on the beam indicated by the second subscript, $g_{12}{}^{(2)}(t)$ refers to observations in which photon ω_1 follows photon ω_2. The final term in (5.9) is the contribution of observations in which the two photons are produced in the *same* cascade emission; it can be much larger than the first term, which arises from photons produced in *different* cascade emissions, if the excitation rate R is sufficiently small. There is of course no similar contribution to $g_{21}{}^{(2)}(t)$ given by (5.7) because, from the nature of a single cascade emission, photon ω_2 must precede photon ω_1.

For comparison with experiment, the single-atom degrees of coherence must be generalised to obtain the properties of the light emitted by a distribution of atoms, following the procedure of §4.3. We assume a large average number $\bar{N}$ of radiating atoms and ignore transit-time effects so that (4.49) is the appropriate kind of connection between the single- and many-atom degrees of coherence. In the present calculation, the degrees of first-order coherence can be set equal to zero for times of the order of the radiative lifetimes where the single-atom degrees of second-order coherence (5.7)–(5.9) show their main time dependence. In any case, the lack of spatial coherence associated with the large angles of collection in the experiments removes the second contribution on the right of (4.49). The many-atom degree of second-order coherence is therefore

$$G_{ij}{}^{(2)}(t) = (g_{ij}{}^{(2)}(t)/\bar{N}) + 1 \tag{5.10}$$

and for $\bar{N} \gg 1$ we find approximately

$$\bar{G}_{22}^{(2)}(t) = \bar{G}_{21}^{(2)}(t) = \bar{G}_{11}^{(2)}(t) = 1 \qquad (5.11)$$

$$\bar{G}_{12}^{(2)}(t) = (2\gamma_1/R\bar{N}) \exp(-2\gamma_1 t) + 1. \qquad (5.12)$$

The classical inequalities (2.6) and (2.9) are no longer violated, but for R sufficiently small that $R\bar{N}$ is smaller than or comparable to $2\gamma_1$, the first term in (5.12) produces a serious violation of (2.8) for times t smaller than about $1/2\gamma_1$.

Figure 11 shows Clauser's experimental results for essentially the right-hand side of (2.8) divided by the left-hand side. The full curves are the theoretical predictions from (5.11) and (5.12) with the numerical values

$$2\gamma_1 = 1 \cdot 19 \times 10^8 \text{ s}^{-1} \quad \text{and} \quad 2\gamma_1/R\bar{N} = 1 \cdot 42. \qquad (5.13)$$

The decay rate for the $7^3S_1 \to 6^3P_1$ transition of Hg is the average of two values given by Mosburg and Wilke (1978), but the other numerical value in (5.13), not available from Clauser's paper, is chosen to give the best fit between theory and experiment. It is seen that quantum theory and experiment are in close agreement with a result that is inconsistent with the classical predictions for the correlations of two light beams, however generated.

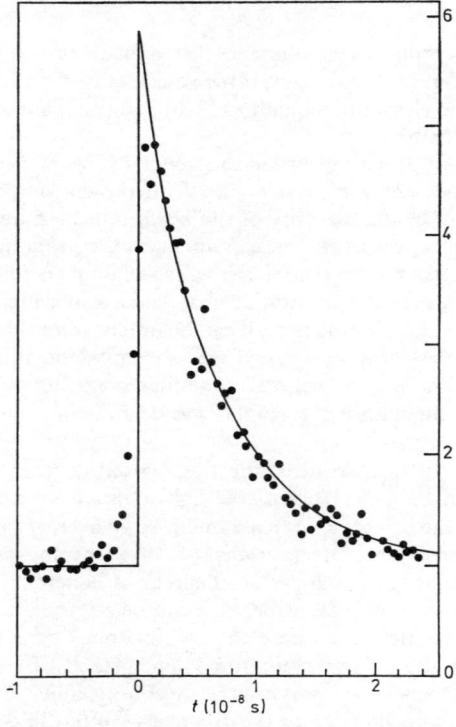

Figure 11. Experimental points from Clauser (1974) and theoretical curves from (5.11) and (5.12) for the two-beam degree of second-order coherence. The function plotted is $[\bar{G}_{12}^{(2)}(t)]^2$ for $t > 0$ and $[\bar{G}_{21}^{(2)}(-t)]^2$ for $t < 0$. The values of parameters are are given in the text.

The experiment in fact measured individually all four degrees of second-order coherence appearing in (5.11) and (5.12) and was unable to detect any time dependence in $\bar{G}_{11}{}^{(2)}(t)$ and $\bar{G}_{22}{}^{(2)}(t)$, in agreement with (5.11). Now $1/2\gamma_1$ in (5.12) is the radiative lifetime of level 1 and $R\bar{N}$ is the rate of atomic excitation, so that the mean number of simultaneously radiating atoms is

$$\bar{n} = R\bar{N}/2\gamma_1 \approx 0\cdot 7. \tag{5.14}$$

Thus the first term in (5.12) is of the same order as the anomalous coincidence term that appears in the classical theory of the degree of second-order coherence, for example the final term in (2.18). The presence of any such term in the single-beam degrees of second-order coherence would have been easily resolved in Clauser's experiment, but no such contribution was found.

In summary, the studies of two-photon cascade emission provide very clear evidence of a violation of the classical inequality (2.8). The results also disagree with the predictions of the kind of classical model described in §2 that lead to the degree of second-order coherence (2.18); nevertheless, the observed degrees of second-order coherence of the individual light beams do lie within the range allowed by classical theory and it is conceivable that some other classical model could account for this aspect of the results (Mandel 1977). Finally, it should be mentioned that the observation of correlations between photons emitted in cascade forms a standard technique for determining radiative lifetimes of the intermediate states (Camhy-Val and Dumont 1970).

5.2. Non-linear optical processes

Various non-linear optical processes produce pairs of different photons in a combined emission and thus lead to non-classical double-beam coincidence statistics of the same kind as in the cascade emission treated above. The first experiment to detect any kind of non-classical behaviour in the degree of second-order coherence, by Burnham and Weinberg (1970), was of this variety. They produced pairs of photons, one each of the frequencies ω_1 and ω_2, by spontaneous parametric splitting of incident photons at the sum frequency $\omega_1 + \omega_2$. The intensity correlations between the beams at frequencies ω_1 and ω_2 were measured. With only one elementary splitting occurring on average during a detection period, the light is effectively close to a state in which each of the two beams has only a single photon. The expressions given in (3.21) and (3.22) then indicate a very large violation of the classical inequality (2.8). In the experiments of Burnham and Weinberg, the observed coincidence rate from the two photons produced in a single splitting event was 100 times the rate of coincidence between photons generated in different events. The experimental value for the right-hand side of (2.8) therefore exceeded that for the left-hand side by a factor of 10^4. A very complete theory of the experiment has been given by Mollow (1973); although the nature of the coincidence effect is similar in outline to that for the linear cascade emission considered in §5.1, the calculation of the non-linear parametric emission is quite different in detail.

Straightforward non-linear two-photon emission also produces light with non-classical statistical properties. Consider the steady-state photon distribution for light emitted by a two-photon transition between two atomic energy levels whose populations are maintained at fixed values. If no light is present initially and no single-photon transitions take place, the steady-state photon distribution is (Simaan and Loudon 1975b)

$$P_{n_1, n_2} = \delta_{n_1, n_2} \bar{n}_1{}^{n_1}/(1+\bar{n}_1)^{1+n_1} \qquad (5.15)$$

where $\bar{n}_1 = \bar{n}_2$ is the mean photon number in each mode. This is a two-photon version of the chaotic distribution of (3.17), and it leads to

$$G_{11}{}^{(2)}(0) = G_{22}{}^{(2)}(0) = 2 \qquad G_{12}{}^{(2)}(0) = 2 + (1/\bar{n}_1) \qquad (5.16)$$

in clear violation of (2.8), particularly for small numbers of photons. Any realistic theory of two-photon emission would of course need a more complete model of the atomic excitation and emission processes, but the above results show the origin of the non-classical effects in a very simple case.

Non-linear emission processes are favourable for observation of non-classical double-beam statistics since the beams of interest exist only as a result of the non-linearity. Their resulting intrinsic weakness produces the large reciprocal photon numbers that usually determine the size of the quantum effects. By contrast, the non-linear processes described in §4.4 for generating non-classical single-beam statistics need to produce changes in the initial statistics of powerful light beams, often subjected to other interactions that tend to mask the rather small quantum effects.

6. Conclusions

The main concern of the review is with the observations of kinds of light whose degrees of second-order coherence violate various inequalities that must hold in the classical theory of light. The inequalities are valid for any theory in which the beam intensity is describable by a single-valued function of position and time, whose magnitude is assumed to be measurable without any consequent change in its value. The domains of validity of the inequalities are taken to characterise the classical theory.

The possible ranges of values of the degrees of second-order coherence are wider in the quantum theory as a result of its inherent distinction between the parameters that describe the state of a system and the operators that describe the process of measurement. Thus, although a classical measurement of the intensity has a clear correspondence with a quantum measurement of the number of photons in a light beam, the latter observation changes the measured photon number for all initial states that are not eigenstates of the photon destruction operator. A subsequent measurement of the photon number then registers a different count in general, in accordance with the usual nature of the measurement process in quantum mechanics. Thus a correlation of two measurements of the photon number made on the same light field does not in general correspond to the analogous correlation of two measurements of the classical intensity.

Similar remarks apply of course to the correlations of arbitrary numbers of classical intensity or photon-number measurements. The present review is restricted to the degree of second-order coherence, this being the lowest order in which the non-classical effects should be seen, and there has in fact been little or no work on the degrees of third- or higher-order coherence in the context of discrepancies between classical and quantum theory.

The main classical inequalities that the degree of second-order coherence should satisfy are (2.6), (2.8) and (2.9). None of these can be established for the quantum degree of second-order coherence, and it has been shown how the theories of various non-linear and linear optical experiments predict the generation of light that can violate all of the classical inequalities. More concretely, there are experimental observations of light that do not satisfy the single-beam inequality (2.9) (Kimble *et al* 1977, 1978, Dagenais and Mandel 1978) and the double-beam inequality (2.8) (Burnham and Weinberg 1970, Clauser 1974). However, the results of all of these experiments are in very good agreement with the predictions of quantum theory.

Antibunched light, defined as light that violates the single-beam inequality (2.6), has not strictly been observed directly. Its detection in resonance fluorescence requires observation of a single radiating atom, and it is similarly difficult to make photon correlation measurements in the non-linear optical processes where antibunched light is predicted. There can, however, be little doubt that the quantum description is also correct for this aspect of the degree of second coherence. The resonance fluorescence measurements of Mandel and co-workers provide extremely strong evidence for the quantum theory of the resonance fluorescence process and the single-atom degree of second-order coherence extracted in their analysis of the measured data displays the antibunching violation of (2.6).

Although the measured statistical properties of light sometimes lie outside the domain of classical theory, it remains a noteworthy feature of this area of study that the vast majority of photon-counting correlation measurements are made on varieties of light whose properties can equally well be described in terms of classical intensity fluctuations.

Acknowledgments

This article is an expanded version of a talk given at the London Mathematical Society Durham Symposium on the mathematical theory of non-linear problems in quantum mechanics and quantum optics in July 1978. I thank E A Power, the symposium organiser, and also L Allen, J H Eberly and P L Knight for much encouragement and advice. I am grateful to T P Hughes for advice on the mercury spectrum, to J F Clauser for providing photographs of his data and to M Dagenais and L Mandel for permission to reproduce their figures.

Appendix. Solutions of the optical Bloch equations

Consider an atom, with ground state $|1\rangle$ and an excited state $|2\rangle$ separated by energy $\hbar\omega_0$, in interaction with a single-mode beam of light having frequency ω and creation and destruction operators $\hat{a}^\dagger$ and $\hat{a}$. The interaction-representation Hamiltonian of the coupled system is

$$\hat{H} = \hbar\omega_0 \hat{\pi}^\dagger \hat{\pi} + \hbar\omega \hat{a}^\dagger \hat{a} + \hbar\{g \exp(\mathrm{i}\Delta t)\hat{\pi}^\dagger \hat{a} + g^* \exp(-\mathrm{i}\Delta t)\hat{a}^\dagger \hat{\pi}\} \quad (A1.1)$$

where

$$\hat{\pi}^\dagger = |2\rangle\langle 1| \quad \hat{\pi} = |1\rangle\langle 2| \quad (A1.2)$$

$$\Delta = \omega_0 - \omega \quad (A1.3)$$

is the detuning, and g contains the electric dipole matrix element of the transition.

Let σ_{ij} be the 2×2 atomic density matrix in the interaction representation and define ρ_{ij} by

$$\rho_{11}=\sigma_{11} \qquad \rho_{12}=\exp(i\Delta t)\sigma_{12}$$
$$\rho_{21}=\exp(-i\Delta t)\sigma_{21} \qquad \rho_{22}=\sigma_{22}. \qquad (A1.4)$$

The equations of motion for the ρ_{ij} as the atom is driven by the light beam and allowed to relax via dissipative processes take the forms (Allen and Eberly 1975)

$$d\rho_{22}/dt = -ig\alpha\rho_{12}+ig^*\alpha^*\rho_{21}-2\gamma\rho_{22} \qquad (A1.5)$$

$$d\rho_{21}/dt = -(\gamma'+i\Delta)\rho_{21}+ig\alpha(\rho_{22}-\rho_{11}) \qquad (A1.6)$$

and the other components are determined at all times by

$$\rho_{11}+\rho_{22}=1 \qquad \rho_{12}=\rho_{21}^*. \qquad (A1.7)$$

In these equations 2γ is the spontaneous radiative decay rate of the excited state (equal to the Einstein A coefficient) and γ' is the sum of γ and the rate of collisions that interrupt the phase of the wavefunction but do not cause atomic transitions. It is assumed that the incident light beam is in a coherent state with complex amplitude α and mean photon number $|\alpha|^2$.

The steady-state solutions of the Bloch equations (A1.5) and (A1.6) are easily found to be

$$\rho_{22}(\infty)=\frac{\gamma'|g\alpha|^2}{\gamma(\gamma'^2+\Delta^2)+2\gamma'|g\alpha|^2} \qquad (A1.8)$$

$$\rho_{21}(\infty)=-\frac{ig\alpha\gamma(\gamma'-i\Delta)}{\gamma(\gamma'^2+\Delta^2)+2\gamma'|g\alpha|^2} \qquad (A1.9)$$

where the combination

$$\Omega=2|g\alpha| \qquad (A1.10)$$

is the Rabi frequency. The short-time solutions are also easily found but they depend on the initial conditions. If we assume

$$\rho_{22}(0)=\rho_{12}(0)=\rho_{21}(0)=0 \qquad (A1.11)$$

then the initial time dependence of ρ_{22} is

$$\rho_{22}(t)=|g\alpha|^2 t^2. \qquad (A1.12)$$

The solutions of the Bloch equations and some of their special cases were first considered by Torrey (1949). More recently Kimble and Mandel (1976) have given a general solution, with graphs of $\rho_{22}(t)$ for ranges of values of the parameters. Many possible experiments of interest are covered by special cases in which the Bloch equations have simple solutions, and we summarise two of these. Consider first a driving field that is sufficiently weak for $|g\alpha|$ to be smaller than the other characteristic frequencies in the equations. The solution to lowest order in $|g\alpha|$ for the boundary condition (A1.11) is

$$\rho_{22}(t)=|g\alpha|^2\left(\frac{\gamma'/\gamma}{\Delta^2+\gamma'^2}+\frac{(2\gamma-\gamma')/\gamma}{\Delta^2+(2\gamma-\gamma')^2}\exp(-2\gamma t)\right.$$

$$\left.-\frac{2[\gamma'(2\gamma-\gamma')+\Delta^2]\cos\Delta t-4\Delta(\gamma-\gamma')\sin\Delta t}{(\Delta^2+\gamma'^2)[\Delta^2+(2\gamma-\gamma')^2]}\exp(-\gamma't)\right)$$

(weak beam). (A1.13)

The off-diagonal density matrix element is needed in the text for a less stringent initial condition in which $\rho_{21}(0)$ is not zero, and the corresponding solution is

$$\rho_{21}(t) = \frac{ig\alpha}{i\Delta + \gamma'} \{\exp[-(i\Delta + \gamma')t] - 1\} + \rho_{21}(0) \exp[-(i\Delta + \gamma')t] \quad \text{(weak beam)}.$$
(A1.14)

The density matrix in the weak-field limit behaves differently for different relative sizes of γ and γ'. At one extreme, where there is no collision broadening and $\gamma' = \gamma$, (A1.13) reduces to

$$\rho_{22}(t) = \frac{|g\alpha|^2}{\Delta^2 + \gamma^2} [1 + \exp(-2\gamma t) - 2\cos\Delta t \exp(-\gamma t)] \quad (\gamma' = \dot{\gamma}) \quad \text{(A1.15)}$$

a well-known result (see, for example, Smith (1978a) where the $\gamma' = \gamma$ case is treated in some detail). At the opposite extreme where $\gamma' \gg \gamma$ and assuming $t \gg 1/\gamma'$, (A1.13) reduces to

$$\rho_{22}(t) = \frac{|g\alpha|^2 \gamma'/\gamma}{\Delta^2 + \gamma'^2} [1 - \exp(-2\gamma t)] \quad (\gamma' \gg \gamma) \quad \text{(A1.16)}$$

the short-time limit (A1.12) does not hold for the times considered here. The same form of solution (A1.16) is obtained from the rate equation

$$d\rho_{22}/dt = R - 2\gamma\rho_{22} \quad \text{(A1.17)}$$

where the excited-state population is determined by the competition between a pumping rate

$$R = 2|g\alpha|^2 \gamma'/(\Delta^2 + \gamma'^2) \quad \text{(A1.18)}$$

and the radiative decay rate 2γ. The same 'rate equation' time dependence as in (A1.16) is again obtained if (A1.13) is integrated over the detuning,

$$\int \rho_{22}(t) d\Delta = (\pi|g\alpha|^2/\gamma)[1 - \exp(-2\gamma t)] \quad \text{(A1.19)}$$

corresponding to illumination of the atom by a broad-band source. These results apply to a homogeneously collision-broadened transition, but a very similar rate-equation limit is obtained for a Doppler-broadened transition (see (24) of Smith (1978b)).

The other special case for which the Bloch equations can be solved fairly easily is that of zero detuning, $\Delta = 0$, but general values of γ, γ' and $|g\alpha|$. The required result for the same initial conditions (A1.11) as before is

$$\rho_{22}(t) = \frac{|g\alpha|^2}{\gamma\gamma' + 2|g\alpha|^2} \left(1 - \frac{2\gamma + \gamma' + \lambda}{2\lambda} \exp[-(2\gamma + \gamma' - \lambda)t/2] \right.$$

$$\left. + \frac{2\gamma + \gamma' - \lambda}{2\lambda} \exp[-(2\gamma + \gamma' + \lambda)t/2]\right) \quad (\Delta = 0) \quad \text{(A1.20)}$$

where

$$\lambda = [(2\gamma - \gamma')^2 - 16|g\alpha|^2]^{1/2}. \quad \text{(A1.21)}$$

It is easy to check that the small $|g\alpha|$ limit of this expression agrees with the zero-detuning limit of (A1.13). The $\gamma' = \gamma$ case of this result has been given by Carmichael and Walls (1976a, b), Allen et al (1977) and Smith (1978a).

References

Agarwal G S 1970 *Phys. Rev.* A **1** 1445–59
—— 1978 *Phys. Rev.* A **18** 1490–506
Agarwal G S, Brown A C, Narducci L M and Vetri G 1977 *Phys. Rev.* A **15** 1613–24
Allen L, Allen B and Knight P L 1977 *Opt. Commun.* **20** 150–4
Allen L and Eberly J H 1975 *Optical Resonance and Two-Level Atoms* (New York: Wiley)
Baltes H P, Quattropani A and Schwendimann P 1979 *J. Phys. A: Math. Gen.* **12** L35–7
Bandilla A 1977 *Opt. Commun.* **23** 299–302
Bandilla A and Ritze H H 1975 *Phys. Lett.* **55A** 285–6
—— 1976 *Opt. Commun.* **19** 169–71
Burnham D C and Weinberg D L 1970 *Phys. Rev. Lett.* **25** 84–7
Camhy-Val C and Dumont A M 1970 *Astron. Astrophys.* **6** 27–50
Carmichael H J, Drummond P, Meystre P and Walls D F 1978 *J. Phys. A: Math. Gen.* **11** L121–6
Carmichael H J and Walls D F 1975 *J. Phys. B: Atom. Molec. Phys.* **8** L77–81
—— 1976a *J. Phys. B: Atom. Molec. Phys.* **9** L43–6
—— 1976b *J. Phys. B: Atom. Molec. Phys.* **9** 1199–219
Chandra N and Prakash H 1970 *Phys. Rev.* A **1** 1696–8
Chaturvedi S, Drummond P and Walls D F 1977 *J. Phys. A: Math. Gen.* **10** L187–92
Clauser J F 1974 *Phys. Rev.* D **9** 853–60
Clauser J F and Shimony A 1978 *Rep. Prog. Phys.* **41** 1881–927
Cohen-Tannoudji C and Reynaud S 1977 *J. Phys. B: Atom. Molec. Phys.* **10** 345–63
Dagenais M and Mandel L 1978 *Phys. Rev.* A **18** 2217–28
Dewael P 1975 *J. Phys. A: Math. Gen.* **8** 1614–9
Drummond P, McNeil K J and Walls D F 1979 *Opt. Commun.* **28** 255–8
Durrant A V 1977 *Am. J. Phys.* **45** 752–7
Every I M 1973 *J. Phys. A: Math., Nucl. Gen.* **6** 1375–82
—— 1975a *J. Phys. A: Math. Gen.* **8** 133–41
—— 1975b *J. Phys. A: Math. Gen.* **8** L69–72
Glauber R J 1963a *Phys. Rev.* **130** 2529–39
—— 1963b *Phys. Rev.* **131** 2766–88
Hanbury Brown R 1974 *The Intensity Interferometer* (London: Taylor and Francis)
Hardy G H, Littlewood J E and Pólya G 1951 *Inequalities* (Cambridge: Cambridge University Press)
Helstrom C W 1979 *Opt. Commun.* **28** 363–4
Hildred G P and Hall A G 1978 *J. Phys. A: Math. Gen.* **11** L209–12
Huber D L 1969 *Phys. Rev.* **178** 93–102
Jakeman E, Pike E R, Pusey P N and Vaughan J M 1977 *J. Phys. A: Math. Gen.* **10** L257–9
Kimble H J, Dagenais M and Mandel L 1977 *Phys. Rev. Lett.* **39** 691–5
—— 1978 *Phys. Rev.* A **18** 201–7
Kimble H J and Mandel L 1976 *Phys. Rev.* A **13** 2123–44
—— 1977 *Phys. Rev.* A **15** 689–99
Knight P L, Molander W A and Stroud C R 1978 *Phys. Rev.* A **17** 1547–9
Kozierowski M and Tanaś R 1977 *Opt. Commun.* **21** 29–31
Krasinski J 1978 *Multiphoton Processes* ed J H Eberly and P Lambropoulos (New York: Wiley) pp277–87
Krasinski J and Dinev S 1976 *Opt. Commun.* **18** 424–6
Lax M 1968 *Phys. Rev.* **172** 350–61
Loudon R 1973 *The Quantum Theory of Light* (Oxford: Clarendon)
McNeil K J and Walls D F 1974 *J. Phys. A: Math., Nucl. Gen.* **7** 617–31
—— 1975 *Phys. Lett.* **51A** 233–4
Mandel L 1974 *IEEE J. Quant. Electron.* **QE-10** 773–4
—— 1976 *Progress in Optics* vol 13, ed E Wolf (Amsterdam: North-Holland) pp27–68
—— 1977 *J. Opt. Soc. Am.* **67** 1101–4
Milonni P W 1976 *Phys. Rep.* **25** 1–81
Mohr U and Paul H 1979 *J. Phys. A: Math. Gen.* **12** L43–6
Mollow B R 1969 *Phys. Rev.* **188** 1969–75
—— 1970 *Phys. Rev.* A **2** 76–80

—— 1973 *Phys. Rev.* A **8** 2684–94
—— 1975 *Phys. Rev.* A **12** 1919–43
Mosburg E R and Wilke M D 1978 *J. Quant. Spectrosc. Radiat. Transfer* **19** 69–81
Mostowski J and Rzazewski K 1978 *Phys. Lett.* **66A** 275–8
Omont A, Smith E W and Cooper J 1972 *Astrophys. J.* **175** 185–99
Paul H, Mohr U and Brunner W 1976 *Opt. Commun.* **17** 145–8
Simaan H D 1978 *J. Phys. A: Math. Gen.* **11** 1799–802
Simaan H D and Loudon R 1975a *J. Phys. A: Math. Gen.* **8** 539–54
—— 1975b *J. Phys. A: Math. Gen.* **8** 1140–58
—— 1978 *J. Phys. A: Math. Gen.* **11** 435–41
Smith R A 1978a *Proc. R. Soc.* A **362** 1–12
—— 1978b *Proc. R. Soc.* A **362** 13–25
Stoler D 1974 *Phys. Rev. Lett.* **33** 1397–400
Swain S 1975 *J. Phys. B: Atom Molec. Phys.* **8** L437–41
Titulaer U M and Glauber R J 1965 *Phys. Rev.* **140** B676–82
Tornau N and Bach A 1974 *Opt. Commun.* **11** 46–9
Torrey H C 1949 *Phys. Rev.* **76** 1059–68
Wagner J, Kurowski P and Martienssen W 1979 *Z. Phys.* **33** 391–402
Walls D F 1977 *Am. J. Phys.* **45** 952–6
—— 1979 *Nature* **280** 451–4
Wódkiewicz K 1978 *Boulder Symp. on the Foundations of Radiation Theory and Quantum Electrodynamics* (New York: Plenum)

Bell's theorem: experimental tests and implications

JOHN F CLAUSER[†] and ABNER SHIMONY[‡][§]

[†] Lawrence Livermore Laboratory—L-437, Magnetic Fusion Energy Division, Livermore, California 94550, USA
[‡] Departments of Physics and Philosophy, Boston University, Boston, Massachusetts 02215, USA

Abstract

Bell's theorem represents a significant advance in understanding the conceptual foundations of quantum mechanics. The theorem shows that essentially all local theories of natural phenomena that are formulated within the framework of realism may be tested using a single experimental arrangement. Moreover, the predictions by these theories must significantly differ from those by quantum mechanics. Experimental results evidently refute the theorem's predictions for these theories and favour those of quantum mechanics. The conclusions are philosophically startling: either one must totally abandon the realistic philosophy of most working scientists, or dramatically revise our concept of space–time.

[§] Work supported in part by the National Science Foundation.

1. Introduction

Realism is a philosophical view, according to which external reality is assumed to exist and have definite properties, whether or not they are observed by someone. So entrenched is this viewpoint in modern thinking that many scientists and philosophers have sought to devise conceptual foundations for quantum mechanics that are clearly consistent with it. One possibility, it has been hoped, is to reinterpret quantum mechanics in terms of a statistical account of an underlying hidden-variables theory in order to bring it within the general framework of classical physics. However, Bell's theorem has recently shown that this cannot be done. The theorem proves that all realistic theories, satisfying a very simple and natural condition called locality, may be tested with a single experiment against quantum mechanics. These two alternatives necessarily lead to significantly different predictions. The theorem has thus inspired various experiments, most of which have yielded results in excellent agreement with quantum mechanics, but in disagreement with the family of local realistic theories. Consequently, it can now be asserted with reasonable confidence that either the thesis of realism or that of locality must be abandoned. Either choice will drastically change our concepts of reality and of space–time.

The historical background for this result is interesting, and represents an extreme irony for Einstein's steadfastly realistic position, coupled with his desire that physics be expressable solely in simple geometric terms. Within the realistic framework, Einstein *et al* (1935, hereafter referred to as EPR) presented a classic argument. As a starting point, they assumed the non-existence of action-at-a-distance and that some of the statistical predictions of quantum mechanics are correct. They considered a system consisting of two spatially separated but quantum-mechanically correlated particles. For this system, they showed that the results of various experiments are predetermined, but that this fact is not part of the quantum-mechanical description of the associated systems. Hence that description is an incomplete one. To complete the description, it is thus necessary to postulate additional 'hidden variables', which presumably will then restore completeness, determinism and causality to the theory.

Many in the physics community rejected their argument, preferring to follow a counter-argument by Bohr (1935), who believed that the whole realistic viewpoint is inapplicable. Many others, however, felt that since both viewpoints lead to the same observable phenomenology, a commitment to either one is only a matter of taste. Hence, the discussion, for the greater part of the subsequent 30 years, was pursued perhaps more at physicists' cocktail parties than in the mainstream of modern research.

Starting in 1965, however, the situation changed dramatically. Using essentially the same postulates as those of EPR, J S Bell showed for a *Gedankenexperiment* of Bohm (a variant of that of EPR) that no deterministic local hidden-variables theory can reproduce all of the statistical predictions by quantum mechanics. Inspired by that work, Clauser *et al* (1969, hereafter referred to as CHSH) added three contributions. First, they showed that his analysis can be extended to cover actual systems, and that experimental tests of this broad class of theories can be performed. Second, they introduced a very reasonable auxiliary assumption which allows tests to be performed with existing technology. Third, they specifically proposed performing such a test by examining the polarisations of photons produced by an atomic cascade, and derived the required conditions for such an experiment.

Curiously, the transition to a consideration of real systems introduced new aspects to the problem. EPR had demonstrated that any ideal system which satisfies a locality condition must be deterministic (at least with respect to the correlated properties). Since that argument applies only to ideal systems, CHSH therefore had postulated determinism explicitly. Yet, it eventually became clear that it is not the deterministic character of these theories that is incompatible with quantum mechanics. Although not stressed, this point was contained in Bell's subsequent papers (1971, 1972)—any non-deterministic (stochastic) theory satisfying a more general locality condition is also incompatible with quantum mechanics. Indeed it is the objectivity of the associated systems and their locality which produces the incompatibility. Thus, the whole realistic philosophy is in question! Bell's (1971) result, however, is in a form that is awkward for an experimental test. To facilitate such tests, Clauser and Horne (1974, hereafter referred to as CH) explicitly characterised this broad class of theories. They then gave a new incompatibility theorem that yields an experimentally testable result and derived the requirements for such a test. Although such an experiment is difficult to perform (and in fact has not yet been performed), they showed that an assumption weaker in certain respects than the one of CHSH allowed the experiments proposed earlier by CHSH to be used as a test for these theories also.

The interpretation of all of the existing results requires at least some auxiliary assumptions, although experiments are possible for which this is not the case. Even though some of the assumptions are very reasonable, this fact allows loopholes still to exist. Experiments now in progress or being planned will be able to eliminate most of these loopholes. However, even now one can assert with reasonable confidence that the experimental evidence to date is contrary to the family of local realistic theories. The construction of a quantum-mechanical world view as an alternative to the point of view of the local realistic theories is beyond the scope of this review.

Section 2 of this review summarises the argument of EPR, appendix 1 discusses various critical evaluations of it, and appendix 2 summarises briefly the history of hidden-variables theories. Section 3 describes the versions of Bell's theorem discussed above as well as some others. Section 4 discusses the requirements for a fully general test and shows why such an experiment is a difficult one to perform. Section 5 is devoted to a description of the cascade-photon experiments proposed by CHSH. First, it discusses the auxiliary assumptions by CHSH and CH. Second, calculations of the quantum-mechanical predictions for these experiments are summarised. Third, there is a discussion of the actual cascade-photon experiments performed so far (Freedman and Clauser 1972, Holt and Pipkin 1973, Clauser 1976, Fry and Thompson 1976). All but the second agree very well with the quantum-mechanical predictions, thus providing significant evidence against the entire family of local realistic theories. Section 5 ends with a critique of the CH and CHSH assumptions. Section 6 summarises and discusses related experiments measuring the polarisation correlation of photons produced in positronium annihilation (Kasday *et al* 1975, Faraci *et al* 1974, Wilson *et al* 1976, Bruno *et al* 1977) and an experiment measuring the spin correlation of proton pairs (Lamehi-Rachti and Mittig 1976). Section 7 is devoted to an evaluation of the experimental results obtained so far and to the prospects for future experiments.

2. The Einstein–Podolsky–Rosen argument

A profound argument for the thesis that a quantum-mechanical description of a physical system is incomplete was presented by EPR in 1935. Their argument rests upon three premises. (i) Some of the quantum-mechanical predictions concerning observations on a certain type of system, consisting of two spatially separated particles, are correct. (ii) A very reasonable criterion for the existence of 'an element of physical reality' is proposed: 'If, without in any way disturbing a system, we can predict with certainty (i.e., with probability equal to unity) the value of a physical quantity, then there exists an element of physical reality corresponding to this physical quantity' (EPR 1935, p777). (iii) There is no action-at-a-distance in nature.

The system which they study consists of two particles, which are prepared in a state such that the sum of their momenta in a given direction (p_1+p_2) and the difference of their positions (x_1-x_2) are both definite. The wavefunction $\delta(x_1-x_2-a)$ quantum mechanically describes this system, for it is an eigenfunction of the operator x_1-x_2 with eigenvalue a, and of the operator p_1+p_2 with eigenvalue 0. By measuring the position of particle 1 one can predict with certainty, according to quantum mechanics, what value will be found if the position of particle 2 is then measured (immediately). In view of premise (iii) the prediction is made without in any way disturbing particle 2, since the two particles are spatially separated. EPR therefore infer that the position of particle 2 has a definite predetermined value, not included in the description by the wavefunction $\delta(x_1-x_2-a)$. By an analogous argument EPR also infer that the momentum of particle 2 has a definite value, contrary to the uncertainty principle. (Of course, the same argument, starting with measurements made upon particle 2, allows them to infer that particle 1 also has both a definite position and a definite momentum.) Hence EPR reach the conclusion that at least in this particular situation the quantum-mechanical description is incomplete. Although they do not use the term 'hidden variables', this expression can be appropriately used to apply to the parts of the complete state which are not comprised in the quantum-mechanical description, and which suffice to fix the outcomes of measurements that are not fully determined quantum mechanically.

In our opinion the reasoning of EPR is impeccable, once an ambiguity in the phrase 'can predict', which occurs in the second premise, is removed. In a narrow sense, one can predict the value of a quantity only *when an experimental arrangement is chosen for determining the value of that quantity*. In a broad sense, one can predict the value of a quantity *if it is possible to choose an experimental arrangement for determining it*. If the narrow sense is accepted, then the argument of EPR clearly does not go through, since the experimental arrangements for measuring the position and momentum of a particle are incompatible. From the standpoint of physical realism the broad sense of 'can predict' is the appropriate one, since from that viewpoint, one conceives a physical system to have a definite set of properties independently of their being observed, but which may of course be explored at the option of the experimenter. In the situation envisaged by EPR one can predict, in the broad sense, both x_2 and p_2. Hence if this sense of the ambiguous phrase is adopted, their argument does go through. An assumption of physical realism clearly underlies the argument by EPR. Bohr's (1935) answer to EPR, defending the completeness of quantum mechanics, consisted essentially of a critique of the realism which they had taken for granted (see appendix 1).

A variant of EPR's argument was given by Bohm (1951), formulated in terms of discrete states. He considered a pair of spatially separated spin-$\frac{1}{2}$ particles produced somehow in a singlet state, for example, by dissociation of the spin-0 system. Various spin components of each of these particles may then be measured independently at the option of the experimenter. The spin part of the state vector is given by:

$$\Psi = \frac{1}{\sqrt{2}} [u_{\hat{n}}^{+}(1) \otimes u_{\hat{n}}^{-}(2) - u_{\hat{n}}^{-}(1) \otimes u_{\hat{n}}^{+}(2)]. \tag{2.1}$$

Here $\sigma \cdot \hat{n} u_{\hat{n}}^{\pm}(1) = \pm u_{\hat{n}}^{\pm}(1)$, so that $u_{\hat{n}}^{\pm}(1)$ quantum mechanically describes a state in which particle 1 has spin 'up' or 'down', respectively, along the direction $\hat{n}$; $u_{\hat{n}}^{\pm}(2)$ has an analogous meaning concerning particle 2. Since the singlet state Ψ is spherically symmetric, $\hat{n}$ can specify any direction. Suppose that one measures the spin of particle 1 along the $\hat{x}$ axis. The outcome is not predetermined by the description Ψ. But from it, one can predict that if particle 1 is found to have its spin parallel to the $\hat{x}$ axis, then particle 2 will be found to have its spin antiparallel to the $\hat{x}$ axis if the $\hat{x}$ component of its spin is also measured. Thus, the experimenter can arrange his apparatus in such a way that he can predict the value of the $\hat{x}$ component of spin of particle 2 presumably without interacting with it (if there is no action-at-a-distance). Likewise, he can arrange the apparatus so that he can predict any other component of the spin of particle 2. The conclusion of the argument is that all components of spin of each particle are definite, which of course is not so in the quantum-mechanical description. Hence, a hidden-variables theory seems to be required.

Some comments are in order concerning EPR's premises in the light of Bell's theorem. If premise (i) is taken to assert that all of the quantum-mechanical predictions are correct, then Bell's theorem has shown it to be inconsistent with premises (ii) and (iii). Actually, in the body of their argument EPR used only a few predictions with probability one, which are atypical in quantum mechanics, whereas the discrepancies which Bell exhibited between local realistic theories and quantum mechanics involved statistical predictions. If it was EPR's intention to aim at a hidden-variables theory which is local and realistic, and which agrees with all the statistical predictions of quantum mechanics—as many readers have understood them—then, of course, Bell's theorem shows mathematically that this aim cannot be achieved. We shall not try to answer the historical question of their intent. Two statements, however, can be made with confidence. First, the argument from their premises is valid, once the above-mentioned ambiguity is cleared up. Second, the physical situation which they envisaged is of immense value for examining the philosophical implications of quantum mechanics and (via Bell's work) for exploring the limitations of the family of local realistic physical theories.

3. Bell's theorem

There is a vast literature concerning the consistency of hidden-variables theories with the algebraic structure of the observables of quantum mechanics. The major results of this literature are summarised in appendix 2, but they are not indispensable for understanding the content and implications of Bell's theorem. Heuristically, however, this literature was very important for Bell's work. In the course of preparing a review article on 'impossibility' proofs of hidden-variables interpretations of quantum mechanics, Bell studied the theories proposed by de Broglie (1928) and Bohm (1952). He noticed, as Bohm had already realised, that in order to reproduce the quantum-theoretic predictions for a system of EPR type, they postulated the existence of non-local interactions between spatially separated particles. Bell was thus led to ask whether the peculiar non-locality exhibited by these models is a generic characteristic of hidden-variables theories that agree with the statistical predictions by quantum mechanics. He proved (Bell 1965) that the answer is positive for the whole class of deterministic hidden-variables theories in the domain of ideal apparatus and systems. Stronger versions of this theorem, which also constrain actual systems, were later proved by Bell himself and by others. These versions state that essentially all realistic local theories of natural phenomena may be tested in a single experimental arrangement against quantum mechanics, and that these two alternatives necessarily lead to observably different predictions.

In this section we review some of these derivations, which we shall refer to collectively as 'Bell's theorem'. Our purpose here is to arrive at versions of Bell's theorem which satisfy the following criteria. (i) The hypotheses seem to be inescapable for anyone who is committed to physical realism and to the non-existence of action-at-a-distance. (ii) Discrepancies with the predictions by quantum mechanics occur in at least one situation which is experimentally realisable. Criterion (i) is, in our opinion, very close to having been achieved, although the hypotheses are violated by some pathological instances of local realistic theories. Criterion (ii) has essentially been achieved; however, the experiment which it specifies is difficult, and has not yet been performed. Additional assumptions, not implicit in locality and realism, have been relied upon to allow easier experiments to be considered. (The assumptions and experiments are discussed in §§5 and 6.) Unfortunately, this fact leaves open various loopholes (discussed in §§5–7). It must be stressed, however, that the existence of these loopholes in no way diminishes the mathematical validity of the versions of Bell's theorem presented in this section.

3.1. Deterministic local hidden-variables theories and Bell (1965)

In his paper of 1965 Bell considered Bohm's *Gedankenexperiment*, described above in §2. That system consists of two spin-$\frac{1}{2}$ particles, prepared in the quantum-mechanical singlet state Ψ given by equation (2.1). Let $A_{\hat{a}}$ be the result of a measurement of the spin component of particle 1 of the pair along the direction $\hat{a}$, and let $B_{\hat{b}}$ be that of particle 2 along direction $\hat{b}$. We take the unit of spin as $\hbar/2$; hence, $A_{\hat{a}}, B_{\hat{b}} = \pm 1$.

The product $A_{\hat{a}} \cdot B_{\hat{b}}$ is a single observable of the two-particle system (even though two distinct operations are needed in order to measure it). It is represented quantum mechanically by a self-adjoint operator on the Hilbert space associated with the system. For this *Gedankenexperiment* one can readily calculate the quantum-mechanical prediction for the expectation value of this observable†:

$$[E(\hat{a}, \hat{b})]_{\Psi} = \langle \Psi | \sigma_1 . \hat{a} \sigma_2 . \hat{b} | \Psi \rangle = - \hat{a} . \hat{b}. \quad (3.1)$$

A special case of equation (3.1) contains the determinism implicit in this idealised system. When the analysers are parallel, we have:

$$[E(\hat{a}, \hat{a})]_{\Psi} = -1 \quad (3.2)$$

for all $\hat{a}$. Thus, one can predict with certainty the result B, by previously obtaining

† The notation of this review is to use the wavefunction or the letters QM as a subscript to denote the quantum-mechanical prediction. We omit the subscript for predictions by the class of theories included by the postulates of Bell's theorem, when this convention does not cause confusion.

the result A (EPR's premise (ii)). Since the quantum-mechanical state Ψ does not determine the result of an individual measurement, this fact (via EPR's argument) suggests that there exists a more complete specification of the state in which this determinism is manifest. We denote this state by the single symbol λ, although it may well have many dimensions, discrete and/or continuous parts, and different parts of it interacting with either apparatus, etc. Presumably the quantum state Ψ is a related partial specification of this state. We thus define a deterministic hidden-variables theory as any physical theory which postulates the existence of states of a system, for which the observables of quantum mechanics always have definite values.

Let Λ be the space of the states λ for an ensemble comprised of a very large number of the observed systems. We make no restrictions as to what type of space this is, nor to its dimensionality, nor do we require linearity for operations with it, but of course we require that a set of Borel subsets of Λ be defined, so that probability measures can be defined upon it. We represent the distribution function for the states λ on the space Λ by the symbol ρ. For this ensemble we take ρ to have norm one:

$$\int_\Lambda d\rho = 1. \tag{3.3}$$

In a deterministic hidden-variables theory the observable $A_{\hat{a}} \cdot B_{\hat{b}}$ has a definite value $(A_{\hat{a}} \cdot B_{\hat{b}})(\lambda)$ for the state λ. For these theories Bell defined locality as follows: *a deterministic hidden-variables theory is local if for all $\hat{a}$ and $\hat{b}$ and all $\lambda \in \Lambda$ we have:*

$$(A_{\hat{a}} \cdot B_{\hat{b}})(\lambda) = A_{\hat{a}}(\lambda) \cdot B_{\hat{b}}(\lambda). \tag{3.4}$$

That is, once the state λ is specified and the particles have separated, measurements of A can depend only upon λ and $\hat{a}$ but not $\hat{b}$. Likewise measurements of B depend only upon λ and $\hat{b}$. Any reasonable physical theory that is realistic and deterministic and that denies the existence of action-at-a-distance is local in this sense. (More general definitions of 'local' will be considered in §3.3.) For such theories the expectation value of $A_{\hat{a}} \cdot B_{\hat{b}}$ is then given by

$$E(a, b) = \int_\Lambda A_{\hat{a}}(\lambda) B_{\hat{b}}(\lambda) \, d\rho. \tag{3.5}$$

Bell's (1965) proof of the theorem consists of showing that if the locality condition (3.4) and the condition (3.2) for partial agreement with quantum mechanics are both satisfied, then the expectation values satisfy a simple inequality. This inequality is then an alternative prediction to that by quantum mechanics for the expectation value of $A_{\hat{a}} \cdot B_{\hat{b}}$. The predictions made by this inequality are quantitatively different from those of equation (3.1).

The demonstration is straightforward. Equation (3.2) can hold if and only if

$$A_{\hat{a}}(\lambda) = -B_{\hat{a}}(\lambda) \tag{3.6}$$

holds for all $\lambda \in \Lambda$. Using equation (3.6) we calculate the following function, which involves three different possible orientations of the analysers:

$$E(\hat{a}, \hat{b}) - E(\hat{a}, \hat{c}) = -\int_\Lambda [A_{\hat{a}}(\lambda) A_{\hat{b}}(\lambda) - A_{\hat{a}}(\lambda) A_{\hat{c}}(\lambda)] \, d\rho$$

$$= -\int_\Lambda A_{\hat{a}}(\lambda) A_{\hat{b}}(\lambda) [1 - A_{\hat{b}}(\lambda) A_{\hat{c}}(\lambda)] \, d\rho.$$

Since $A, B = \pm 1$, this last expression can be written:

$$|E(\hat{a}, \hat{b}) - E(\hat{a}, \hat{c})| \leq \int_\Lambda [1 - A_{\hat{b}}(\lambda) A_{\hat{c}}(\lambda)] \, d\rho.$$

Using equations (3.3), (3.5) and (3.6) we have:

$$|E(\hat{a}, \hat{b}) - E(\hat{a}, \hat{c})| \leq 1 + E(\hat{b}, \hat{c}). \tag{3.7}$$

Inequality (3.7) is the first of a family of inequalities which are collectively called 'Bell's inequalities'.

A simple instance of the disagreement between the predictions of equation (3.1) and inequality (3.7) is provided by taking $\hat{a}$, $\hat{b}$ and $\hat{c}$ to be coplanar, with $\hat{c}$ making an angle of $2\pi/3$ with $\hat{a}$, and $\hat{b}$ making an angle of $\pi/3$ with both $\hat{a}$ and $\hat{c}$. Then:

For these directions:
$$\hat{a}.\hat{b} = \hat{b}.\hat{c} = \tfrac{1}{2} \qquad \hat{a}.\hat{c} = -\tfrac{1}{2}.$$

$$|[E(\hat{a}, \hat{b})]_\Psi - [E(\hat{a}, \hat{c})]_\Psi| = 1 \quad \text{while} \quad 1 + [E(\hat{b}, \hat{c})]_\Psi = \tfrac{1}{2}.$$

These values do not satisfy inequality (3.7). Hence the quantum-mechanical prediction and that by inequality (3.7) are incompatible, at least for some pairs of analyser orientations.

The version of Bell's theorem just proved can be summarised as follows: no deterministic hidden-variables theory satisfying equation (3.2) and the locality condition (3.1) can agree with all of the predictions by quantum mechanics concerning the spins of a pair of spin-$\tfrac{1}{2}$ particles in the singlet state.

3.2. Foreword to the non-idealised case

Any argument whose scope is strictly limited to a discussion of ideal systems is of little value to working physicists, who endeavour to describe systems that can and do occur in practice. The immense heuristic value of Bell's (1965) argument, outlined in §3.1, is that it leads to formulations that provide direct experimental predictions for systems which can actually be produced in a laboratory. By itself, the derivation given in §3.1 is insufficient to do this, because of its reliance upon the existence of a pair of analyser orientations for which there is a perfect correlation. That is, the above proof hinges strongly upon the condition that equation (3.2) hold exactly. Use is made of this equation in three ways. First, it allows the proof to go through mathematically. Second, determinism is derivable from it and does not have to be postulated separately. Finally, for reasons to be discussed, it assumes that the locality postulate is reasonable.

Unfortunately, equation (3.2) cannot hold exactly in an actual experiment. Any real detector will have an efficiency less than 100%, and any real analyser will have some attenuation as well as some leakage into its orthogonal channel. Since we are attempting to deal with not just one but a whole class of theories, it is quite possible that in some of these theories the above imperfections are inherently correlated with the measurement and detection processes in a way that depends upon the state λ. The problems which arise when these three implications cannot be drawn will be considered in turn.

The problem concerning the derivation's mathematical reliance upon equation (3.2) was first solved by CHSH. They demonstrated that a different proof of the theorem follows from the above formalism, without requiring equation (3.2) to hold. They derived a different inequality that is violated by the quantum-mechanical predictions for systems which never achieve the perfect correlation of equation (3.2), but which do achieve a necessary minimum correlation. The inequality which results from their analysis is:

$$|E(a, b) - E(a', b)| + E(a, b') + E(a', b') \leq 2. \tag{3.8}$$

When equation (3.2) does hold, inequality (3.8) implies inequality (3.7) as a special case. Since essentially this same inequality was subsequently derived by Bell (1971) for the more general non-deterministic case presented in §3.4, we will not present the CHSH derivation here.

The second problem—that determinism is no longer derivable—is not a serious one. One needs merely to assume that determinism holds for the theories under consideration. Indeed this was the approach by CHSH. Thereby, they produced a very powerful result, which constrains deterministic local hidden-variables theories for realisable systems. However, it was subsequently noticed by Bell (1971, 1972) and Clauser and Horne (1974) that this assumption is not needed. On the contrary, a weakening of the locality requirement can be made which still allows inequality (3.8) to be derived, but which significantly increases the scope of the theorem. The theorem thus applies to a class of fundamentally stochastic theories, as well as to deterministic theories in which there are hidden variables in the apparatus.

The third problem is a very delicate one, yet one of great importance. In the idealised situation, whenever a particle is observed at one apparatus an associated particle is *always* observed at the other apparatus. The selection of the sub-ensemble of observed particles from among all of those emitted by the source depends only upon the collimator and source geometry and can have no dependence upon the parameters $\hat{a}$ and $\hat{b}$. Hence ρ was defined for the observed particles, and one can then be confident that it is independent of $\hat{a}$ and $\hat{b}$.

In the actual case, on the other hand, observed particles are paired with particles which, for some reason, are not observed at all, i.e. in neither a spin-up nor a spin-down channel.

The sub-ensemble which we used in the idealised case is then further partitioned into four disjoint sub-ensembles, i.e. those for which (*a*) both particles are observed, (*b*) only particle 1 is observed, (*c*) only particle 2 is observed, and (*d*) neither particle is observed. The distribution ρ of the union of these four sets is clearly independent of $\hat{a}$ and $\hat{b}$. However, the mode of partitioning may well depend upon $\hat{a}$ and $\hat{b}$, since the detection and various attenuation processes occur 'downstream' from the analysers. Hence there is no reason to expect that the composition, and thus the distribution, of each sub-ensemble is independent of $\hat{a}$ and $\hat{b}$. (This fact was noticed by Pearle (1970) and Clauser and Horne (1974). The latter contrived a hidden-variables theory in which ρ becomes dependent upon $\hat{a}$ and $\hat{b}$ when sub-ensemble (*a*) alone is considered and which yields exactly the quantum-mechanical predictions for the system.) Thus if we are to use equation (3.3) for a normalisation condition, and to expect that ρ is independent of $\hat{a}$ and $\hat{b}$, the ensemble for which it is defined must also include the unobserved particles. Since their number is unknown and may be very large, it is no longer obvious how to compare the prediction by inequality (3.8) with experiment.

Three approaches to this problem have been pursued. The approach used by CHSH is to introduce an auxiliary assumption, that if a particle passes through a spin analyser, its probability of detection is independent of the analyser's orientation. Unfortunately, this assumption is not contained in the hypotheses of locality, realism or determinism. Moreover, it also has the undesirable feature that it makes the process of 'passage' or 'non-passage' a primitive one, and thereby excludes from consideration theories for which partial passage is appropriate.

A second approach, that used by Bell (1971) (although not specifically stated but clear from the context), is to employ an auxiliary apparatus ('event-ready' detectors) to measure the number of pairs emitted by the source. This possibility is shown schematically in figure 1. For this scheme, one can simply take the ensemble to consist of the particles which actually trigger the 'event-ready' detectors. Whether or not a triggering occurs clearly does not depend upon the analyser orientations. No problem with locality arises from the presence of the signals propagating to the remote apparatuses, since these signals can be simply considered as part of the state λ. Unfortunately, in practice most conceivable 'event-ready' detectors depolarise or destroy the particles. The value of this approach is thus limited.

An altogether different approach was employed by Clauser and Horne (1974). They derived an inequality from the hypotheses of locality and realism in which only ratios of the observed particle detection probabilities appear, and the normalisation condition equation (3.3) is not required for its derivation. The influence of the size

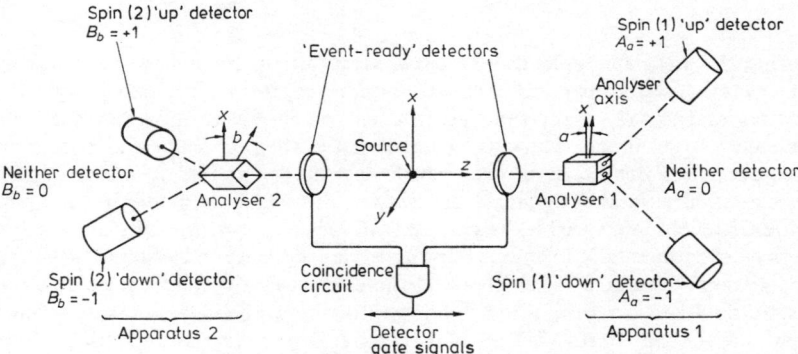

Figure 1. Apparatus configuration used for Bell's 1971 proof. 'Event-ready' detectors signal both arms that a pair of particles has been emitted. For a given gate signal, the result on either arm is assigned the value $+1$ if the corresponding spin-up detector responds, -1 if the spin-down detector responds, and 0 if neither detector responds.

of the ensemble thus vanishes. Their apparatus arrangement does not have the 'event-ready' detectors of figure 1, nor does it have two detectors for each apparatus but only one. It is thus much simpler, and is shown schematically in figure 2.

In the remainder of this section, we will show how these latter two approaches proceed. First, however, we will discuss the aforementioned generalisation of the locality postulate to include inherently stochastic theories.

3.3. Generalisation of the locality concept

Consider either of the experimental configurations for Bohm's *Gedankenexperiment*, described in §3.2. Actually there is nothing in the proof which requires the systems to be spin-$\frac{1}{2}$ particles. They may be any discrete-state quantum-mechanically correlated emissions. (However, not all quantum-mechanically correlated systems are predicted to violate the resulting inequalities. A careful choice is required to find one which is an appropriate test case.) In Bohm's *Gedankenexperiment* the symbols $\hat{a}$ and $\hat{b}$ are taken to represent the orientations of the Stern–Gerlach magnets used for

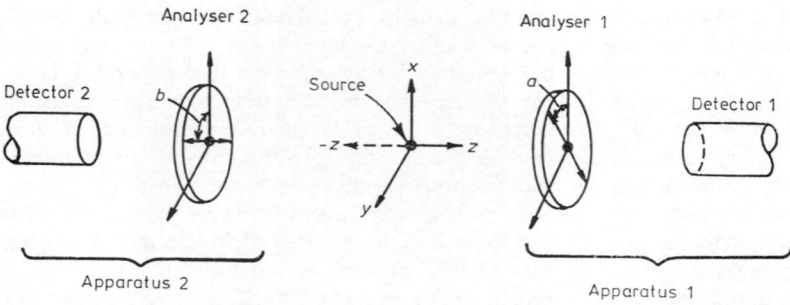

Figure 2. Apparatus configuration used in the proofs by CHSH and by CH. A source emitting particle pairs is viewed by two apparatuses. Each apparatus consists of an analyser and an associated detector. The analysers have parameters a and b respectively, which are externally adjustable. In the above example a and b represent the angles between the analyser axes and a fixed reference axis.

measuring the associated spin components. However, in general, a and b may represent any associated apparatus parameters under control by the experimenter. As before, A_a and B_b represent the measurement outcomes at apparatuses 1 and 2, respectively. Appropriate values will be assigned to these outcomes, as necessary.

The preceding definition of locality will now be generalised to include systems whose evolution is inherently stochastic, as well as to include deterministic systems with additional random variables associated with either apparatus, and that may locally affect their experimental outcomes. Suppose a pair of correlated systems, which have a joint state λ, separate. They then continue to evolve perhaps in an inherently stochastic way, and given λ, a and b, one can define probabilities for any particular outcome at either apparatus. We allow that, given λ, these two probabilities may each depend upon the associated (local) apparatus parameter, a or b respectively, and of course upon λ, but we assume that these probabilities are otherwise independent of each other.

This definition of locality seems very common-sensical. It says that the outcome (or the probability of outcomes) of a measurement performed on one part of a composite system is independent of what aspects of the other component the experimenter chooses to measure. It by no means excludes the possibility of obtaining knowledge concerning system 2 from an examination of system 1. The state λ contains information common to both systems, and a measurement on one of these presumably reveals some of this. Nor does it prevent a measurement performed on one component of a composite system from locally disturbing that component. What it does prescribe, in essence, is that the measured value of a quantity on one system is not causally affected by what one chooses to measure on the other system, since the two systems are well separated (e.g. space-like separated) when the measurements are performed.

3.4. Bell's 1971 proof

We now describe Bell's (1971) proof, using this generalised locality definition. The apparatus configuration appropriate to this proof was discussed in §3.2 and is shown in figure 1. Given that a particle pair was emitted into the associated apparatuses, the results of either measurement can have one of three possible outcomes, to which the following values were assigned by Bell:

$$A_a(\lambda) = \begin{cases} +1, \text{ 'spin-up' detector triggered by particle 1} \\ -1, \text{ 'spin-down' detector triggered by particle 1} \\ 0, \text{ particle 1 not detected} \end{cases} \quad (3.9(a))$$

and

$$B_b(\lambda) = \begin{cases} +1, \text{ 'spin-up' detector triggered by particle 2} \\ -1, \text{ 'spin-down' detector triggered by particle 2} \\ 0, \text{ particle 2 not detected.} \end{cases} \quad (3.9(b))$$

For a given state λ of the emitted composite system, we denote the expectation values for these quantities by the symbols $\bar{A}_a(\lambda)$ and $\bar{B}_b(\lambda)$. In the general case these average values will differ from the values assigned by equations (3.9). Since the values for A and B are bounded by 1, it follows that:

$$|\bar{A}_a(\lambda)| \leq 1 \qquad |\bar{B}_b(\lambda)| \leq 1. \tag{3.10}$$

Using the general definition of locality of §3.3, we can write the expectation value for the product $A_a B_b$ as:

$$E(a, b) = \int_\Lambda \bar{A}_a(\lambda) \bar{B}_b(\lambda) \, d\rho. \tag{3.11}$$

Since we are including in our ensemble only those particles which have previously triggered the 'event-ready' detectors, we are assured that the distribution ρ and the range of integration Λ are independent of a and b. Now consider the expression:

$$E(a, b) - E(a, b') = \int_\Lambda [\bar{A}_a(\lambda)\bar{B}_b(\lambda) - \bar{A}_a(\lambda)\bar{B}_{b'}(\lambda)] \, d\rho$$

where we take a' and b' to be alternative settings for analysers 1 and 2, respectively. This can be rewritten as:

$$E(a, b) - E(a, b') = \int_\Lambda \bar{A}_a(\lambda)\bar{B}_b(\lambda)[1 \pm \bar{A}_{a'}(\lambda)\bar{B}_{b'}(\lambda)] \, d\rho$$
$$- \int_\Lambda \bar{A}_a(\lambda)\bar{B}_{b'}(\lambda)[1 \pm \bar{A}_{a'}(\lambda)\bar{B}_b(\lambda)] \, d\rho.$$

Using inequalities (3.10), we then have:

$$|E(a, b) - E(a, b')| \leq \int_\Lambda [1 \pm \bar{A}_{a'}(\lambda)\bar{B}_{b'}(\lambda)] \, d\rho + \int_\Lambda [1 \pm \bar{A}_{a'}(\lambda)\bar{B}_b(\lambda)] \, d\rho$$

or

$$|E(a, b) - E(a, b')| \leq \pm [E(a', b') + E(a', b)] + 2 \int_\Lambda d\rho.$$

Hence:

$$-2 \leq E(a, b) - E(a, b') + E(a', b) + E(a', b') \leq 2. \tag{3.12}$$

By re-definition of the parameters a, a', b and b' in the central expression of (3.12), the minus sign may be permuted to any one of the four terms. Inequality (3.12) and its permutations are one form of Bell's inequality, and represent a general prediction for the theories covered by the above assumptions.

In order to complete the proof of the theorem, it is sufficient to show that in at least one situation the predictions by quantum mechanics contradict inequality (3.12). The quantum-mechanical prediction $[E(\hat{a}, \hat{b})]_{QM}$ for the two spin-$\frac{1}{2}$ particle example, when due account is taken of imperfections in the analysers, detectors and state preparation, will be of the form:

$$[E(\hat{a}, \hat{b})]_{QM} = C \hat{a} \cdot \hat{b} \tag{3.13}$$

where the coefficient C is bounded by one for actual systems, and is equal to plus or minus one only in the idealised case. Suppose we take $\hat{a}$, $\hat{a}'$, $\hat{b}$ and $\hat{b}'$ to be coplanar vectors as shown in figure 3 with $\phi = \pi/4$, and calculate:

$$[E(\hat{a}, \hat{b}) - E(\hat{a}, \hat{b}') + E(\hat{a}', \hat{b}) + E(\hat{a}', \hat{b}')]_{\text{QM}} = 2\sqrt{2}C.$$

There is a wide range of values for C for which the prediction by inequality (3.12) disagrees with that by equation (3.13). Hence the proof is complete.

3.5. The proof by Clauser and Horne

Clauser and Horne (1974) also proved Bell's theorem for general local realistic theories, including inherently stochastic theories. Their proof is noteworthy in that it

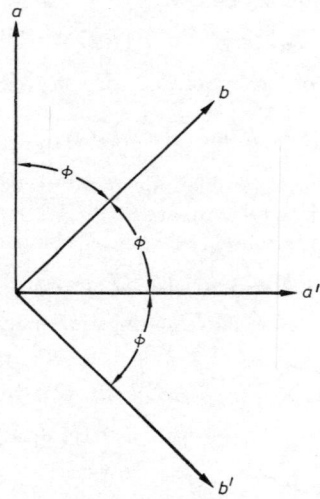

Figure 3. Optimal orientations for a, a', b and b'. If the correlation is of the form $C_1 + C_2 \cos n\phi$, then the maximum violation of the inequalities occurs at $n\phi = \pi/4$.

defines an experiment which might actually be performed and which does not require that auxiliary assumptions be made. The apparatus configuration which they used for the proof was introduced by Clauser et al (1969) and is shown schematically in figure 2. In contrast to the configuration of figure 1, theirs has only one detector in each arm and no 'event-ready' detectors. For each analyser/detector assembly there are only two possible outcomes: 'count' and 'no-count'. The results are thus formulated in terms of probabilities for single and coincidence counts, rather than the expectation values considered in §§3.1 and 3.4.

Suppose that during a period of time, while the adjustable parameters have the values a and b, the source emits, say, N of the two-component systems of interest. For this period, denote by $N_1(a)$ and $N_2(b)$ the number of counts at detectors 1 and 2, respectively, and by $N_{12}(a, b)$ the number of simultaneous counts from the two detectors (coincidence counts). When N is sufficiently large, the probabilities for these results for the whole ensemble (with due allowance for random errors) are given by:

$$p_1(a) = N_1(a)/N$$
$$p_2(b) = N_2(b)/N \qquad (3.14)$$
$$p_{12}(a, b) = N_{12}(a, b)/N.$$

CH derive an inequality which constrains ratios of the probabilities in equations (3.14) rather than their absolute magnitudes. Thereby, the influence of the quantity N vanishes, so that it does not have to be measured.

Their derivation is straightforward. Following the discussion of §3.3, we expect a well-defined probability $p_1(\lambda, a)$ of detecting component 1, given the state λ of the composite system and the parameter a of the first analyser; a probability $p_2(\lambda, b)$ of detecting component 2, given λ and b; and a probability $p_{12}(\lambda, a, b)$ of detecting both components, given λ, a and b. Following our discussion of §3.3, we assume that, given λ, a and b, the probabilities $p_1(\lambda, a)$ and $p_2(\lambda, b)$ are independent. Thus we write the probability of detecting both components as

$$p_{12}(\lambda, a, b) = p_1(\lambda, a)p_2(\lambda, b). \qquad (3.15)$$

The ensemble average probabilities of equations (3.14) are then given by:

$$p_1(a) = \int_\Lambda p_1(\lambda, a)\, d\rho$$
$$p_1(b) = \int_\Lambda p_2(\lambda, b)\, d\rho \qquad (3.16)$$
$$p_{12}(a, b) = \int_\Lambda p_1(\lambda, a)p_2(\lambda, b)\, d\rho.$$

To proceed, CH introduce the following lemma, the proof of which may be found in their paper: if x, x', y, y', X, Y are real numbers such that $0 \leq x, x' \leq X$ and $0 \leq y, y' \leq Y$, then the inequality:

$$-XY \leq xy - xy' + x'y + x'y' - Yx' - Xy \leq 0 \qquad (3.17)$$

holds. Inequality (3.17) and equation (3.15) yield:

$$-1 \leq p_{12}(\lambda, a, b) - p_{12}(\lambda, a, b') + p_{12}(\lambda, a', b) + p_{12}(\lambda, a', b') - p_1(\lambda, a') - p_2(\lambda, b) \leq 0. \qquad (3.18)$$

Integrating inequality (3.18) over λ with distribution ρ, and using equation (3.16), one obtains the result:

$$-1 \leq p_{12}(a, b) - p_{12}(a, b') + p_{12}(a', b) + p_{12}(a', b') - p_1(a') - p_2(b) \leq 0. \qquad (3.19)$$

(Obtaining the left-hand inequality also required the use of equation (3.3), but the right-hand one did not. Since the left-hand inequality requires a measurement of the absolute magnitude of probabilities, the 'event-ready' detectors of figure 1 will be needed to test it.) The right-hand side of inequality (3.19) can be rewritten in the following form:

$$\frac{p_{12}(a, b) - p_{12}(a, b') + p_{12}(a', b) + p_{12}(a', b')}{p_1(a') + p_2(b)} \leq 1. \qquad (3.20(a))$$

As desired, it involves only a quantity that is independent of N. Using equations (3.14), and defining $R(a, b)$ as the rate of coincident detections, and $r_1(a)$ and $r_2(b)$ as the rate of single-particle detections by either apparatus, inequality (3.20(a)) can be rewritten directly in terms of a ratio of observable count rates:

$$\frac{R(a, b) - R(a, b') + R(a', b) + R(a', b')}{r_1(a') + r_2(b)} \leqslant 1. \quad (3.20(b))$$

Inequalities (3.20) are thus a general prediction for any local realistic theory of natural phenomena.

In order to complete the proof of the theorem it suffices to exhibit an instance in which the quantum-mechanical counterpart to inequalities (3.20) fails. This is done in §4, when we discuss the experimental requirements for a valid test of these theories.

3.6. Symmetry considerations

Almost all of the experiments which have been proposed for testing the predictions by Bell's inequalities involve pairs of polarised particles (either photons or massive particles). In these experiments the parameters a and b, considered abstractly in §§3.4 and 3.5, are taken to be orientation angles relative to some reference axis in a fixed plane. In most of these experiments, the method of preparing the pairs of polarised particles attempts to achieve cylindrical symmetry about a normal to the fixed plane and reflection symmetry with respect to planes through this normal. This symmetry is exhibited in the quantum-mechanical predictions for detection rates and correlations:

$[p_1(a)]_{\rm QM}$ and $[r_1(a)]_{\rm QM}$ are independent of a.
$[p_2(b)]_{\rm QM}$ and $[r_2(b)]_{\rm QM}$ are independent of b.
$[p_{12}(b)]_{\rm QM}$, $[R(a, b)]_{\rm QM}$ and $[E(a, b)]_{\rm QM}$ are functions only of $|a-b|$.

We now assume that the corresponding predictions for local realistic theories exhibit the same symmetries:

$$p_1(a) \equiv p_1 \text{ and } r_1(a) \equiv r_1 \text{ are independent of } b$$
$$p_2(b) \equiv p_2 \text{ and } r_2(b) \equiv r_2 \text{ are independent of } b \quad (3.21)$$
$$p_{12}(a, b) \equiv p_{12}(|a-b|), \; R(a, b) \equiv R(|a-b|) \text{ and } E(a, b) \equiv E(|a-b|).$$

We must emphasise two points concerning equation (3.21). First, these symmetry relations do not simply follow from the corresponding quantum-mechanical symmetry relations or from the symmetry of the experimental arrangement, for one does not know what symmetry-breaking factors may lurk at the level of the hidden variables. Second, no harm is done in assuming equations (3.21), since they are susceptible to experimental verification.

Now suppose that we take a, a', b and b' so that:

$$|a-b| = |a'-b| = |a'-b'| = \tfrac{1}{3}|a-b'| = \phi$$

as in figure 3. With the use of equation (3.21), inequalities (3.12) and (3.20) simplify to

$$|3E(\phi) - E(3\phi)| \leqslant 2 \quad (3.22)$$

and

$$S(\phi) \leqslant 1 \quad (3.23)$$

where we have defined:

$$S(\phi) \equiv \frac{3p_{12}(\phi) - p_{12}(3\phi)}{p_1 + p_2} \quad (3.24(a))$$

in terms of probabilities, or equivalently in terms of count rates:

$$S(\phi) \equiv \frac{3R(\phi) - R(3\phi)}{r_1 + r_2}. \quad (3.24(b))$$

3.7. The proof by Wigner, Belinfante and Holt

A simple method of proving Bell's theorem for deterministic local hidden-variables theories was invented independently by Wigner (1970) and Belinfante (1973), and extended by Holt (1973). The method consists of subdividing the space Λ of states of a two-component system into subspaces corresponding to various possible values of the observables of interest, and then performing some easy calculations on the measures of these subspaces. Rather than duplicate their proofs, which are readily available, we show how their method can be used to derive the inequality of CH.

Consider the apparatus configuration of figure 2. Assume that the detection or non-detection of component 1 is completely determined by the parameter a of the first analyser and the state of the composite system, but is independent of the parameter b of the other analyser, and so forth for component 2. As such, the discussion is for the restricted situation in which determinism applies. Under this assumption, we can exhaustively subdivide the space Λ into 16 mutually disjoint subspaces Λ (ij; kl), where each letter can take on the value 0 or 1, with 1 denoting detection and 0 non-detection; with i and j referring to the results if the parameter of the first analyser is chosen respectively to be a or a'; and with k and l referring to the results if the parameter of the second analyser is chosen respectively to be b or b'. For example, Λ (10; 01) is the subspace in which component 1 will be detected if its associated parameter is chosen to be a but will not if the parameter is chosen to be a', while component 2 will not be detected if its associated parameter is chosen to be b but will be detected if that parameter is chosen to be b'. (Note that there is no question of simultaneously examining detection or non-detection for two different values of a parameter. Indeed, such observations are mutually exclusive. Rather, the subspace is defined in terms of what will happen if any one of the various experiments is performed. Since the theories are assumed to be deterministic, these values are all determined once a, b, λ and the apparatus configuration are specified.) If a probability measure ρ is assumed to be given on Λ (determined presumably by the way in which the composite system is prepared), then $\rho(ij; kl)$ is defined to be the probability that the composite state is in $\Lambda(ij; kl)$. Clearly, all $\rho(ij; kl)$ are non-negative. Because the 16 subspaces are disjoint and exhaustive, we have:

$$\sum_{ijkl} \rho(ij; kl) = 1. \quad (3.25)$$

We now define $p_1(a)$ to be the probability that component 1 will be detected if its parameter is chosen to be a; $p_2(b)$ to be the probability that component 2 will be detected if its parameter is chosen to be b; and $p_{12}(a, b)$ to be the probability of joint detection of both components if the two parameters are chosen respectively to be a and b. Analogous definitions are given for the other values of the parameters. Then we have:

$$p_{12}(a, b) = \rho(11; 11) + \rho(11; 10) + \rho(10; 11) + \rho(10; 10)$$
$$p_{12}(a, b') = \rho(11; 11) + \rho(11; 01) + \rho(10; 11) + \rho(10; 01)$$
$$p_{12}(a', b) = \rho(11; 11) + \rho(11; 10) + \rho(01; 11) + \rho(01; 10)$$
$$p_{12}(a', b') = \rho(11; 11) + \rho(11; 01) + \rho(01; 11) + \rho(01; 01) \quad (3.26)$$
$$p_1(a') = \rho(11; 11) + \rho(11; 10) + \rho(11; 01) + \rho(11; 00)$$
$$+ \rho(01; 11) + \rho(01; 10) + \rho(01; 01) + \rho(01; 00)$$
$$p_2(b) = \rho(11; 11) + \rho(11; 10) + \rho(10; 11) + \rho(10; 10)$$
$$+ \rho(01; 11) + \rho(01; 10) + \rho(00; 11) + \rho(00; 10).$$

It follows that:

$$p_{12}(a, b) - p_{12}(a, b') + p_{12}(a', b) + p_{12}(a', b') - p_1(a') - p_2(b)$$
$$= -\rho(11; 01) - \rho(11; 00) - \rho(10; 11) - \rho(10; 01) - \rho(01; 10)$$
$$- \rho(01; 00) - \rho(00; 11) - \rho(00; 10). \quad (3.27)$$

Consequently, we recover inequality (3.19) derived by Clauser and Horne for the more general stochastic case:

$$-1 \leq p_{12}(a, b) - p_{12}(a, b') + p_{12}(a', b) + p_{12}(a', b') - p_1(a') - p_2(b) \leq 0.$$

The demonstration of the incompatibility between this inequality and quantum mechanics is thus the same as that of § 3.5, and hence the theorem is proved.

3.8. *Stapp's proof*

Stapp's version of Bell's theorem (1971, 1977) appears to be very general, for it dispenses with all assumptions about the state of the system and about probability measures on the space of states. The proof was generalised by Eberhard (1977) to include realisable systems. Stapp considered a long series of N occurrences of Bohm's *Gedankenexperiment*. In each occurrence a pair of spin-$\frac{1}{2}$ particles is produced in the singlet state in a space–time region S_0. The particles propagate in opposite directions along a given axis. Particle 1 proceeds to a space–time region S_1, where it is deflected 'up' or 'down' by a Stern–Gerlach magnet oriented in either the $\hat{a}$ or the $\hat{a}'$ direction, and particle 2 proceeds to the region S_2 where it is deflected up or down by a magnet oriented in either the $\hat{b}$ or the $\hat{b}'$ direction. S_1 and S_2 are supposed to have space-like separation, and the choice for orienting the first magnet along $\hat{a}$ or $\hat{a}'$ is made when particle 1 is in S_1, and similarly with the choice for orienting the second magnet. Let the number 1 or -1 be recorded for a particle entering the field of a Stern–Gerlach magnet accordingly as it is deflected 'up' or 'down'. Let $r_{\alpha j}(\hat{a}, \hat{b})$ (where $\alpha = 1, 2$ and $j = 1, \ldots, N$) be the number recorded for the αth particle of the jth pair if the two magnets are oriented in the $\hat{a}$ and $\hat{b}$ directions respectively, and let $r_{\alpha j}(\hat{a}, \hat{b}')$, $r_{\alpha j}(\hat{a}', \hat{b})$ and $r_{\alpha j}(\hat{a}', \hat{b}')$ have analogous meanings. Clearly the orientations $\hat{a}$ and $\hat{a}'$ are mutually exclusive, as are $\hat{b}$ and $\hat{b}'$. Although only one of the four possible pairs of orientations $(\hat{a}, \hat{b})$, $(\hat{a}, \hat{b}')$, $(\hat{a}', \hat{b})$, $(\hat{a}', \hat{b}')$ can occur in the real world, Stapp made an assumption of 'counterfactual definiteness', that $r_{\alpha j}(\hat{a}, \hat{b})$, $r_{\alpha j}(\hat{a}, \hat{b}')$, etc, are all definite numbers. In addition, he made an assumption of individual locality, that:

$$r_{1j}(\hat{a}, \hat{b}) = r_{1j}(\hat{a}, \hat{b}') \qquad (3.28(a))$$

$$r_{1j}(\hat{a}', \hat{b}) = r_{1j}(\hat{a}', \hat{b}') \qquad (3.28(b))$$

$$r_{2j}(\hat{a}, \hat{b}) = r_{2j}(\hat{a}', \hat{b}) \qquad (3.28(c))$$

$$r_{2j}(\hat{a}, \hat{b}') = r_{2j}(\hat{a}', \hat{b}'). \qquad (3.28(d))$$

Stapp then showed that the $8N$ numbers $r_{\alpha j}(\hat{a}, \hat{b})$, etc, must disagree with some of the statistical predictions of quantum mechanics. Some critics of Stapp have argued that his assumption of counterfactual definiteness is understandable only from the standpoint of a deterministic local hidden-variables theory. Stapp (1978, §4) has replied, however, that his assumption requires no commitment to determinism, but only to the possibility of speaking (as is commonly done in the sciences) of possible worlds as well as the actual one. He makes the explicit assumption that each of the four choices $(\hat{a}, \hat{b})$, $(\hat{a}, \hat{b}')$, $(\hat{a}', \hat{b})$ and $(\hat{a}', \hat{b}')$ is made in some possible world. It may nevertheless be objected that Stapp has not given a reason for demanding the existence of a quadruple of possible worlds which mesh together as in equations $(3.28(a))$–(d). The combination of no action-at-a-distance with the idea of possible worlds only seems to require four pairs of possible worlds, one pair meshing as in equation $(3.28(a))$, one as in equation $(3.28(b))$, etc. It is not obvious why these four relations need to govern a cluster of four possible worlds unless determinism is supposed. An answer to this objection is provided by Stapp (1978), in which the following equivalence theorem is proved.

Let I be the set of individual outcomes $r_{\alpha j}(c, d)$, where c is $\hat{a}$ or $\hat{a}'$, d is $\hat{b}$ or $\hat{b}'$, α is 1 or 2, and j is $1, \ldots, N$. Let $P(I)$ be the set of probabilities

$$P = (\{r_1 | \hat{a}\}, \{r_2 | \hat{b}\}, \{r_1, r_2 | \hat{a}, \hat{b}\})$$

determined by the appropriate frequencies in I:

$$\{r_1 | \hat{a}\} = N(r_1, \hat{a})/N$$

(where $N(r_1, \hat{a})$ is the number of j such that $r_1 = r_{1j}(\hat{a}, \hat{b}) = r_{1j}(\hat{a}, \hat{b}')$ by individual locality),

$$\{r_1, r_2 | \hat{a}, \hat{b}\} = N(r_1, r_2, \hat{a}, \hat{b})$$

(where $N(r_1, r_2, \hat{a}, \hat{b})$ is the number of j such that $r_1 = r_{1j}(\hat{a}, \hat{b})$ and $r_2 = r_{2j}(\hat{a}, \hat{b})$), etc.

Let L_P be the set of P which satisfy the following probabilistic locality conditions: there exists a discrete set of λ, a probability weight function ρ defined on this set, and probabilities $p_1(\lambda, \hat{a}, r_1)$, $p_2(\lambda, \hat{b}, r_2)$ for the outcomes r_1 and r_2 respectively (given λ and given $\hat{a}$ or $\hat{b}$), such that:

$$\{r_1, r_2 | \hat{a}, \hat{b}\} = \sum_\lambda \rho(\lambda) p_1(\lambda, \hat{a}, r_1) p_2(\lambda, \hat{b}, r_2)$$

$$\{r_1 | \hat{a}\} = \sum_\lambda \rho(\lambda) p_1(\lambda, \hat{a}, r_1) \qquad (3.29)$$

$$\{r_2 | \hat{b}\} = \sum_\lambda \rho(\lambda) p_1(\lambda, \hat{b}, r_2).$$

Finally, let L be the set of I which satisfy the individual locality conditions $(3.28(a))$–(d). Then the equivalence theorem asserts (i) if $I \in L$, then $P(I) \in L_P$, (ii) if $P \in L_P$, then there is an $I \in L$ such that $P(I)$ is approximately equal to P.

Note that (3.29) is essentially the CH probabilistic locality condition, except that a sum over discrete values of λ is used instead of an integral over the space Λ; but since the integral can always be approximated by a sum, this difference is not crucial. Because of this equivalence theorem, the theorem of CH and of Bell that no P which belongs to L_P can agree statistically with quantum mechanics entails that no I which belongs to L can agree statistically with quantum mechanics and, conversely, the theorem of Stapp that no I belonging to L can agree statistically with quantum mechanics implies the theorem of CH and Bell. Stapp's equivalence theorem, therefore, shows that, contrary to appearances, his proof of Bell's theorem and those of CH and Bell (1971) are of equal strength. J S Bell (personal communication) has remarked that part (ii) of Stapp's equivalence theorem is an example of the possibility of simulating a stochastic process with a deterministic one.

3.9. Other versions of Bell's theorem

Several other versions of Bell's theorem have been discovered. The proofs are mathematically correct, but with hypotheses in some respects problematic, either from a philosophical point of view or from their inherent restriction to idealised systems.

A very general derivation of Bell's theorem has been presented by Bell (1976). It was critically evaluated by Shimony *et al* (1976), who challenged one of the premises. Bell (1977) replied to this criticism. If we retain the notation of §3.8, we can express the essential assumption of Bell (1976) in the following way: the complete state of region S_1 is independent of the choice between $\hat{b}$ and $\hat{b}'$ in S_2, and likewise the complete state of S_2 is independent of the choice between $\hat{a}$ and $\hat{a}'$ in S_1. Shimony *et al* (1976) criticised this assumption on the ground that the backward light cones of S_1 and S_2 overlap in a region S, and it is possible that a factor in S affecting the choice between $\hat{b}$ and $\hat{b}'$ leaves some trace in S_1. Bell's reply (1977) to this objection stresses the spontaneity of the experimenter's choice between $\hat{b}$ and $\hat{b}'$ and between $\hat{a}$ and $\hat{a}'$, but this answer seems to us to depend upon too strong a commitment to indeterminism for his argument to be fully general (see also Shimony 1978).

The proof by d'Espagnat (1975) has the virtue of staying quite close to the original ideas of EPR by reasoning in terms of the intrinsic properties of the system. We shall not try to summarise his argument, partly because of its length and partly because of a premise which is impossible to be realised experimentally. D'Espagnat assumes (as Bell did in 1965) that one has a system like a pair of spin-$\frac{1}{2}$ particles in the singlet state, such that one can measure an observable of one of the pair and then infer with absolute certainty the value of a corresponding observable of the other pair (equation (3.2)). The same criticism can also be made of the arguments of Gutkowski and Masotto (1974), Selleri (1978) and Schiavulli (1977) but it should be noted that they derive a number of generalisations of Bell's inequalities which have not been obtained elsewhere.

4. Considerations regarding a general experimental test

Following Bell's (1965) results, many readers believed that local realistic theories were *ipso facto* discredited, because quantum mechanics has been so abundantly confirmed in a variety of experimental situations. Indeed, some of the most striking confirmations of quantum mechanics, such as the spectrum of helium, concerned correlated pairs of particles. However, upon careful examination, one finds that situations exhibiting the disagreement discovered by Bell are rather rare, and none had ever been experimentally realised. Moreover, the reasoning of the previous sections indicates that the treatment of correlated but spatially separated systems may well be the point of greatest vulnerability of quantum mechanics. In view of the consequences of Bell's theorem it is thus important to design experiments to test explicitly the predictions made for local realistic theories via Bell's theorem.

Starting with the simple configurations specified in §3, the first problem is to find a suitable system whose quantum-mechanical predictions directly violate the predictions in the theorem, but additionally one that is accessible with available technology. In fact, this has not yet been done! (Possible avenues in this direction are discussed in §7.) In the present section we examine the requirements for a fully general test, and see why the problem is difficult. Since the presence of auxiliary counters is required by the apparatus configuration of figure 1, and usually these depolarise or destroy the emissions, we will confine our discussion to the apparatus configuration of figure 2.

We thus compare the quantum-mechanical predictions for this configuration with those by inequality (3.23). The left-hand side of inequality (3.19) is not considered here, since it cannot be expressed in terms of ratios of observable probabilities. It will, however, become useful for the discussion of §5.

4.1. Requirements for a general experimental test

Consider an experiment, with a configuration similar to that of figure 2, whose quantum-mechanical predictions take the following form:

$$[p_{12}(\phi)]_{QM} = \tfrac{1}{4}\, \eta_1 \eta_2 f_1 g [\epsilon_+^1 \epsilon_+^2 + \epsilon_-^1 \epsilon_-^2\, F \cos(n\phi)]$$
$$[p_1]_{QM} = \tfrac{1}{2}\, \eta_1 f_1 \epsilon_+^1 \qquad (4.1)$$
$$[p_2]_{QM} = \tfrac{1}{2}\, \eta_2 f_2 \epsilon_+^2.$$

This general form is characteristic of the quantum-mechanical predictions for the actual experiments of interest (see, for example, equation (5.15)). In these expressions η_i represents the effective quantum efficiency of detector $i (i = 1, 2)$, and

$$\epsilon_+^i \equiv \epsilon_M^i + \epsilon_m^i \qquad \epsilon_-^i \equiv \epsilon_M^i - \epsilon_m^i. \qquad (4.2)$$

The terms ϵ_M^i and ϵ_m^i are the maximum and minimum transmissions of the analysers relative to the pertinent orthogonal basis. The functions f_1 and f_2 are the collimator efficiencies, i.e. the probability that an appropriate emission enters apparatus 1 or 2. Typically, these are simply proportional to the collimator acceptance solid angles. The function g is the conditional probability that, given emission 1 enters apparatus 1, then emission 2 will enter apparatus 2. The function F is a measure of the initial-state purity and the inherent quantum-mechanical correlation of the two emissions. For the actual cascade-photon experiments (see §5), these functions depend on the collimator solid angles. The values of n are 1 or 2 depending upon whether the experiment is performed with fermions or bosons.

Inserting equation (4.1) into the definition of $S(\phi)$, equation (3.24(a)), we find the quantum-mechanical prediction for this function to be given by:

$$S_{QM}(\phi) = \tfrac{1}{4}\, \eta g\, \{2\epsilon_+ + [3 \cos(n\phi) - \cos(3n\phi)]\, F(\epsilon_-^2/\epsilon_+)\}. \qquad (4.3)$$

Here for simplicity we have taken $\eta \equiv \eta_1 = \eta_2$, $f_1 = f_2$, $\epsilon_+ = \epsilon_+^1 = \epsilon_+^2$ and $\epsilon_- \equiv \epsilon_-^1 = \epsilon_-^2$. Selecting the optimum value $\phi = \pi/4n$, one finds that the condition for a violation of inequality (3.23) is given by:

$$\eta g \epsilon_+ [\sqrt{2}\,(\epsilon_-/\epsilon_+)^2\, F + 1] > 2. \tag{4.4}$$

Thus, a correlation experiment with values in the domain specified by inequality (4.4) is capable of distinguishing between the prediction, inequality (3.23), and that of quantum theory, equation (4.3). Although such experiments are apparently possible, there is at present no existing experimental result in this domain, and thus none in violation of any inequality which does not require additional assumptions for its derivation.

For a direct test of inequality (4.4) the requirements are stringent, which accounts for the fact that, so far, no such experiment has been attempted.

(i) A source must emit pairs of discrete-state systems, which can be detected with high efficiency.

(ii) Quantum mechanics must predict strong correlations of the relevant observables of each pair (polarisations in the experiments so far). Correspondingly, the ensemble of pairs must have high quantum-mechanical purity.

(iii) The analysers must be capable of allowing systems in certain states to pass with great efficiency, while simultaneously rejecting nearly all of those in orthogonal states.

(iv) The collimators (and filters if these are necessary to remove unwanted emissions, etc) must have very high transmittances and not depolarise the emissions.

(v) The source must produce the systems via a two-body decay. A three- (or more) body decay cannot be used, because the resulting angular correlation will make $g \ll 1$.

(vi) Another requirement should be added in order to achieve an airtight argument against locality: the parameters a and b must be rapidly changed while the emissions are in flight. A detection event should be space-like separated from the corresponding parameter change event at the far apparatus (see §7). This suggestion was first made by Bohm and Aharonov (1957).

For a practical experiment, it is of course also necessary for the counting rate to be sufficiently high to make the required integration time reasonable.

4.2. Three important experimental cases

Let us examine how the failure of any of these parameters to approximate the ideal case prevents a violation of inequality (3.23) from arising. Figure 4 shows the prediction by equation (4.3) for three important cases of interest, along with the prediction by inequality (3.23).

Case I, nearly ideal. In the domain of nearly ideal apparatus, we have $g \approx \epsilon_+ \approx \epsilon_- \approx \eta \approx 1$. For these conditions we find a violation of inequality (3.23) for a wide range of ϕ, with a maximum violation at $n\phi = \pi/4$.

Case II, poor detector efficiencies or co-focusing. When $g \ll 1$ holds, because of imperfect collimator alignment and/or a weak angular correlation inherent in a three-body decay, or when $\eta \ll 1$ holds, because the detector efficiencies are low, then the amplitude of $S(\phi)$ contracts in amplitude about a value close to zero. The quantum-mechanical predictions enter a domain where no violation of inequality (3.23) occurs.

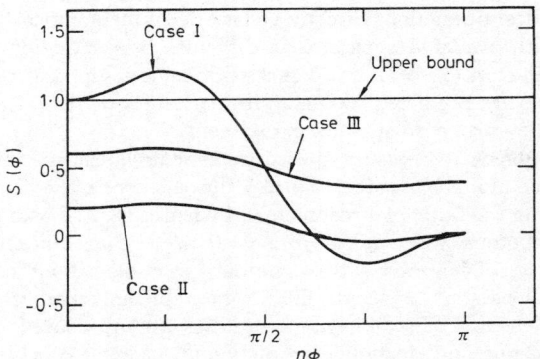

Figure 4. Typical dependence of $S(\phi)$ upon $n\phi$, for cases I–III. Upper bound for $S(\phi)$ set by inequality (3.23) is $+1$. Case I experiments (nearly ideal) have $\eta \approx g \approx F \approx \epsilon_+ \approx \epsilon_- \approx 1$. Case II experiments have nearly ideal parameters $F \approx \epsilon_- \approx \epsilon_+ \approx 1$, but have $\eta \ll 1$ and/or $g \ll 1$. Case III experiments have nearly ideal parameters $\eta \approx g \approx 1$, but have $F \ll 1$ and/or $\epsilon_-/\epsilon_+ \ll 1$.

This case is typical of the low-energy cascade-photon experiments to be described in §5.

Case III, weak correlation. The third case occurs when the predicted correlation is weak. The correlation coefficient F and/or the parameter ϵ_-/ϵ_+ may be much less than unity. This will occur, for example, if the emissions are only weakly correlated, if the initial state is impure, if the emissions suffer significant depolarisation in passing through the apparatus, or if the analyser efficiencies are low. The curve $S(\phi)$ then contracts in amplitude symmetrically about a value slightly less than $+\frac{1}{2}$, and again no violation of inequality (3.23) occurs. This case is typical of the positronium annihilation and the proton–proton S-wave scattering experiments to be described in §6.

The manner in which the amplitude of $S(\phi)$ contracts is of more importance than it may seem. To perform a test of the local realistic theories in the domain of case II and III experiments requires a credible auxiliary assumption that $S(\phi)$ can be rescaled somehow to an amplitude sufficient to violate the inequalities. Case II experiments (discussed in §5) are more favourable in this respect than are those of case III. For the former, a replacement of p_1 and p_2 (singles rates) with carefully selected coincidence rates can provide this rescaling at the small price of accepting only a very mild auxiliary assumption. On the other hand, rescaling case III experiments (see §6) requires one to assume a certain *ad hoc* modification of the basic correlation coefficient F. However, the measurement of this coefficient is in many respects a primary objective of the experiment. Any such assumption must then be scrutinised very carefully, for it inherently becomes the weak point of the experiment.

5. Cascade-photon experiments

The essential problem in testing the predictions in Bell's theorem against those by quantum mechanics is to find experimentally realisable situations in which the quantum-mechanical predictions directly violate Bell's inequalities. In §4 we showed that to do so with available apparatus is difficult. The situation is not hopeless, however. Clauser et al (1969) showed that with a mild supplementary assumption, actual experiments are predicted by quantum mechanics to yield a violation of Bell's inequality, and they proposed such an experiment.

Their suggestion is to measure the correlation in linear polarisation of photon pairs emitted in an atomic cascade. Figure 5 shows a schematic diagram of a typical apparatus for doing this, that of Freedman and Clauser (1972), who reported the first such test. The photons were emitted in a $J=0 \to J=1 \to J=0$ atomic cascade. The decaying atoms were viewed by two symmetrically placed optical systems, each consisting of two lenses, a wavelength filter, a rotatable and removable polariser, and a single-photon detector. The following quantities were measured: $R(\phi)$, the coincidence rate for two-photon detection as a function of the angle ϕ between the planes of linear polarisation, defined by the orientations of the inserted polarisers; R_1, the

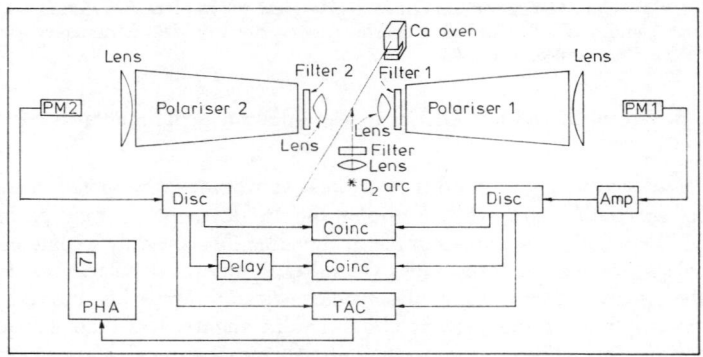

Figure 5. Schematic diagram of apparatus and associated electronics of the experiment by Freedman and Clauser. Scalers (not shown) monitored the outputs of the discriminators and coincidence circuits (figure after Freedman and Clauser).

coincidence rate with polariser 2 removed; R_2, the coincidence rate with polariser 1 removed; R_0, the coincidence rate with both polarisers removed.

The details of this experiment along with other similar ones will be discussed in §5.3. First, however, we describe the auxiliary assumption(s) which render this a reasonable test, and present the resulting inequalities. Then we describe the quantum-mechanical predictions for this and similar arrangements.

5.1. Predictions by local realistic theories

5.1.1. Assumptions for cascade-photon experiments.

The initial assumption by CHSH is, given that a pair of photons *emerges* from the polarisers, the probability of their joint detection is independent of the polariser orientations a and b. Clauser and Horne (1974) showed that an alternative assumption leads to the same results. Their assumption is that for every pair of emissions (i.e. for each value of λ), the probability of a count with a polariser in place is less than or equal to the corresponding probability with the polariser removed. The assumption of CH is stronger than that of CHSH in so far as it is stated for each value of λ, whereas CHSH make an assertion only for the total sub-ensemble of photons which pass through the polarisers. On the other hand, the assumption of CH is more general, in that the processes 'passage' and 'non-passage' through a polariser (which are not observable, and which are inappropriate for many possible theories) are not considered primitive. Furthermore, CH only assume an inequality, which is weaker than the equality of CHSH. Both assumptions, in our opinion, are physically plausible, but each gives a certain loophole to those who wish to defend local hidden-variables theories in spite of the experimental evidence which will be presented below.

Let us discuss the consequences of the CH assumption. We denote by the symbol ∞ an apparatus configuration in which the analyser is absent. Let $p_1(\lambda, \infty)$ denote the probability of a count from detector 1 when analyser 1 is absent and the state of the emission is λ. A similar probability $p_2(\lambda, \infty)$ may be defined for apparatus 2. Thus, the assumption is that:

$$0 \leqslant p_1(\lambda, a) \leqslant p_1(\lambda, \infty) \leqslant 1$$
$$0 \leqslant p_2(\lambda, b) \leqslant p_2(\lambda, \infty) \leqslant 1 \quad (5.1)$$

for every λ, and for all values of a and b. Inequalities (5.1) and (3.17) and arguments similar to those which led from (3.16) to (3.19) yield immediately the result:

$$-p_{12}(\infty, \infty) \leqslant p_{12}(a, b) - p_{12}(a, b') + p_{12}(a', b) + p_{12}(a', b') - p_{12}(a', \infty) - p_{12}(\infty, b) \leqslant 0.$$
$$(5.2)$$

Note that all terms in inequality (5.2) are joint probabilities for coincident counts at the two detectors. Inequality (3.19), in contrast, contains the two terms p_1 and p_2, which are probabilities of a count at a single detector. Furthermore, both the upper and lower limits in inequality (5.2) can be written as a ratio of probabilities, so that both can be tested without the need for the 'event-ready' detectors of figure 1.

5.1.2. *Additional symmetries*. Again we can invoke a rotational invariance argument similar to that of §3.6; thus we require:

(i) $p_{12}(a, \infty)$ is independent of a, and likewise $R_1(a) \equiv R_1$
(ii) $p_{12}(\infty, b)$ is independent of b, and likewise $R_2(b) \equiv R_2$
(iii) $p_{12}(a, b) \equiv p_{12}(\phi)$, and likewise $R(a, b) \equiv R(\phi)$, where $\phi = |a-b|$.

These conditions are not always satisfied, and they obviously fail when each of the particles has a definite linear polarisation. However, for all of the actual experiments to be described in this section, the conditions are at least satisfied by the quantum-mechanical predictions, and more importantly no experimental deviations from them have been detected. It is noteworthy that this set of conditions is frequently satisfied, even in situations where some of those of §3.6 are not. For example, in many of the cascade-photon experiments, the singles rate r_2 contains an extraneous contribution from excitation to the intermediate state of the cascade by channels not involving the first level of the cascade. Such excitation may result in the emission of polarised light at the wavelength of the second photon of the cascade, but no coincidences.

With these conditions, inequality (5.2) becomes:

$$-p_{12}(\infty, \infty) \leqslant 3p_{12}(\phi) - p_{12}(3\phi) - p_{12}(a', \infty) - p_{12}(\infty, b) \leqslant 0 \quad (5.3)$$

for all a' and b.

Since the emission rates in all of the various experiments were held constant, and in most cases monitored by an auxiliary apparatus, we can write the ratios of probabilities as ratios of count rates:

$$p_{12}(\phi)/p_{12}(\infty, \infty) = R(\phi)/R_0$$
$$p_{12}(a, \infty)/p_{12}(\infty, \infty) = R_1/R_0 \qquad (5.4)$$
$$p_{12}(\infty, b)/p_{12}(\infty, \infty) = R_2/R_0.$$

Inserting equations (5.4) into inequality (5.3), we can write this form of Bell's inequality in terms of coincidence rates:

$$-R_0 \leqslant 3R(\phi) - R(3\phi) - R_1 - R_2 \leqslant 0. \qquad (5.5)$$

Inequality (5.5) was first derived by Clauser et al (1969), but by using their alternative auxiliary assumption.

Freedman (1972) showed that inequality (5.5) can be further contracted to a form which is very convenient for comparison with experimental results. If we take the optimal value for upper-limit violation by cascade-photon experiments $\phi = \pi/8$, then inequality (5.5) becomes:

$$-R_0 \leqslant 3R(\pi/8) - R(3\pi/8) - R_1 - R_2 \leqslant 0.$$

On the other hand, if we take the optimal value for lower-limit violation $\phi = 3\pi/8$, using the fact that $9\pi/8$ represents the same angle as $\pi/8$, it becomes:

$$-R_0 \leqslant 3R(3\pi/8) - R(\pi/8) - R_1 - R_2 \leqslant 0.$$

Dividing both inequalities by R_0, and subtracting the second inequality from the preceding one, we obtain the simple inequality:

$$|R(\pi/8) - R(3\pi/8)|/R_0 \leqslant \tfrac{1}{4}. \qquad (5.6)$$

Inequality (5.6) has the advantage that it can be checked by measuring the frequency of joint detection of photons with the polarisers in only two different relative orientations, and it dispenses with the need to measure rates with only one polariser removed.

5.2. Quantum-mechanical predictions for a $J = 0 \to 1 \to 0$ two-photon correlation

5.2.1. An idealised case. Even if ideal polarisation analysers and photo-detectors are assumed, the violation or non-violation of inequality (5.6) depends upon the quantum state in which the photon pairs are prepared. It is instructive to demonstrate that a violation does occur with perfect apparatus if the photons are propagating in opposite directions from the source along the $\hat{z}$ axis, with total angular momentum 0 and total parity $+1$. Their state is an ideal limit of ones which can actually be prepared in a laboratory. The polarisation part of the two-photon wavefunction is:

$$\Psi_0 = \frac{1}{\sqrt{2}}\left[\begin{pmatrix}1\\0\\0\end{pmatrix} \otimes \begin{pmatrix}1\\0\\0\end{pmatrix} + \begin{pmatrix}0\\1\\0\end{pmatrix} \otimes \begin{pmatrix}0\\1\\0\end{pmatrix}\right] \qquad (5.7)$$

where $\begin{pmatrix}1\\0\\0\end{pmatrix}$ represents polarisation along the $\hat{x}$ axis and $\begin{pmatrix}0\\1\\0\end{pmatrix}$ represents polarisation along the $\hat{y}$ axis, and where the first of two juxtaposed column vectors refers to photon 1 and the second to photon 2. A projection operator for linear polarisation along an axis, lying in the xy plane and making an angle θ with the $\hat{x}$ axis is:

$$Q(\theta) \equiv \begin{bmatrix} \cos^2\theta & \cos\theta\sin\theta & 0 \\ \cos\theta\sin\theta & \sin^2\theta & 0 \\ 0 & 0 & 0 \end{bmatrix} \quad (5.8)$$

as one can check by noting that the vector $\begin{pmatrix}\cos\theta\\\sin\theta\\0\end{pmatrix}$, which represents linear polarisation in this direction, is an eigenvector of $Q(\theta)$ with eigenvalue 1. Similarly the vector $\begin{pmatrix}-\sin\theta\\\cos\theta\\0\end{pmatrix}$, representing linear polarisation perpendicular to this direction, is an eigenvector of $Q(\theta)$ with eigenvalue 0, as is $\begin{pmatrix}0\\0\\1\end{pmatrix}$ which represents polarisation along the z axis (which of course is excluded by transversality). Consequently, the quantum-mechanical prediction for this case is:

$$[R(\phi)/R_0]_{\Psi_0} = \langle \Psi_0 | Q(a) \otimes Q(b) | \Psi_0 \rangle = \tfrac{1}{4}(1 + \cos 2\phi) \quad (5.9)$$

where, as before, we have taken $\phi = |a - b|$. From this result we find that the quantum-mechanical predictions

$$[R(\pi/8)/R_0 - R(3\pi/8)/R_0]_{\Psi_0} = \tfrac{1}{4}\sqrt{2} \quad (5.10)$$

violate inequality (5.5).

5.2.2. *Quantum-mechanical predictions for $J = 0 \to 1 \to 0$ cascade, ideal analysers, and finite solid-angle detectors.* Consider a $J = 0 \to J = 1 \to J = 0$ atomic cascade in which no angular momentum is exchanged with the nucleus, and in which both transitions are electric dipole. Since the atom is both initially and finally in states with zero total angular momentum, and since there is a parity change in each transition, the emitted photon pair has zero total angular momentum and even parity. We can therefore exactly write the angular wavefunction of the photon pair as:

$$\Psi = \frac{1}{\sqrt{3}} [Y_{1,1}^1(\hat{\eta}_1) Y_{1,-1}^1(\hat{\eta}_2) - Y_{1,0}^1(\hat{\eta}_1) Y_{1,0}^1(\hat{\eta}_2) + Y_{1,-1}^1(\hat{\eta}_1) Y_{1,1}^1(\hat{\eta}_2)] \quad (5.11)$$

where $\hat{\eta}_1$ and $\hat{\eta}_2$ are variable directions of propagation of the first and second photons, and where Y_{jm}^1 is the spherical vector function for total angular momentum j, magnetic quantum number m, and parity -1 (see, for example, Akhiezer and Beretstetskii (1965) for notation). Now suppose that the lenses which make the photons impinge normally upon the polarisation analysers collect light in cones of half-angle ξ. The wavefunction of a photon pair which emerges from the pair of lenses can be represented as $D(\xi)\Psi$, where $D(\xi)$ is an operator which is exhibited in the appendix to Shimony (1971). An argument is outlined in that paper that if ξ is infinitesimal, then $D(\xi)\Psi$ is equal (except for normalisation) to the ideal two-photon polarisation vector Ψ_0 of equation (5.7). This is a reasonable result, since there is no orbital angular momentum if the two photons propagate along a straight line. Therefore the fact that the photon pair has total angular momentum 0 implies that it has zero spin angular momentum, as in the state Ψ_0. Of course, a finite value of ξ is essential in an actual experiment in order to obtain a non-vanishing count rate. The quantum-mechanical prediction for the coincidence rates with the polarisation state $D(\xi)\Psi$ is then:

$$[R(\phi)/R_0]_{D(\xi)\Psi} = \langle D(\xi)\Psi | Q(a) \otimes Q(b) | D(\xi)\Psi \rangle = \tfrac{1}{4} + \tfrac{1}{4} F_1(\xi) \cos(2\phi) \quad (5.12)$$

where $F_1(\xi)$ is a monotonically decreasing function, has the value 1 for $\xi=0$, and diminishes to 0·9876 at $\xi=30°$. Equation (5.12) shows a somewhat weaker polarisation correlation than one finds in equation (5.9), as a result of the admixture of orbital angular momentum states when light is collected in a non-zero solid angle. However, the diminution of correlation is small, even for fairly large values of ξ, and it is evident that inequality (5.5) will be violated by the probabilities of equation (5.12), with $\xi=30°$.

5.2.3. Quantum-mechanical correlation for $J=0 \to 1 \to 0$ cascade in an actual experiment. In an actual experiment one does not have ideal linear polarisation analysers, and equation (5.12) must be corrected in order to take into account the inefficiency of actual analysers. We let $\epsilon_M{}^j$ be the maximum transmittance of the jth analyser ($j=1, 2$) and $\epsilon_m{}^j$ be the minimum transmittance. (The former is 1 and the latter is 0 for an ideal analyser, but values for the analysers will be given in the summaries below of the experiments which have actually been performed.) Then equation (5.9) must be replaced by the following:

$$[R(a,b)/R_0]_{QM} = \epsilon_M{}^1 \epsilon_M{}^2 \langle D(\xi)\Psi | Q(a) \otimes Q(b) | D(\xi)\Psi \rangle$$
$$+ \epsilon_M{}^1 \epsilon_m{}^2 \langle D(\xi)\Psi | Q(a) \otimes \bar{Q}(b) | D(\xi)\Psi \rangle$$
$$+ \epsilon_m{}^1 \epsilon_M{}^2 \langle D(\xi)\Psi | \bar{Q}(a) \otimes Q(b) | D(\xi)\Psi \rangle \quad (5.13)$$
$$+ \epsilon_m{}^1 \epsilon_m{}^2 \langle D(\xi)\Psi | \bar{Q}(a) \otimes \bar{Q}(b) | D(\xi)\Psi \rangle$$

where
$$\bar{Q}(a) = 1 - Q(a) \qquad \bar{Q}(b) = 1 - Q(b). \qquad (5.14)$$

We thus find:

$$[R(\phi)/R_0]_{QM} = \tfrac{1}{4}(\epsilon_M{}^1 + \epsilon_m{}^1)(\epsilon_M{}^2 + \epsilon_m{}^2) + \tfrac{1}{4}(\epsilon_M{}^1 - \epsilon_m{}^1)(\epsilon_M{}^2 - \epsilon_m{}^2) F_1(\xi) \cos 2\phi \quad (5.15)$$

(see Clauser *et al* 1969, Horne 1970, Shimony 1971). Again, the quantum-mechanical counterpart of inequality (5.5) is violated, if suitable values of the transmittances are used.

5.2.4. Other cascades. If the photon pair is obtained from a $J=1 \to J=1 \to J=0$ cascade with equal populations in the initial Zeeman sublevels and no coherence among them (so that the density matrix of the initial level is $\tfrac{1}{3} I$), but the preceding experimental arrangement is otherwise unchanged, then the quantum-mechanical prediction for the probability of joint detection is the same as the right-hand side of equation (5.15), except that $F_1(\xi)$ is replaced by $-F_2(\xi)$, where $F_2(0)=1$. The function $F_2(\xi)$ decreases monotonically more rapidly than $F_1(\xi)$ (Clauser *et al* 1969, Horne 1970, Holt 1973). A systematic survey of other possible cascades has been made by Fry (1973).

5.3. Description of experiments

So far, there have been four experiments of the type just described. Three of these have agreed with the quantum-mechanical predictions, and one has agreed with the predictions by local realistic theories via Bell's theorem.

5.3.1. Experiment by Freedman and Clauser (1972).
Freedman and Clauser (1972, see also Freedman 1972) observed the 5513 Å and 4227 Å pairs produced by the $4p^2\ ^1S_0 \to 4p4s\ ^1P_1 \to 4s^2\ ^1S_0$ cascade in calcium. Their arrangement is shown schematically in figure 5. Calcium atoms in a beam from an oven were excited by resonance absorption to the $3d4p\ ^1P_1$ level, from which a considerable fraction decayed to the $4p^2\ ^1S_0$ state at the top of the cascade. No precaution was necessary for eliminating isotopes with non-zero nuclear spin, since 99·855% of naturally occurring calcium has zero nuclear spin. Pile-of-plates polarisation analysers were used, with transmittances $\epsilon_M^1 = 0.97 \pm 0.01$, $\epsilon_m^1 = 0.038 \pm 0.004$, $\epsilon_M^2 = 0.96 \pm 0.01$, $\epsilon_m^2 = 0.037 \pm 0.004$. Each analyser could be rotated by angular increments of $\pi/8$, and the plates could be folded out of the optical path on hinged frames. The half-angle ξ subtended by the primary lenses was 30°. Coincidence counting was done for 100 s periods; periods during which all plates were removed alternated with periods during which all were inserted. In each run the ratios $R(\pi/8)/R_0$ and $R(3\pi/8)/R_0$ were determined. Corrections were made for accidental coincidences, but even without this correction, the results still significantly violated inequality (5.6). The average ratios for roughly 200 h of running time are:

$$[R(\pi/8)/R_0]_{\text{expt}} = 0.400 \pm 0.007 \qquad [R(3\pi/8)/R_0]_{\text{expt}} = 0.100 \pm 0.003$$

and therefore:

$$[R(\pi/8)/R_0 - R(3\pi/8)/R_0]_{\text{expt}} = 0.300 \pm 0.008$$

in clear disagreement with inequality (5.6). The quantum-mechanical predictions are obtained from equation (5.15) (with allowances for uncertainties in the measurement of the transmittances and the subtended angle):

$$[R(\pi/8)/R_0 - R(3\pi/8)/R_0]_{\text{QM}} = (0.401 \pm 0.005) - (0.100 \pm 0.005) = 0.301 \pm 0.007.$$

The agreement between the experimental results with the quantum-mechanical predictions is excellent. Agreement is also found for other values of the angle ϕ, as well as for measurements made with only one or the other polariser removed.

5.3.2. Experiment by Holt and Pipkin (1973).
Holt and Pipkin (1973, see also Holt 1973) observed 5676 Å and 4047 Å photon pairs produced by the $9^1P_1 \to 7^3S_1 \to 6^3P_0$ cascade in the zero nuclear-spin isotope ^{198}Hg (see figure 6 for a partial level diagram of mercury). Atoms were excited to the 9^1P_1 level by a 100 eV electron beam. The density matrix of the 9^1P_1 level was found to be approximately $\frac{1}{3}I$ by measurements of the polarisation of the 5676 Å photons, so that equation (5.15) with $F_1(\xi)$ replaced by $-F_2(\xi)$ is used to calculate the quantum-mechanical predictions for the coincidence counting rates. Calcite prisms were employed as polarisation analysers, with measured transmittances:

$$\epsilon_M^1 = 0.910 \pm 0.001 \qquad \epsilon_M^2 = 0.880 \pm 0.001$$
$$\epsilon_m^1 < 10^{-4} \qquad \epsilon_m^2 < 10^{-4}.$$

The half-angle ξ was taken to be 13° ($F_2(13°) = 0.9509$). The quantum-mechanical prediction is:

$$[R(3\pi/8)/R_0 - R(\pi/8)/R_0]_{\text{QM}} = 0.333 - 0.067 = 0.266$$

which only marginally exceeds the value $\frac{1}{4}$ allowed by inequality (5.6). The experimental result in 154·5 h of coincidence counting, however, is:

$$[R(3\pi/8)/R_0 - R(\pi/8)/R_0]_{\text{expt}} = 0.316 \pm 0.011 - 0.099 \pm 0.009 = 0.216 \pm 0.013$$

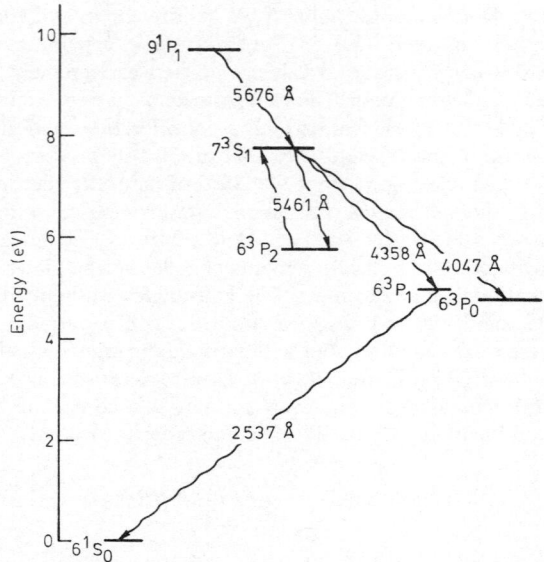

Figure 6. Partial level scheme for atomic mercury. Experiments by Holt and Pipkin, and by Clauser, excited the atoms to the 9^1P_1 level by electron bombardment, and observed photons emitted by the $9^1P_1 \to 7^3S_1 \to 6^3P_0$ cascade. The experiment by Fry and Thompson excited atoms to the 6^3P_2 (metastable) level by electron bombardment. Downstream, the atoms were excited by a tunable dye laser to the 7^3S_1 level, and photons were observed from the $7^3S_1 \to 6^3P_1 \to 6^1S_0$ cascade.

in good agreement with inequality (5.6) but in sharp disagreement with the quantum-mechanical prediction. Since this result is very surprising, Holt and Pipkin took great care to check possible sources of systematic error: the contamination of the source by isotopes with non-zero nuclear spin, perturbation by external magnetic or electric fields, coherent multiple scattering of the photons (radiation trapping), polarisation sensitivity of the photomultipliers, and spurious counts from residual radioactivity and/or cosmic rays, etc.

One such systematic error was found in the form of stresses in the walls of the Pyrex bulb used to contain the electron gun and mercury vapour. Estimates of the optical activity of these walls were then made, and the results were corrected correspondingly. (The values presented above include this correction.) It is noteworthy, however, that only the retardation sum for both windows was measured, for light entering the cell from one side and exiting through the opposite side. On the other hand, in the present experiment in which light exits from both windows, the relevant quantity is the retardation difference.

It is also noteworthy that in the subsequent experiment by Clauser (§5.3.3), a correlation was first measured which agreed with the results of Holt and Pipkin. Stresses were then found in one lens which were due to an improper mounting. (These were too feeble to be detected by a simple visual check using crossed Polaroids.) The stresses were removed, the experiment was re-performed, and excellent agreement with quantum mechanics was then obtained. On the other hand, Holt and Pipkin did not repeat their experiment when they discovered the stresses in their bulb.

A second criticism is that Holt and Pipkin took the solid-angle limit to be that imposed by a field stop placed outside the collimating lenses. It is possible that lens aberrations may have allowed a larger solid angle than they recognised. A ray-tracing calculation was in fact performed to assure that this was not the case. However, a solid stop ahead of the lens would have given one greater confidence that this did not, in fact, occur.

5.3.3. Experiment by Clauser (1976). Clauser (1976) repeated the experiment of Holt and Pipkin, using the same cascade and same excitation mechanism, though with a source consisting mainly of the zero-spin isotope ^{202}Hg. (The depolarisation effect due to some residual non-zero nuclear spin isotopes was calculated, using some results of Fry (1973).) Pile-of-plates polarisers were used with transmittances:

$$\epsilon_M^1 = 0.965 \qquad \epsilon_m^1 = 0.011 \qquad \epsilon_M^2 = 0.972 \qquad \epsilon_m^2 = 0.008$$

and the half-angle ξ taken to be $18.6°$. The quantum-mechanical prediction is:

$$[R(3\pi/8)/R_0 - R(\pi/8)/R_0]_{QM} = 0.2841.$$

The experimental result, from 412 h of integration, is:

$$[R(3\pi/8)/R_0 - R(\pi/8)/R_0]_{expt} = 0.2885 \pm 0.0093$$

in excellent agreement with the quantum-mechanical prediction, but in sharp disagreement with inequality (5.6).

5.3.4. Experiment by Fry and Thompson (1976). Fry and Thompson (1976) observed the 4358 Å and 2537 Å photon pairs emitted by the $7^3S_1 \to 6^3P_1 \to 6^1S_0$ cascade in the zero nuclear-spin isotope ^{200}Hg. Their experiment is shown schematically in figure 7.

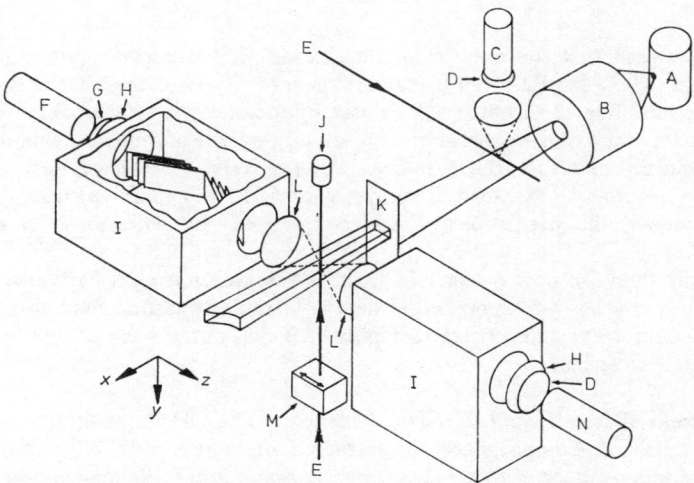

Figure 7. Schematic diagram of the experimental arrangement of Fry and Thompson. Polariser plate arrangement is also indicated. Actual polarisers have 14 plates. A, Hg oven; B, solenoid electron gun; C, RCA 8575; D, 4358 Å filter; E, 5461 Å laser beam; F, Amperex 56 DUVP/03; G, 2537 Å filter; H, focusing lens; I, pile-of-plates polariser; J, laser beam trap; K, atomic beam defining slit; L, light collecting lens; M, crystal polariser; N, RCA 8850 (figure after Fry and Thompson).

An atomic beam consisting of natural mercury was used as a source of ground-state (6^1S_0) atoms. The excitation of these to the 7^3S_1 level occurred in two steps at different locations along the beam. First, the atoms were excited by electron bombardment to the metastable 6^3P_2 level. Downstream, where all rapidly decaying states had vanished, a single isotope was excited to the 7^3S_1 level by resonant absorption of 5461 Å radiation from a narrow-bandwidth tunable dye laser. The technique provided a high data accumulation rate, since only the cascade of interest was excited. Photons were collected over a half-angle ξ of $19\cdot9° \pm 0\cdot3°$, and pile-of-plates analysers were used, with transmittances:

$$\epsilon_M{}^1 = 0\cdot98 \pm 0\cdot01 \qquad \epsilon_m{}^1 = 0\cdot02 \pm 0\cdot005 \qquad \epsilon_M{}^2 = 0\cdot97 \pm 0\cdot01 \qquad \epsilon_m{}^2 = 0\cdot02 \pm 0\cdot005.$$

The density matrix of the 7^3S_1 level was ascertained by polarisation measurements of the 4358 Å photons; it was found to be diagonal even though the Zeeman sublevels were not equally populated. The quantum-mechanical prediction is:

$$[R(3\pi/8)/R_0 - R(\pi/8)/R_0]_{QM} = 0\cdot294 \pm 0\cdot007.$$

The experimental result is:

$$[R(3\pi/8)/R_0 - R(\pi/8)/R_0]_{expt} = 0\cdot296 \pm 0\cdot014$$

in excellent agreement with the quantum-mechanical prediction, but again in sharp disagreement with inequality (5.5). Because of the high pumping rate attainable with the dye laser, it was possible to gather the data in a remarkably short period of 80 min which, of course, diminished the probability of errors due to variations in the operation of the apparatus, and facilitated checking for systematic errors.

5.4. Are the auxiliary assumptions for cascade-photon experiments necessary and reasonable?

We have seen that the data from the cascade-photon experiments are *sufficient* to refute the whole family of local realistic theories, if either the CHSH or the CH auxiliary assumption is accepted. Both assumptions are very reasonable. Yet both are conceivably false. One may ask the question: are the experimental data, by themselves, sufficient to refute the theories? Alternatively, is at least some auxiliary assumption *necessary*? The answer was given by CH, who contrived a local hidden-variables model, the predictions of which agree exactly with those of quantum mechanics.

One may then ask how reasonable are these assumptions. In particular do they disagree with any known experimental data? A similar question may also be asked about the counter-example. It seems highly artificial, but are any of its implications experimentally testable?

5.4.1. Critique of the CH and CHSH assumptions.
The CHSH assumption (§5.1) is, given that a pair of photons emerges from the polarisers, the probability of their joint detection is independent of the polariser orientation a and b. It may appear that the assumption can be established experimentally by measuring detection rates when a controlled flux of photons of known polarisation impinges on each detector. From the standpoint of local realistic theories, however, these measurements are irrelevant, since the distribution ρ when the fluxes are thus controlled is almost certain to be different from that governing the ensemble in the correlation experiments. We thus

see no way of directly testing this assumption, and thus no experiments with which it disagrees.

It is noteworthy, however, that there exists an important hidden-variables theory—the semiclassical radiation theory—which correctly predicts a large body of atomic physics data, but which denies both the CHSH assumption as well as a presupposition of it. The presupposition is that one can speak unequivocally of a photon's passage or non-passage through the polarisation analysers. In the semiclassical radiation theory, however, a photon partially passes its respective polariser and departs with a reduced (classical) amplitude. Furthermore, this amplitude depends upon the polariser's orientation and thereby determines the probability of the photon's subsequent detection (in violation of the CHSH assumption that all photons have the same detection probability, independent of either polariser's orientation). Nonetheless, the predictions for this theory are consistent with those by inequality (5.2), and the theory is refuted by the cascade-photon experiments (Clauser 1972). Evidently an alternative assumption is possible which allows inequality (5.2) to constrain theories denying this presupposition.

Such an assumption was provided by CH. This assumption (§5.1) is that for every pair of emissions, the probability of a count with the polariser in place is less than or equal to the corresponding probability with the polariser removed. This assumption appears reasonable because the insertion of a polarisation analyser imposes an obstacle between the source of the emissions and the detector, and it is natural to believe that an obstacle cannot increase the probability of detection. To be sure, we know of situations in which the insertion of an additional optical element (apparently an obstacle) does increase the probability of detection, e.g. the insertion of a diagonally oriented linear polariser between two crossed polarisers. However, the situation appropriate to the CH assumption is quite different from the one just mentioned, since no polarising elements follow the inserted polarisers. Moreover, if the third polariser is a two-channel device, such as a Wollaston prism, the increased detection rate observed in one channel occurs at the expense of the detection rate in the orthogonal channel. The sum of the rates from both channels actually decreases when the second polariser is inserted. Correspondingly, if the third polariser is replaced by a polarisation-insensitive detector, in a closer parallel to the situation of the cascade-photon experiments, then the detection rate is always reduced when a polariser is inserted ahead of this detector.

These considerations, unfortunately, are by no means sufficient to prove the CH assumption, since these observations concern ensemble-average probabilities. The CH assumption requires that the probability be diminished upon the insertion of a polariser *for all* λ.

5.4.2. The counter-example by Clauser and Horne. Clauser and Horne (1974) produced a local hidden-variables model whose predictions agree exactly with those by quantum mechanics. In their model the rate at which photons jointly pass through the polarisation analysers is in agreement with Bell's inequalities, but the joint detection rate agrees with the quantum-mechanical predictions. The model requires that the detected photon pairs be selected in a very special manner from among those which pass through the analysers, and that those which have not passed through a polariser have a different detection probability from those which have. Although the selection is done entirely locally, it does have the appearance of being highly artificial and, indeed, almost conspiratorial against the experimenter.

The model applies only as long as the net detector efficiencies are smaller than a certain maximum value p_{max} which depends on the analyser efficiencies ϵ_+ and ϵ_-: that is (using the notation of §4) when:

$$[\eta f]_{expt} \leqslant p_{max}. \tag{5.16}$$

With the conditions holding for the experiment by Freedman and Clauser, these values are $[\eta f]_{expt} \approx 0.004$ and $p_{max} \approx 0.4$. This comparison can be improved somewhat by reference to some experimental results by Clauser†. Inequality (5.16) then becomes $\eta_{expt} \leqslant p_{max}$. For the experiment by Freedman and Clauser the value $\eta_{expt} \approx 0.06$ holds.

Despite our caution concerning the CHSH and CH assumptions, we regard the experimental refutation which relies upon them to be compelling. It is striking that only a highly artificial model has so far been found which is local and yet yields quantum-mechanical detection rates in the cascade-photon experiments, and even this model can be excluded by rather modest improvements in the apparatus. There is also some hope for a theorem to the effect that any model consistent with the experimental data will have anomalous features as does the CH model.

6. Positronium annihilation and proton–proton scattering experiments

6.1. Historical background

Two experiments testing predictions based on Bell's theorem have been performed using the high-energy photons produced by positronium annihilation. The historical background of these experiments is interesting. Wu and Shaknov (1950) determined the parity of the ground state of positronium by a method suggested by Wheeler that consisted of measuring the polarisation correlation of γ rays produced by positronium annihilation. The photons Compton-scattered, and two-photon coincidences were observed as a function of azimuthal scattering angles, a and b. Two relative angles 0 and $\pi/2$ were employed. From the ratio of these two coincidence rates they were able to infer that the parity of the ground state is negative. Bohm and Aharonov (1957), with different motivations, showed that these data are explained by quantum mechanics if the polarisation state of the photon pair is assumed to be:

$$\Psi_1 = \frac{1}{\sqrt{2}} \left[\begin{pmatrix} 1 \\ 0 \\ 0 \end{pmatrix} \otimes \begin{pmatrix} 0 \\ 1 \\ 0 \end{pmatrix} - \begin{pmatrix} 0 \\ 1 \\ 0 \end{pmatrix} \otimes \begin{pmatrix} 1 \\ 0 \\ 0 \end{pmatrix} \right] \tag{6.1}$$

† Clauser (1974) noticed that the parameters of existing experimental results were inappropriate to determine whether or not transmission and reflection of a photon at a dielectric surface (similar to one of the surfaces in a pile-of-plates polariser) are, in fact, mutually exclusive possibilities. He thus performed an experiment which confirmed that they are. This behaviour is in marked contrast to that of the semiclassical radiation theory, in which a photon is simultaneously transmitted and reflected by the surface. He also performed a variation of this experiment (unpublished) in which the dielectric surface was replaced by a fine mesh mirror ($\approx 50\%$ transmission), and again photons were observed to be either transmitted or reflected but not both simultaneously. One can conclude from this result that, at least for the purposes of local realistic theories, the simultaneous emission of a photon into any two different solid-angle elements does not occur. Since the probabilities relevant to the CH counter-example are conditional upon the photons actually entering the collimator, it follows that the solid-angle parameter f can be dropped for the purposes of inequality (5.16).

but are incompatible with the assumption that the ensemble of photon pairs can be described by a mixture of states, each of which is a product of two single-photon polarisation states. They therefore concluded that the data of Wu and Shaknov confirm the existence of states of two-particle systems which are 'non-separable', even though the particles are spatially remote from each other (see appendix 1).

Clauser et al (1969) investigated the possibility of using the arrangement of Wu and Shaknov, perhaps with some variation, for the further purpose of checking whether the observed frequencies can violate Bell's inequalities. It is, indeed, easy to show that if efficient linear polarisation analysers existed for 0·5 MeV photons (with transmittances $\epsilon_M - \epsilon_m$ greater than ~ 0.83), then the quantum-mechanical values for the coincidence rates, with the joint polarisation state given by equation (6.1), can violate the inequalities (3.12) or (3.20). Unfortunately no such analysers exist, and Compton scattering does no more than give a scattering distribution, described by the Klein–Nishina formula, that is dependent upon the direction of linear polarisation. They concluded that no variant of the Wu–Shaknov experiment can provide a test of the predictions based on Bell's theorem (see Horne 1970).

6.2. The experiment by Kasday, Ullman and Wu

It was argued by Kasday et al (1970, 1975, hereafter referred to as KUW; see also Kasday 1971) that such photon pairs can be used to test the predictions based on Bell's theorem if one accepts two auxiliary assumptions: (i) in principle, ideal linear polarisers can be constructed for high-energy photons; (ii) the results, which would be obtained in an experiment using ideal analysers, and those obtained in a Compton scattering experiment, are correctly related by quantum theory.

Their experimental arrangement (a variant of that of Wu and Shaknov) is shown schematically in figure 8. Positrons were emitted by a ^{64}Cu source, stopped and

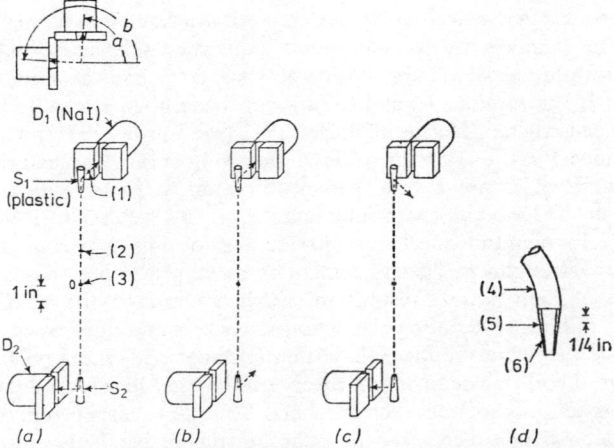

Figure 8. Schematic diagram of the experimental arrangement of KUW. The lead collimator is not shown. (a) Four-fold coincidence event; (b) (c), three-fold coincidence events; (d) detail of scatterer. a, b are the azimuthal angles of the scattered photons. (1) Scattered γ with energy E, absorbed by D_1, (2) annihilation γ, (3) positron source and absorber, (4) light pipe, (5) plastic scatterer, (6) MgO-coated aluminium light reflector (figure after Kasday et al).

annihilated in copper at the place labelled by 0. The annihilation γ-rays were emitted in all directions; the vertical direction was selected by a lead collimator. The scatterers were plastic scintillators. Lead slits selected a narrow range of acceptance azimuthal angle about the angles a and b. The top slit-detector assembly was then rotated to vary the relative azimuthal angle. Accepted coincidence events had a fourfold coincidence among the two scatterers and two detectors, as well as a sum-energy requirement that the total energy deposited in each scatterer plus detector equals the annihilation energy. It is noteworthy that this is the only experiment which employs the arrangement of figure 1. (Here the ensemble consists of the pairs jointly scattered by the scintillators.)

KUW applied assumptions (i) and (ii) as follows. Imagine two ideal linear polarisation analysers in the plane perpendicular to the direction of propagation of the selected annihilation photons (the vertical direction in figure 8), which are respectively oriented in the directions a and b of the two slits. If the state λ of the photon pair is given, a deterministic hidden-variables theory will determine whether each photon will pass through its respective analyser. This is their use of assumption (i). If photon 1 will pass its ideal analyser, then linear polarisation in the a direction is assigned to it; and if photon 2 will correspondingly pass its analyser, then it is assigned linear polarisation in the direction b. KUW then use assumption (ii) to assert that the angular scattering distribution of each respective photon is given by the Klein–Nishina formula (a distribution which is dependent upon the photon's initial linear polarisation). With the Klein–Nishina formula one can calculate the probability that the scattered photons will enter the respective acceptance slits. Quantum mechanics makes a definite prediction for this joint probability. Deterministic local hidden-variables theories together with assumptions (i) and (ii) also imply an inequality governing this probability which will disagree with the quantum-mechanical predictions. The experimental data of KUW are in good quantitative agreement with the quantum-mechanical predictions.

This experiment is less decisive, in our opinion, as a refutation of the family of local realistic theories than are the cascade-photon experiments discussed in §5, because it relies upon assumptions which are considerably stronger than the assumption needed by the latter. If assumptions (i) and (ii) are not made, then a local hidden-variables model can be constructed (Horne 1970, Bell 1971 (see Kasday 1971)) which yields the same predictions for the experiment as those by quantum mechanics. This consideration, by itself, is not a fully sufficient reason to prefer the cascade-photon experiments, since a local hidden-variables model, albeit a much more artificial one, also exists which yields quantum-mechanical predictions for those experiments (see §5.4).

The relative strengths of the supplementary assumptions provides a better reason for preference. There is one respect in which assumption (ii) of KUW is quite unconvincing. The only definite polarisation states acknowledged by quantum theory are the various modes of elliptic polarisation (circular and linear polarisation being special cases of these). Since quantum theory can be used to calculate the relationship required for assumption (ii) between ideal and Compton polarimeters only when the state of a photon is one recognised by quantum theory itself, this assumption presupposes that photons which enter the Compton polarimeters are in a quantum-mechanically describable state. Such a supposition is strongly in conflict with the postulates of Bell's theorem. The state λ presumably is not such a state, and moreover there is no prescription within quantum theory for calculating the results of an experiment for these more general states.

In a general hidden-variables scheme, the state λ of the photons clearly cannot be represented as one of definite linear polarisation (the special case in which it can is the hypothesis studied by Furry). KUW's decomposition of the state λ of a photon into linear and/or circular polarisation basis states is undoubtedly not possible in general.

Indeed, even in a quantum-mechanical treatment of the problem, the photons are acknowledged not to be in a state of definite polarisation. Quantum mechanically, neither photon's polarisation is in a definite state, but each is in what is known as an 'improper mixture' of such states (see d'Espagnat (1976) for a discussion of improper mixtures). In quantum theory, the only correct procedure for handling such systems is to perform calculations for the composite two-photon state. Thus we see that the 'marriage' between quantum mechanics and a general local realistic theory required by assumption (ii) results in a fatally incorrect handling of both theories.

6.3. The experiments by Faraci et al, Wilson et al and Bruno et al

An experiment very similar to that of KUW (but with ^{22}Na as a source) was performed by Faraci *et al* (1974) with very different results. Their data disagree sharply with the quantum-mechanical predictions based upon the polarisation state of equation (6.1), and are at the extreme limit permitted by Bell's inequalities (given the assumptions of KUW). Their data also showed a variation in correlation strength which depends upon the source-to-scatterer distances. Since their paper is quite condensed, it is difficult to conjecture whether or not a systematic error is responsible for these results. KUW, however, present various criticisms of this work as well as a clarification of various misinterpretations of their own work by these authors.

Wilson *et al* (1976) repeated the experiment using ^{64}Cu as a source. In contrast with the results of Faraci *et al*, they found complete agreement with the quantum-mechanical predictions, and no significant variation of the correlation strength when the scatterer positions were changed.

Bruno *et al* (1977) also repeated the experiment using ^{22}Na as a source, but used alternatively Cu and Plexiglass as the annihilator. To discriminate against multiple scattering events they imposed a sum-energy restriction as did KUW and Wilson *et al* (but not Faraci *et al*), and also varied the scatterer sizes. Residual triplet-positronium contribution was ascertained by the use of the different annihilator materials. Again, no violation of the quantum-mechanical prediction was observed, for any of various source–scatterer distances.

6.4. Proton–proton scattering experiment by Lamehi-Rachti and Mittig

The only test of the predictions in Bell's theorem which has been performed so far not using photons is that of Lamehi-Rachti and Mittig (1976). They measured the spin correlations in proton pairs prepared by low-energy S-wave scattering. The scattering geometry is shown in figure 9. Protons from the Saclay tandem accelerator were scattered by a target containing hydrogen. The incident and recoil protons each entered analysers at $\theta_{lab} = 45°$ ($\theta_{cm} = 90°$). The protons were scattered by a carbon foil, and detected at positions labelled L_1 or R_1, and L_2 or R_2 in the figure. Coincidences were sought between detectors on opposite arms, as they varied the azimuthal angle of the detector pair of one arm.

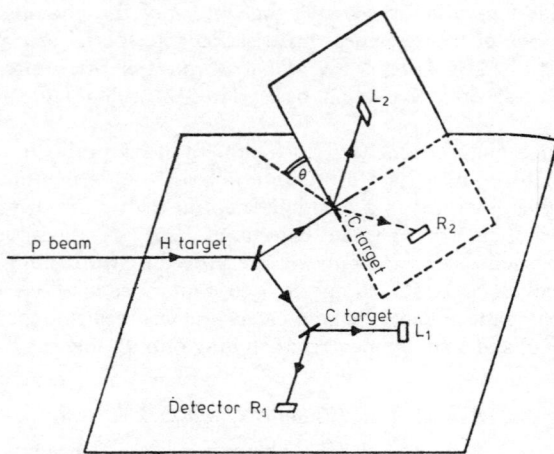

Figure 9. Proton–proton scattering geometry for the experiment by Lamehi-Rachti and Mittig (after Lamehi-Rachti and Mittig 1976).

Auxiliary assumptions similar to those required for the positronium experiments allow them to compare the data with the predictions for local realistic theories. It should be noted that their geometry requires an additional assumption not necessary for the positronium experiments. Since the analysers are only sensitive to the transverse components of the spin, and since $\theta_{cm} = 90°$, the correlation is of the form:

$$E(a, b) = C \cos a \cos b$$

and cannot violate inequalities (3.12) or (3.19) no matter what value $C \leqslant 1$ has. They thus assume that the quantum-mechanically predicted rotational invariance of the S-wave scattering (supposing negligible triplet contribution) allows them to decompose the correlation into a rotationally invariant part (singlet) and a non-rotationally invariant part (triplet). They then extrapolate the results back to a form which violates Bell's inequalities, and rely upon other experimental evidence to set an upper limit to the triplet scattering contribution. An arrangement in which one of the protons is electrostatically deflected through 90°, or magnetically precessed through 90°, would have eliminated the need for this last assumption.

They obtain good agreement with the quantum-mechanical predictions. If one accepts their assumptions, then Bell's inequalities are violated. However, even more reliance on quantum mechanics is needed than for the positronium experiments, and the criticisms of those experiments apply here more acutely.

7. Evaluation of the experimental results and prospects for future experiments

7.1. Two problems

There are two very different problems involved in evaluating the experiments so far performed for testing the predictions in Bell's theorem. The first is to determine the significance of the anomalous results of Holt and Pipkin and Faraci *et al*. The second is to determine what possibilities remain open if only the experiments which favour quantum mechanics are accepted as veridical, and the anomalous results are attributed to spurious effects. In this section we discuss both of these problems.

7.1.1. Significance of the experimental discrepancies. The probability is extremely high, in our opinion, that the results contradicting the predictions by quantum mechanics were due to systematic errors. This opinion is *not* based on a conservative acknowledgment of the great success of quantum mechanics in the atomic domain. Rather, it is based upon the consideration that quantum mechanics predicts strong correlations, whereas Bell's theorem sets a limit upon such correlations. Virtually any conceivable systematic error will wash out a strong correlation so as to produce results in accordance with Bell's theorem, rather than speciously strengthen a weak correlation. We also note that the predictions by quantum mechanics are quantitatively precise. Therefore, in order to maintain that a local realistic theory governs nature, one must invoke experimental errors not only to explain a violation of the inequalities in seven out of nine experiments, but also to explain a very close quantitative agreement with the quantum-mechanical predictions in these seven. In view of the delicacy of these experiments we are not surprised that two anomalous results were obtained among nine. Experience with the experimental techniques and an awareness of the probable systematic errors, one expects, will lead to greater uniformity of results in later repetitions of the experiments. The results of the more recent experiments already indicate this to be so.

7.1.2. Loopholes with auxiliary assumptions. The assumptions for cascade-photon experiments are criticised in §5.4 and those for the positronium and the proton–proton scattering experiments in §§6.2 and 6.4. The opinion is advanced that those for the former are considerably weaker than those for the latter; hence, the cascade-photon experiments are to be preferred. Evidently, none of these assumptions can be directly tested, and thus neither argument is at present fully conclusive.

On the other hand, an indirect test of the assumptions of CH and CHSH may become possible. The counter-example for the cascade-photon experiments (in contrast to that for the positronium and the proton–proton scattering experiments) exploits minor technological imperfections in the apparatus. Indeed, improvements in the polariser efficiencies and/or the photomultiplier quantum efficiencies can make this counter-example obsolete. There is, to our knowledge, nothing fundamentally restricting significant improvements in either of these.

The cascade-photon experiments performed so far were all done on a very small budget (in comparison with modern large-scale experimentation). They were designed simply for testing inequality (5.2), and the various arrangements were sufficient to that end. Now suppose that a theorem (a strengthening of the one conjectured in §5.4) can be proved that the model of CH is essentially the only local hidden-variables model which reproduces quantum-mechanical data in the cascade-photon experiments. Since only a modest improvement in some of these parameters is sufficient to rule out this counter-example, the added expense of a significantly improved apparatus, in our opinion, would be justified.

7.2. Experiments without auxiliary assumptions about detector efficiencies

Even though the experimental results concerning local realistic theories appear highly convincing, it is still desirable to have an experiment for which auxiliary assumptions are not required. It was shown in §4 that the requirements for such a scheme are demanding. Experiments using photons for this purpose do not appear to be feasible in the foreseeable future, since there seems to be no way of resolving the dilemma that highly efficient polarisation analysers can be achieved only for low-energy photons, while highly efficient detectors can be made only for high-energy photons. Furthermore, the two-body decay requirement is problematic with low-energy photons. Charged particles are evidently unusable, since an elegant argument by Bohr (see Mott and Massey 1965) indicates that magnetic state selection of their spin components violates the uncertainty principle. Hence, most schemes under consideration involve using either neutral particles and/or discrete states other than those associated with spin components. For example, Bell (1971) and Clauser et al (see Fehrs 1973) were inspired by a paper of Inglis (1961) to consider the charge-conjugation correlations shown in the decay of neutral kaon pairs produced by proton–antiproton annihilation. It was concluded that the exponentially decaying envelope of the correlations precludes the observation of a direct violation of Bell's inequalities in this system.

There is hope that the requirements for efficient analysis and detection can be achieved by observing the dissociation fragments of a metastable molecule, with a pair of Stern–Gerlach magnets as analysers. The latter have virtually 100% transmission, and proper design of the magnetic fields can minimise spurious spin-flips (Majorana transitions) during propagation of the decay fragments. Alkali metal and halogen atoms, if used as the decay fragments, can be detected individually by ionisation or electron attachment at a hot surface with nearly 100% efficiency. The parameters a and b can be taken to be the amplitudes of suitable resonant radio-frequency fields, applied in such a way as to coherently rotate the particle spins. Such an experiment holds promise of testing local realistic theories without any auxiliary assumptions, and with no loopholes other than the possibility of communication between the analysers (see §7.3).

7.3. Preventing communication between the analysers

Both the special and general theories of relativity preclude the existence of action-at-a-distance. This fact is, of course, the primary motivation for the various locality postulates considered above. However, in all of the experiments described so far, action-at-a-distance in the relativistic sense is not precluded, since the analysers are always kept at fixed orientations for periods of several seconds. Thus, there is ample time for information about the orientation of one analyser to be transmitted by some unknown mechanism (consistent with relativity theory) to the other apparatus (and/or other particle) thereby influencing its results. It is thus conceivable that such a mechanism is instrumental in producing quantum-mechanical coincidence counting rates in the above experiments. To test this possibility requires an experiment in which the parameters a and b are adjusted with great rapidity while the correlated particles are in flight. If the event consisting of the adjustment of the parameter a of the first analyser is wholly space-like separated from the detection event of particle 2, and similarly concerning adjustment of parameter b and the detection of particle 1, then no signal with subluminal speed can convey information about the orientation of one analyser to the other apparatus in time to affect the probability of detecting the respective particles. In other words, if the parameters a and b are adjusted with sufficient rapidity, then the non-occurrence of action-at-a-distance implies locality. For photons the required rapid adjustment of the analyser orientations can be accomplished, for example, by using modern electro-optical devices such as high-speed Pockell's cells. Aspect (1976) proposed the use of acousto-optical devices for basically the same purpose.

However, even with such devices it is impossible to block the loophole completely. Since the backward light cones of the detection and adjustment events overlap, it may be claimed that events in the overlap region are responsible for determining the choices of the parameters a and b as well as the observed results. In this way the quantum-mechanical coincidence counting rates can still be accounted for without any direct causal connection between opposite sides of the experiment, and hence without introducing action-at-a-distance. Such an argument, however, seems unacceptable on methodological grounds, for it could be used to justify an *ad hoc* dismissal of any disagreeable data in almost any conceivable scientific experiment.

7.4. Conclusion

Although further experimental investigations of the family of theories governed by Bell's theorem are desirable, we are tentatively convinced that no theory of this kind can correctly describe the physical world. Nonetheless, we find this conclusion disturbing, since the philosophical point of view which most working scientists have found natural, at least until quite recently, requires a local realistic theory. Because of the evidence in favour of quantum mechanics from the experiments based upon Bell's theorem, we are forced either to abandon the strong version of EPR's criterion of reality—which is tantamount to abandoning a realistic view of the physical world (perhaps an unheard tree falling in the forest makes no sound after all)—or else to accept some kind of action-at-a-distance. Either option is radical, and a comprehensive study of their philosophical consequences remains to be made.

Appendix 1. Criticism of EPR argument by Bohr, Furry and Schrödinger

The argument of EPR is powerful, since their conclusion surely follows from their plausible premises. Most of the community of physicists rejected EPR's conclusion, however, because of a reply by Bohr (1935), which essentially consisted of a subtle analysis of their premise (ii). His argument is that when the phrase 'without in any way disturbing the system' is properly understood it is incorrect to say that system 2 is not disturbed by the experimentalist's option to measure a rather than a' on system 1.

'Of course there is, in a case like that just considered, no question of a mechanical disturbance of the system under investigation during the last critical stage of the measuring procedure. But even at this stage there is essentially the question of *an influence on the very conditions which define the possible types of predictions regarding the future behaviour of the system*. Since these conditions constitute an inherent element of the description of any phenomenon to which the term "physical reality" can be properly attached, we see that the argumentation of the mentioned authors does not justify their conclusion that quantum-mechanical description is essentially incomplete.'

It is beyond the scope of the present review to analyse Bohr's claim that the term 'reality' can be used unambiguously in microphysics only when the experimental arrangement is specified. We are not convinced that Bohr ever succeeded in giving a coherent statement of his philosophical position (see, for example, Shimony 1971, Stein 1972, Hooker 1972). We must admit, however, in consideration of the experimental evidence presented in this review against EPR's conclusion, that Bohr's position remains as one of the few feasible options concerning the foundations of quantum mechanics.

An early important reaction to the argument by EPR was to question premise (i). Furry (1936) and Schrödinger (1935) independently considered the possibility that, after systems 1 and 2 become spatially separated and cease effectively to interact, their joint wavefunction no longer has the form (2.1), but rather becomes a mixture of simple product states, each having the form:

$$\Psi'_{\hat{n}} = u_{\hat{n}}^{\pm}(1) \otimes u_{\hat{n}}^{\mp}(2). \tag{A1.1}$$

For each element of the mixture, both 1 and 2 are then in definite quantum states. This possibility is sometimes called 'Furry's hypothesis', but that nomenclature is inappropriate. What Furry did was to show that for any choice of $u_{\hat{n}}(1)$ and $u_{\hat{n}}(2)$, there exist in principle pairs of observables, M of 1 and S of 2, such that the statistical predictions for joint measurements of M and S based upon mixtures of the $\Psi'_{\hat{n}}$ are different from those based upon equation (2.1). Since Furry believed quantum mechanics to be correct, he concluded that a state like that of equation (2.1) does not automatically evolve into a mixture of the $\Psi'_{\hat{n}}$ when 1 and 2 separate from each other (see the conclusion of §4 of his paper). It is more appropriate to call this scheme 'Schrödinger's hypothesis', since he explicitly stated that it may be true. Strong evidence against this hypothesis was presented in 1957 by Bohm and Aharonov (see §6.1). More recent experimental evidence confirming their conclusions has been discussed by Kasday (1971) and Clauser (1972, 1977). It is noteworthy that this hypothesis is such a natural one that many physicists apparently believe it to be a resolution of the EPR 'paradox' without recognising the theoretical and experimental evidence against it.

Appendix 2. Hidden-variables theories

A theory which asserts that the quantum-mechanical description of a physical system is incomplete and requires supplementation in order to specify completely the state of the system is commonly called a *hidden-variables theory*†. Those properties of the system which are proposed as supplements to the quantum-mechanical description are commonly called *hidden variables* or sometimes *hidden parameters*. The history of hidden-variables theories is quite intricate, because the various proponents and opponents have made different assumptions about the conditions of adequacy which a hidden-variables theory should satisfy. We shall review here only as much of this history as is needed to provide the background for Bell's theorem. Other reviews of this subject (some including discussions of Bell's theorem) may be found in Bell (1966), Capasso *et al* (1970), Belinfante (1973), Jammer (1974) and d'Espagnat (1976).

In 1926–7 deBroglie wrote several papers proposing an interpretation of the wavefunction very different from that of Bohr. He supposed that the wavefunction

† The literature is not always consistent on this point, and many authors have included a hypothesis of determinism in their definition of a hidden-variables theory. In this review hidden-variables theories for which determinism holds are referred to as deterministic hidden-variables theories.

associated with a particle is a physically real field propagating in physical space in accordance with the Schrödinger equation. He also supposed that the particle always has a definite position and a definite momentum. Thus his interpretation was actually a hidden-variables theory. Finally, he assumed an intimate coupling between the particle and the field described by the function Ψ, so that the latter can be considered a 'guiding wave' or 'pilot wave' for the particle. This coupling then accounts for interference and diffraction phenomena. Several serious difficulties were found in deBroglie's theory (see deBroglie 1960), especially concerning many-particle systems and the S-wave state of a particle. As a result, deBroglie set aside his investigations of this kind until he was re-encouraged by the work of Bohm in 1952.

We shall not discuss in detail the various other models considered so far (see, for example, Madelüng 1926, deBroglie 1953, Bohm and Vigier 1954, Freistadt 1957, Andrade e Silva and Lochak 1969). Some of these models assume that in addition to the potential recognised in classical mechanics the particle is subject to a 'quantum potential' $h^2\nabla^2 R/2mR$, where R is the amplitude of the wavefunction. Other models assume that the wavefunction describes an averaged or smoothed state of a fluid medium, subject to random fluctuations which are not taken into account by the wavefunction, but which are nevertheless important for understanding the statistical behaviour of particles moving in the medium. For the most part, the advocates of these models do not claim that they are anything but tentative descriptions of the subquantum level of the physical world. Their significance lies in providing existence proofs that a theory can be deterministic in character, and nevertheless agree with many of the statistical predictions by quantum mechanics. We have found, in our discussion of Bell's theorem in §3, that a theory can achieve complete agreement with quantum mechanics only if it is non-local.

Leaving aside the locality problem, we may ask how is it possible that so many mathematically consistent hidden-variables theories have been devised when various theorems have claimed that the structure of the class of quantum-mechanical observables precludes such theories? We now discuss two such theorems.

A2.1. Von Neumann's theorem

The most famous theorem of this type is due to von Neumann (1932). Let $\mathcal{O}$ be the class of observables, and suppose that every self-adjoint operator on a Hilbert space $\mathcal{H}$ of dimension greater than 1 represents a member of $\mathcal{O}$ (but it is not excluded that $\mathcal{O}$ has other members). A state is specified by defining an expectation value Exp(A) on every $A \in \mathcal{O}$, and it is assumed that Exp satisfies the following conditions.

(i) Exp(1) = 1, where 1 is the observable which, by definition, always has the value unity.

(ii) For each $A \in \mathcal{O}$ and each real number r, Exp(rA) = rExp(A).

(iii) If A is non-negative, then Exp(A) $\geq$ 0.

(iv) If $A, B, C, \ldots$, are arbitrary observables, then there is an observable $A + B + C + \ldots$ (which does not depend upon the choice Exp) such that:

$$\text{Exp}(A+B+C+\ldots) = \text{Exp}(A) + \text{Exp}(B) + \text{Exp}(C) + \ldots.$$

The theorem asserts that there exists a self-adjoint operator A on $\mathcal{H}$ such that Exp(A^2) $\neq$ [Exp(A)]^2, i.e. the state defined by Exp is not dispersion-free over the quantum-mechanical observables.

The intuitive meaning of the conclusion of von Neumann's theorem is that no state of the system—not even a state different from those recognised by quantum mechanics—can assign definite values simultaneously to all quantum-mechanical observables. Von Neumann's theorem is mathematically correct, but its physical significance is doubtful. The fact that it was often cited over three decades as a proof for the completeness of quantum mechanics is a kind of historical aberration, and has wrought much confusion. Its crucial weakness is the supposition that any possible state of the system must satisfy condition (iv), even when $A, B, C, \ldots$, are non-commuting operators and therefore represent observables which, according to quantum mechanics, cannot be simultaneously measured. The actual procedure for measuring $A+B$, when A and B do not commute, is different from the procedures for measuring A and B separately and does not presuppose any information about the value of either A or B. Consequently, the fact that the additivity of condition (iv) is satisfied by quantum-mechanical states is a peculiarity of quantum mechanics, and there is no reason to suppose that it is satisfied by non-quantum-mechanical states. This criticism of the physical significance of von Neumann's theorem was made by Siegel (1966) and Bell (1966). (In the two-dimensional Hilbert space of a spin-$\frac{1}{2}$ particle, Bell also constructed a family of dispersion-free states which are physically reasonable, even though they violate condition (iv). Pearle (1965) and Kochen and Specker (1967) independently constructed similar models.) There is evidence[†] that Einstein was critical of condition (iv) as early as 1938 and therefore did not consider von Neumann's theorem to be an obstacle to the 'completion' of quantum mechanics as demanded by the argument of EPR. An interesting survey by Jammer (1974, pp272-7) shows that others were critical of condition (iv), but not with complete clarity.

A2.2. Gleason's theorem

In 1957 Gleason proved a theorem which is free from the unphysical condition (iv) of von Neumann's theorem, and which has frequently been considered to be a decisive proof of the impossibility of any consistent hidden-variables theory. We shall not state the theorem itself but rather shall state a corollary[‡] which can be compared directly with von Neumann's theorem.

Let $\mathcal{O}$ be a class of observables containing all those represented by the self-adjoint operators on a Hilbert space of dimension greater than 2, and let Exp be a real-valued function over $\mathcal{O}$, which satisfies the following conditions:

1, 2, 3, as in von Neumann's theorem.

4'. If $A, B, C, \ldots$, are commuting self-adjoint operators on $\mathcal{H}$, then

$$\mathrm{Exp}(A+B+C+\ldots) = \mathrm{Exp}(A) + \mathrm{Exp}(B) + \mathrm{Exp}(C) + \ldots.$$

Then Exp is not dispersion-free.

This corollary is weaker than von Neumann's theorem in one respect: it does not apply to a Hilbert space of dimension 2, and therefore it permits the models of Bell, Kochen and Specker, and Pearle. But it is much stronger in one crucial respect: it requires additivity only over commuting operators, for which the values of $A+B+C+$

---

† Professor P G Bergmann was an assistant to Einstein at that time, and he reported Einstein's criticism to one of us. We regret that we have no evidence concerning Einstein's opinion of von Neumann's argument in 1935, when the paper of EPR was written.

‡ There are several direct proofs of essentially this corollary which do not rely upon the main theorem of Gleason: Bell (1966), Kochen and Specker (1967) and Belinfante (1973). The proof given by Jammer (1974, pp298-9) is the same as that of Bell, who is not credited.

... can in principle be determined by summing the results of simultaneous measurements of $A, B, C, \ldots,$.

The conditions for this corollary are physically plausible, and its conclusion seems to be strong enough to preclude all non-trivial hidden-variables theories (i.e. all which apply to a system which is quantum-mechanically represented by a Hilbert space of dimension greater than 2). However, Bell (1966) pointed out the possibility of a family of non-trivial hidden-variables theories which do not satisfy all the conditions of the corollary and therefore are not bound by its conclusion. Suppose A, B and C are self-adjoint operators such that A commutes with both B and C, but B and C do not commute with each other. Therefore A can in principle be measured simultaneously with B, or it can be measured simultaneously with C, and different experimental arrangements are required for the two measurements. In the corollary it was assumed that when the state of the system is fully specified, the function $\text{Exp}(X)$ has a definite value for each observable X, however X is measured. 'It was tacitly assumed that measurement of an observable must yield the same value independently of what other measurements may be made simultaneously' (Bell 1966, p451). But there is no *a priori* reason that this assumption should be true. 'The result of an observation may reasonably depend not only on the state of the system (including hidden variables) but also on the complete description of the apparatus' (Bell 1966, p451). It is physically reasonable, therefore, to consider hidden-variables theories in which the expectation values have the form $\text{Exp}(X; \mathscr{C})$, where $\mathscr{C}$ indicates the 'context' of the measurement of X, i.e. all the quantities measured simultaneously with X. Gleason's theorem and its corollary do not preclude the possibility that such a 'contextual' hidden-variables theory can be dispersion-free, so that the result of measuring any observable is precisely determined by the state of the system (including hidden variables) together with the 'context' of the measurement.

Bell's proposal of a new family of hidden-variables theories sheds light on models like the one given by Bohm in 1952. This model antedated Gleason's work, but Bohm (1952, p187) defended it against von Neumann's impossibility theorem in the following way:

'the so-called "observables" are ... not properties belonging to the observed system alone, but instead potentialities whose precise development depends just as much on the observing apparatus as on the observed system. In fact, when we measure the momentum "observable", the final result is determined by hidden parameters in the momentum-measuring device as well as by hidden parameters in the observed electron.'

This passage does not propose the consideration of contextual hidden-variables theories as explicitly as Bell does, but it can be construed retrospectively as implicitly agreeing with Bell.

The next step in the history of hidden-variables theories was taken by Bell, once he was convinced that impossibility theorems like that of Gleason do not establish *a priori* the inconsistency of hidden-variables models. By taking these models seriously, he was free to examine whether they shared any physically unreasonable properties in spite of their mathematical consistency, and to inquire whether such properties are inevitable in any hidden-variables theory which agrees with the predictions by quantum mechanics. In this way he was heuristically led to the study of locality.

Acknowledgments

We wish to thank Professor Michael A Horne for many valuable suggestions for this review. One of us (AS) wishes to thank the Department of Theoretical Physics of the University of Geneva for its hospitality.

References

Akhiezer A I and Berestetskii V B 1965 *Quantum Electrodynamics* (New York: Interscience)
Andrade e Silva J L and Lochak G 1969 *Quanta* (New York: McGraw-Hill)
Aspect A 1976 *Phys. Rev.* D **14** 1944–51
Belinfante F J 1973 *A Survey of Hidden-Variables Theories* (Oxford: Pergamon)
Bell J S 1965 *Physics* **1** 195–200
—— 1966 *Rev. Mod. Phys.* **38** 447–52
—— 1971 *Foundations of Quantum Mechanics* ed B d'Espagnat (New York: Academic) pp171–81
—— 1972 *Science* **177** 880–1
—— 1976 *Communication at the 6th Gift Conf., Jaca, June 1975* Res. Th 2053–CERN
—— 1977 *Epistemological Lett.* **15** 79–84
Bohm D 1951 *Quantum Theory* (Englewood Cliffs, NJ: Prentice Hall) pp614–22
—— 1952 *Phys. Rev.* **85** 169–93
Bohm D and Aharonov Y 1957 *Phys. Rev.* **108** 1070–6
Bohm D and Vigier J P 1954 *Phys. Rev.* **96** 208–16
Bohr N 1935 *Phys. Rev.* **48** 696–702
de Broglie L 1926 *C. R. Acad. Sci., Paris* **183** 447–8
—— 1927 *C. R. Acad. Sci., Paris* **184** 273
—— 1928 *J. Phys. Radium* **8** 225–41
—— 1953 *La Physique Quantique: Restera-t-elle Indeterministe?* (Paris: Gauthier Villars)
—— 1960 *Non-Linear Wave Mechanics* (Amsterdam: Elsevier)
Bruno M, d'Agostino M and Maroni C 1977 *Nuobo Cim.* **40** B142–52
Capasso V, Fortunato D and Selleri F 1970 *Riv. Nuovo Cim.* **2** 149–99
Clauser J F 1972 *Phys. Rev.* A **6** 49–54
—— 1974 *Phys. Rev.* D **9** 853–60
—— 1976 *Phys. Rev. Lett.* **36** 1223–6
—— 1977 *Nuovo Cim.* **33** 740–6
Clauser J F and Horne M A 1974 *Phys. Rev.* D **10** 526–35
Clauser J F, Horne M A, Shimony A and Holt R A 1969 *Phys. Rev. Lett.* **23** 880–4
Colodny R G (ed) 1972 *Paradigms and Paradoxes* (Pittsburgh, Pa.: University of Pittsburgh Press)
Eberhard P 1977 *Nuovo Cim.* **38** B75
Einstein A, Podolsky B and Rosen N 1935 *Phys. Rev.* **47** 777–80
d'Espagnat B (ed) 1971 *Foundations of Quantum Mechanics. Proceedings of the International School of Physics 'Enrico Fermi'* Course XLIX (New York: Academic)
—— 1975 *Phys. Rev.* D **11** 1424–35
—— 1976 *Conceptual Foundations of Quantum Mechanics* (Reading, Mass.: Benjamin) 2nd edn
Faraci G, Gutkowski S, Notarrigo S and Pennisi A R 1974 *Lett. Nuovo Cim.* **9** 607–11
Fehrs M H 1973 *PhD Thesis* Boston University
Freedman S J 1972 *Lawrence Berkeley Lab. Rep. No* LBL 391
Freedman S J and Clauser J F 1972 *Phys. Rev. Lett.* **28** 938–41
Freistadt H 1957 *Nuovo Cim. Suppl.* **5** 1–70
Fry E S 1973 *Phys. Rev.* A **8** 1219–32
Fry E S and Thompson R C 1976 *Phys. Rev. Lett.* **37** 465–8
Furry W H 1936 *Phys. Rev.* **49** 393–9
Gleason A M 1957 *J. Math. Mech.* **6** 885–93
Gutkowski D and Masotto G 1974 *Nuovo Cim.* **22** B121–9
Holt R A 1973 *PhD Thesis* Harvard University

Holt R A and Pipkin F M 1973 *Preprint* Harvard University
Hooker C A 1972 *Paradigms and Paradoxes* ed R G Colodny (Pittsburgh, Pa.: University of Pittsburgh Press) pp67–302
Horne M A 1970 *PhD Thesis* Boston University
Inglis D R 1961 *Rev. Mod. Phys.* **33** 1–7
Jammer M 1974 *The Philosophy of Quantum Mechanics* (New York: Wiley)
Kasday L R 1971 *Foundations of Quantum Mechanics* ed B d'Espagnat (New York: Academic) pp195–210
Kasday L R, Ullman J D and Wu C S 1970 *Bull. Am. Phys. Soc.* **15** 586
—— 1975 *Nuovo Cim.* **25** B633–61
Kochen S and Specker E 1967 *J. Math. Mech.* **17** 59–87
Lamehi-Rachti M and Mittig W 1976 *Phys. Rev.* **14** 2543–55
Madelüng E 1926 *Z. Phys.* **40** 322–6
Mott N F and Massey H S W 1965 *The Theory of Atomic Collisions* (Oxford: Oxford University Press) pp214–9
von Neumann J 1932 *Mathematische Grundlagen der Quantenmechanik* (Berlin: Springer-Verlag) (Engl. trans. 1955 *Mathematical Foundations of Quantum Mechanics* (Princeton, NJ: Princeton University Press) pp307–25)
Pearle P 1965 *Preprint* Harvard University
—— 1970 *Phys. Rev.* D **2** 1418–25
Schiavulli L 1977 *Preprint* Universitá di Bari
Schrödinger E 1935 *Proc. Camb. Phil. Soc.* **31** 555–63
Selleri F 1978 *Foundations of Physics* **8** 103–16
Shimony A 1971 *Foundations of Quantum Mechanics* ed B d'Espagnat (New York: Academic) pp182–94, 470–80
—— 1978 *Epistemological Lett.* **18** 1–3
Shimony A, Horne M A and Clauser J F 1976 *Epistemological Lett.* **13** 1–8
Siegel A 1966 *Differential Space, Quantum Systems, and Prediction* ed N Wiener, A Siegel, B Rankin and W T Martin (Cambridge, Mass.: MIT Press)
Stapp H P 1971 *Phys. Rev.* D **3** 1303–20
—— 1978 *Whiteheadian Approach to Quantum Theory and the Generalised Bell's Theorem* Foundations of Physics to be published
Stein H 1972 *Paradigms and Paradoxes* ed R G Colodny (Pittsburgh, Pa.: University of Pittsburgh Press) pp367–438
Wigner E P 1970 *Am. J. Phys.* **38** 1005–9
Wilson A R, Lowe J and Butt D K 1976 *J. Phys. G: Nucl. Phys.* **2** 613–24
Wu C S and Shaknov I 1950 *Phys. Rev.* **77** 136

Errata in the paper of Clauser and Shimony

p.1888, 6 lines from bottom: insert "almost" before "all".

p.1889, line 16: substitute "3.4" for "3.1".

p.1899, 12 lines from bottom: right hand side of the equation should be
$N(r_1, r_2, \hat{a}, \hat{b})/N$.

p.1901, last line: insert) after ϵ_- .

p.1902, line 11: substitute "3.23" for "4.4".

p.1923, line 15: substitute "$\hbar$" for "h".

p.1924, last line: on p.297 Jammer does give credit to Bell.

Additional References and Comments

I. Collections and surveys:

Wheeler, J.A. and Zurek, W.H., Eds., 1983, <u>Quantum Theory and Measurement</u> (Princeton, Princeton U. Press). Section I of this comprehensive anthology contains the paper of Einstein, Podolsky, and Rosen and comments by Bohr and Schrödinger. Section III contains nine papers on hidden variables theories and Bell's theorem.

Epistemological Letters, Written Symposium on Hidden Variables (distributed by Institut de la Méthode, Bienne, Switzerland). About thirty-five mimeographed pamphlets, containing contributions and discussions on the question of hidden variables theories, were distributed between 1975 and 1984. Many, but not all, of the contributions have appeared in more standard publications.

Pipkin, F.M., 1978, Adv. in Atom. and Mol. Phys. <u>14</u>, 281-340.

Selleri, F and Tarozzi, G., 1981, Riv. Nuov. Cim. <u>4</u>, no.2, 1-53.
The discussion of the debate between Bell and Shimony, Clauser, and Horne concerning "local beables" fails to take into account the final paper in the exchange, that of Shimony in Issue 18 of Epistemological Letters.

Bell, J.S., 1983, Found. Phys. <u>12</u>, 989-999. This is a valuable historical document, presenting the background and the heuristics of Bell's theorem in a way that is not available elsewhere.

II. New derivations of inequalities of Bell's type:

Garuccio, A. and Selleri, F., 1980, Found. Phys. <u>10</u>, 209-216.

Stapp, H.P., 1980, Found. Phys. 10, 767-795. A defense is given of the author's derivation in Phys. Rev. D 10, 1303 (1971) against the criticisms in Sect. 3.8 of the paper of Clauser and Shimony. The latter are not persuaded.

Mermin, N.D., 1980, Phys. Rev. D 22, 356-361.

Mermin, N.D. and Schwarz, G., 1982, Found. Phys. 12, 101-135.

Ogren, M., 1983, Phys. Rev. D 27, 1766-1773.

Garg, A. and Mermin, N.D., 1984, Found. Phys. 14, 1-39.

III. Analysis of the locality condition:

Fine, A. 1980, in PSA 1980, vol. 2, Eds. A. Asquith and R. Giere (East Lansing, Philosophy of Science Association), 535, with critical comments by A. Shimony, ibid., 572.

Jarrett, J., 1984 or 1985, Nous. Bell's locality condition is shown to be equivalent to the conjunction of two conditions. When the first is violated, a message can be sent superluminally; quantum mechanics satisfies this condition. When the second is violated, as it is by quantum mechanics, superluminal communication does not result.

Shimony, A., 1985, in Proceedings of the Third Oxford Conference on Quantum Gravity, Eds. Isham, C.J. and Penrose, R. (forthcoming). This paper modifies one of the alternatives listed in Sect. 7.4 of the paper of Clauser and Shimony by exploring the possibility of peaceful coexistence between quantum mechanical non-locality and relativistic space-time structure.

Bohm, D. and Hiley, B., 1984, in Foundations of Quantum Mechanics in the Light of New Technology, Eds. Kamefuchi, S. et al. (Tokyo, Physical Society of Japan), 231-232.

Bohm, D. and Hiley, B., 1984, Found. Phys. 14, 255-274.

Eberhard, P., 1978, Nuov. Cim. B 46, 392-419. This is one of several papers showing that quantum mechanical correlations cannot be used for superluminal communication. See also Ghirardi, G.C., Rimini, A., and Weber, T., 1980, Lett. Nuov. Cim. 27, 293-298; Page, D., 1982, Phys. Lett. A 91, 57-60.

IV. Recent experiments:

Aspect, A., Dalibard, J., and Roger, G., 1982, Phys. Rev. Lett. 49, 1804-1807. This is the most important of all the recent experimental results concerning Bell's theorem. The authors tested a Bell-type inequality using an apparatus which effectively chose the orientations of the polarisation analysers for each of a pair of photons while the photons were in flight. Consequently, it seems difficult to see how the locality condition of Bell can be violated if locality holds in the sense of relativity theory, which prohibits communication about the orientation of one polarisation analyser to the other in this experimental arrangement. Thus, any hidden variables theory which preserves relativistic locality must imply a Bell-type inequality for this experimental arrangement, and the results violating the inequality constitute a refutation of relativistically local hidden variables theories. There is, however, a loophole in the argument, which has been analysed in detail by A. Zeilinger. The devices which effect the change of orientation of the polarisation analysers are periodically driven acoustical wave generators, and one could maintain that the hidden variables on one side of the apparatus might become aware of the periodicity of the device on the other side. The independent drifting of the two generators makes this loophole implausible.

Aspect, A, Grangier, P., and Roger, G, 1982, Phys. Rev. Lett. 49, 91-94. This is the only polarisation correlation test of a Bell-type inequality which makes use of two-channel polarisation analysers for each photon. Their result violates the inequality by more than forty standard deviations and is in excellent agreement with the quantum mechanical predictions. Although the analysers are not oriented while the photons are in flight, as in the experiment of Aspect, Dalibard, and Roger, the evidence against local hidden variables theories which it provides is impressive. Rapisarda, V, 1982, Ist. Naz. Fis. Nucl. Catania, Italy, 14 pp (unnumbered report) also proposed an experiment using two-channel analysers for each photon.

V. Proposals for further experiments:

Lo, T.K. and Shimony, A, 1981, Phys. Rev. A 23, 3003-3012. This paper proposes the measurement of spin correlations of daughter atoms from Na_2 dissociation, using Stern-Gerlach magnets as analysers. The purpose of the experiment is to block a loophole in the optical correlation experiments arising from the fact that efficient detectors are unavailable for low-energy photons. As a result, the local model of Clauser and Horne, Phys. Rev. D 10, 526-535 (1974) is able to account for the coincidence counting rates in actual experiments. In principle both spin analysis and detection can be efficient for atoms, thereby preventing the applicability of this model. The proposed experiment was criticised by Santos, E., 1984, Phys. Rev. A 30, 2128-2129, on the ground that dissociations of distinct molecules may be strongly correlated. Shimony, A., 1984, Phys. Rev. A 30,

130-2131, answers this and other criticisms but notes some difficulties pointed out by D. Prichard, which require some modifications of the proposed experiment.

Shimony, A., 1984, in Foundations of Quantum Mechanics in the Light of New Technology, Eds. Kamefuchi, S. et al. (Tokyo, Physical Society of Japan), 225-230. Most of this paper is an analysis of non-locality, but in the last section an experimental test is proposed for a non-local hidden variables theory of J.-P. Vigier, 1982, Astr. Nachr. 303, 55.

VI. Significance of Bell's theorem and related experiments.

Stapp, H.P., 1977, Nuov. Cim. B 40, 191-205. This paper strongly emphasises the experimental confirmation of non-locality.

Fine, A, 1982 a,b, Phys. Rev. Lett. 48, 291-295, and J.Math. Phys. 23, 1306-1310. These papers argue that the essential content of Bell's theorem is the non-existence of certain joint probability distributions, and that locality is a peripheral issue. His argument was criticised by Garg, A. and Mermin, N.D., 1982, Phys. Rev. Lett. 49, 242, and by Shimony, A., 1984, Br. J. Phil. Sci. 35, 25-45.

Marshall, T. and Santos, E., 1985, "Local realist model for the coincidence rates in atomic-cascade experiments," Phys. Lett. (forthcoming). This paper constructs a family of local hidden variables models which agree with the predictions of photon polarisation correlation experiments by violating the "no enhancement" assumption of Clauser and Horne (discussed in Sect. 5.1.1 of Clauser and Shimony).

Franson, J.D. (1985 or thereafter), "Bell's theorem and delayed determinism," Johns Hopkins Applied Physics Laboratory preprint (unpublished). This paper argues that the experiment of Aspect, Dalibard, and Roger discussed above does not rule out a class of local theories in which the event is not detected until some time after its occurrence. If, however, results contrary to a Bell-type inequality are obtained when the intensity of the source is considerably diminished, then any initial plausibility of this class of models would largely be lost.

d'Espagnat, B., 1984, Phys. Reports 110, 202-264. An extensive discussion of both physical and philosophical aspects of Bell's theorem.

APR 0 3 1989